Advances in Postharvest and Analytical Technology of Horticulture Crops

Monika Thakur • Tarun Belwal
Editors

Advances in Postharvest and Analytical Technology of Horticulture Crops

Springer

Editors
Monika Thakur
Amity Institute of Food Technology
Amity University
Noida, Uttar Pradesh, India

Tarun Belwal
Vegetable and Fruit Improvement Center
Texas A&M University
College Station, TX, USA

ISBN 978-981-97-7246-9 ISBN 978-981-97-7247-6 (eBook)
https://doi.org/10.1007/978-981-97-7247-6

This Springer imprint is published by the registered company Springer Nature Singapore Pte Ltd.
The registered company address is: 152 Beach Road, #21-01/04 Gateway East, Singapore 189721, Singapore

Preface

In recent years, the field of horticulture has witnessed remarkable advancements in postharvest techniques and analytical technologies, transforming the way we handle, preserve, and understand horticultural crops. The book "*Advances in Postharvest and Analytical Technology of Horticulture Crops*" is a comprehensive compilation of cutting-edge research and innovation that showcases the strides made in this dynamic and critical domain.

Horticulture plays a pivotal role in our food supply, dietary health, and overall well-being. As demands for high-quality, nutritious, and visually appealing produce continue to rise, there has been a growing emphasis on optimizing postharvest handling techniques and leveraging advanced analytical methods to ensure the freshness, safety, and nutritional value of horticultural crops. This book serves as a valuable resource for researchers, practitioners, and enthusiasts alike who are committed to enhancing the entire postharvest process.

Within the pages of this book, readers will explore a diverse range of topics, including the latest advancements in postharvest storage, packaging, transportation, and quality assessment of horticultural products. From emerging technologies such as controlled atmosphere storage and smart packaging to innovative approaches in quality control and shelf-life prediction, each chapter offers insights into how modern science is reshaping the horticultural landscape.

Additionally, the book delves into the realm of analytical technology, shedding light on the sophisticated tools and methodologies used for the chemical, physical, and sensory analysis of horticultural crops. Rapid developments in spectroscopy, chromatography, imaging techniques, and sensor technology have revolutionized our ability to understand the intrinsic properties of fruits, vegetables, and ornamental plants. These advancements not only enable us to maintain product integrity but also empower us to address consumer preferences and market demands with precision.

The contributors to this book are esteemed researchers and experts who have dedicated their careers to advancing the science of horticulture. Their collective wisdom and ground breaking research findings are encapsulated in each chapter, offering readers a comprehensive overview of the current state and future directions of postharvest and analytical technology. As we embark on this journey through *Advances in Postharvest and Analytical Technology of Horticulture Crops*, we encourage readers to explore the challenges and the new technologies designed to overcome them. It is our hope that this book serves as a source of inspiration, fostering collaboration and sparking new ideas that will continue to drive progress in the vital field of horticulture.

Noida, Uttar Pradesh, India Monika Thakur
College Station, TX, USA Tarun Belwal

Acknowledgments

The comprehensive overview of Introduction to Horticulture Crops with its advancements in postharvest and analytical technologies, as well as the different challenges faced in these areas, can be found in *Advances in Postharvest and Analytical Technology of Horticulture Crops*. We sincerely thank all of the authors who have kindly agreed to share their work on various horticultural crop features and contributed chapters. We could not have brought this volume without the meticulous work they put in. Sincere gratitude is extended to Springer Publishing for all of their help and collaboration during the volume's release. We express our sincere gratitude to the Springer production team for their tireless work in getting the book published.

Finally, we would want to express our gratitude to our family for their incredible support and encouragement.

Monika Thakur
Tarun Belwal

Contents

Part III Advances in Analytical Technology for Horticulture Crops

Editors and Contributors

About the Editors

Monika Thakur is working as professor, Amity Institute of Food Technology, Amity University, Uttar Pradesh, India. She is currently engaged in the development of functional foods, extraction of bioactive components, and applied microbiology. She was awarded PhD in 2006 by Himachal Pradesh University. She is Double Gold Medalist (MSc and MPhil) and received several prestigious awards, including Post-Graduate Studies Himachal Pradesh (2004), Himalayan Research Scholar Association (2007), Wiley Research Academy (2017), Young Scientist Award, IITT (2020), Best Speaker Award (2020, 2021), Young Researcher Award (2021), and Best Paper Presentation (2023 Nutricon Malaysia). Apart from her teaching assignments, she has been involved in various research and extension activities. She has guided more than 30 dissertations, and 3 students have completed PhD under her guidance. She is a member of editorial board and reviewer in journals of national and international repute. She has published more than 57 publications, 37 book chapters, 10 books, and a compendium; holds 3 patents and 3 copyrights; has presented more than 70 papers; has established 2 industrial consultancies; has conducted 3 technology transfers; and has sanctioned DST and MSME projects. She has also organized various training programs, faculty development programs, international conferences, and national seminars. Her research interests include plant sciences, phytochemicals, functional foods and nutraceuticals, mycology, food microbiology and food safety, development of functional food products, and related areas.

Tarun Belwal is currently working at Texas A&M University, Texas, USA. Before this, Dr Belwal worked at the Department of Drug Science and Technology, University of Turin, Italy, and at the College of Biosystems Engineering and Food Science, Zhejiang University, China. Dr Belwal completed his graduation in pharmaceutical science (BPharma) from HNB Garhwal University and his master's in biotechnology (MTech) from VIT University, Vellore, before receiving his PhD

(Biotechnology) from Kumaun University, Uttarakhand, India. During his PhD, he worked at GB Pant National Institute of Himalayan Environment (NIHE), Almora, Uttarakhand, on the utilization of natural products as nutraceutical agents and examined their functional activities, development of the plant *in vitro* cell culture, and advanced extraction techniques, for which he was awarded the prestigious Governor Award for Best Research by the Uttarakhand Government for 2 years. He has completed a project as principal investigator (2018–2020), funded by CPDSF, China. His excellent research credentials reflect over 100 peer-reviewed scientific publications (>25 as first/corresponding author), 15 book chapters, and 6 edited books with over 6600 citations. Dr. Belwal has guided several PhDs, postgraduates, and undergraduate students and jointly shares the research outcomes. He also collaborated with several leading global researchers, which led to fruitful collaborative outcomes. Dr Belwal is currently serving as associate editor of *Phytomedicine* and *Heliyon* journals (Elsevier). He is also an editorial board member of *Food Chemistry* (Elsevier). He has also been listed in the top 2% global research scientist category for the years 2021 and 2022, as analyzed and published by Stanford University. Dr Belwal has interdisciplinary research, managerial, and collaborative skills, and experiences to secure the goal of integrating research into the ground. His research interests include phytomedicine, phytopharmacology, analytical chemistry, development and scale-up bioprocess, nutraceuticals for health use, nano-chemistries for agro-food science, biosensors, ultrasonic sonochemistry, and related areas.

Contributors

Himani Agrawal Department of Functional Metabolomics of Plants, Institute of Bioorganic Chemistry, Polish Academy of Science, Poznan, Poland

P. K. Anagha Division of Fruits and Horticultural Technology, ICAR-Indian Agricultural Research Institute, New Delhi, India

Bindvi Arora Division of Food Science and Postharvest Technology, ICAR-Indian Agricultural Research Institute, New Delhi, India

Saiteja Atluri Amity Institute of Organic Agriculture, Amity University, Noida, India

Priya Awasth Department of Post-Harvest Technology, College of Horticulture, Banda University of Agriculture and Technology, Banda, India

Darshayeeta Baruah Amity Institute of Food Technology, Amity University, Noida, Uttar Pradesh, India

Kanchan Bhatt Department of Food Science and Technology, Dr. YS Parmar University of Horticulture and Forestry, Nauni, Solan, Himachal Pradesh, India

Suma H. Budnimath Department of Horticulture, College of Agriculture, Padannakkad, Kerala Agricultural University, Kasaragod, Kerala, India

Vishal Chugh Department of Basic and Social Sciences, College of Horticulture, Banda University of Agriculture and Technology, Banda, India

K. C. Dileep Department of Food Science and Technology, Dr. YS Parmar University of Horticulture and Forestry, Nauni, Solan, Himachal Pradesh, India

Neeru Dubey Ernst and Young, New Delhi, India

D. S. Gaikwad Amity Institute of Organic Agriculture, Amity University, Noida, Uttar Pradesh, India

Shubham Gangwar Department of Post-Harvest Technology, College of Horticulture, Banda University of Agriculture and Technology, Banda, India

Mahesh Gupta CSIR-Institute of Himalayan Bioresource Technology, Palampur, Himachal Pradesh, India

Hardeep Department of Horticulture, Maharana Pratap Horticultural University, Karnal, Haryana, India

M. L. Jat Department of Horticulture, CCS Haryana Agricultural University, Hisar, Haryana, India

Mohan Lal Jat Department of Horticulture, CCS Haryana Agricultural University, Hisar, Haryana, India

R. K. Jat Department of Fruit Science, College of Horticulture, SDAU, Jagudan, Gujarat, India

Alka Joshi Division of Food Science and Postharvest Technology, ICAR-Indian Agricultural Research Institute, New Delhi, India

Naresh Kumar Agriculture University Jodhpur, Jodhpur, Rajasthan, India

Rakesh Kumar College of Agriculture, Agricultural University Jodhpur, Jodhpur, Rajasthan, India

Simple Kumar Amity International Centre for Post-Harvest Technology and Cold Chain Management, Amity University, Noida, Uttar Pradesh, India

Sonu Kumar Department of Horticulture, CCS Haryana Agricultural University, Hisar, Haryana, India

Sunil Kumar Krishi Vigyan Kendra, Kolhapur, India

S. G. Lekshmi Division of Food Science and Postharvest Technology, ICAR-Indian Agricultural Research Institute, New Delhi, India

Rajan Mahendra Division of Food Science and Post-Harvest Technology, ICAR-Indian Agriculture Research Institute, PUSA, New Delhi, India

Guneshori Maisnam Amity Institute of Organic Agriculture, Amity University, Noida, Uttar Pradesh, India

Girish N. Mathad Department of Tropical Agriculture and International Cooperation, National Pingtung University of Science and Technology, Pingtung, Taiwan

Vigya Mishra Department of Post-Harvest Technology, College of Horticulture, Banda University of Agriculture and Technology, Banda, India

Rajesh Mor Department of Horticulture, Maharana Pratap Horticultural University, Karnal, Haryana, India

Ajay Narola Division of Food Science and Postharvest Technology, ICAR-Indian Agricultural Research Institute, New Delhi, India

Swarajya Laxmi Nayak ICAR-Directorate of Medicinal and Aromatic Plants Research, Anand, Gujarat, India

Rongsenyangla Ozukum Adisan Rural Platform India Private Limited, New Delhi, India

Prity Pant Food Technology, William Carey University Shillong, Shillong, Meghalaya, India

Kshitij Parmar Amity Institute of Organic Agriculture, Amity University, Noida, Uttar Pradesh, India

H. R. Raghavendra ICAR-Indian Agricultural Research Institute, Gogamukh, Assam, India

Vinita Rajput Krishi Vigyan Kendra, Sirsa, India

Naleeni Ramawat College of Agriculture, Agricultural University Jodhpur, Jodhpur, Rajasthan, India

Priyanka Rana Department of Food Science and Technology, Dr. YS Parmar University of Horticulture and Forestry, Nauni, Solan, Himachal Pradesh, India

Agrim Rattan Atmospheric Sciences, IIT Delhi, New Delhi, India

Rishita CCS Haryana Agricultural University, Hisar, India

Shruti Sethi Division of Food Science and Postharvest Technology, ICAR-Indian Agricultural Research Institute, New Delhi, India

Niharika Shanker Amity Institute of Food Technology, Amity University Uttar Pradesh, Noida, India

Abha Sharma Amity Institute of Organic Agriculture, Amity University, Noida, Uttar Pradesh, India

Abhay Sharma Horticulture Research Station, Seobagh, Kullu, HP, India

Anshu Sharma Department of Food Science and Technology, Dr. YS Parmar University of Horticulture and Forestry, Nauni, Solan, Himachal Pradesh, India

Jitendra Singh Shivran Department of Horticulture, College of Agriculture, GBPUAT, Pantnagar, Uttarakhand, India

P. M. Sindhu Division of Food Science and Postharvest Technology, ICAR-Indian Agricultural Research Institute, New Delhi, India

Akhilesh K. Singh Department of Plant Protection, College of Horticulture, Banda University of Agriculture and Technology, Banda, India

Saiatluri Teja Amity Institute of Organic Agriculture, Amity University Uttar Pradesh, Noida, UP, India

Deepshikha Thakur Amity Institute of Organic Agriculture, Amity University Uttar Pradesh, Noida, UP, India

Manisha Thakur Department of Food Science and Technology, Dr. YS Parmar University of Horticulture and Forestry, Nauni, Solan, Himachal Pradesh, India

Monika Thakur Department of Food Science and Technology, Dr. YS Parmar University of Horticulture and Forestry, Nauni, Solan, Himachal Pradesh, India

Vijay Yadav Tokala The Postharvest Education Foundation, La Pine, OR, USA

Ajay Tomar Amity Institute of Organic Agriculture, Amity University, Noida, Uttar Pradesh, India

Sumit Bhausaheb Uhre ICAR-Central Institute of Post-Harvest Engineering & Technology (CIPHET), Ludhiana, Punjab, India

VikonoKsh Division of Food Science and Postharvest Technology, ICAR-Indian Agricultural Research Institute, New Delhi, India

Shalini Singh Visen Amity Food and Agriculture Foundation, Amity University Uttar Pradesh, Noida, India

Sagar Wadkar National Cooperative Union of India (NCUI), New Delhi, India

Abbreviations

%	Percent
&	and
/	per
μCT	Microfocus X-ray computed tomography
μg	microgram
μl	microliter
μm	micrometer
AACs	Aptamer-affinity columns
ACIAR	Australian Centre for International Agricultural Research
ASC	Added sugar content
ASE	Accelerated solvent extraction
BSA	Bovine serum albumins
CA	Controlled atmosphere
CDC	Centers for Disease Control
CE-MS	Capillary electrophoresis-mass spectrometry
CFU/ml	Colony-forming unit per ml
CHO	Carbohydrates
cm	Centimeter
conc.	Concentrated
COP	Coefficient of Performance
CREMNET	Crop and Resource Management Network
dil.	Dilute
DNA	Deoxy ribonucleic acid
e.g.	For example
EAE	Enzyme-assisted extraction
EC	Evaporative cooling
Ed.	Edition
ed.	Editor
eds.	Editors
EFF	Enhanced Freshness Formulation
EFSA	European Food Safety Authority

EIA	Assurance enzyme immunoassay
ELISA	Enzyme-linked immunosorbent assay
et al.	et. alia; and others
etc.	et. cetera
FAO	Food and Agriculture Organization
FAVs	Fruits and vegetables
FBO	Food Business Operator
Fig.	Figure
FIGs	Farmers Interest Groups
FPOs	Farmers Producers' Companies
FSMS	Food Safety Management System
FSSAI	Food Safety Standards Authority of India
g	Gram
GC	Gas Chromatography
GC-MS-MS (QqQ)	Gas chromatography-tandem mass spectrometry (triple quadrupole)
GC-ECD/NPD	Gas chromatography with electron capture detector/nitrogen phosphorus detector
GC-TOF-MS	Gas chromatography-time-of-flight mass spectrometry
GDP	Gross domestic product
GMOs	Genetically modified organisms
GMP	Good manufacturing practices
GRAS	Generally recognized as safe
HACCP	Hazard Analysis Critical Control Points HAZOP Hazard and Operability Analysis
HCs	Horticulture crops
HIS	Hyperspectral imaging
HPLC-PDA	HPLC with photodiode array detector
HPLC-UV	HPLC with ultraviolet detector
HPP	High pressure processing
hrs.	Hours
i.e.	That is
IACs	Immunoaffinity columns
in vitro	Within glass
in vivo	Within living
IR	Infrared
IRRI	International Rice Research Institute
Kg	Kilogram
l	Liter
LAR	Locally acquired resistance
LD	Longissimus Dorsi
LDL	Low density lipoprotein
LLE	Liquid-liquid extraction
LSR	Liquid-to-solid ratio
m	Meter

MAE	Microwave-assisted extraction
MALDI-TOF	Matrix-assisted laser desorption ionization-time of flight
mg	Milligram
MH	Million hectares
min	Minutes
MIPs	Molecularly Imprinted Polymers
MIR	Mid-infrared
MIRs	Mid-infrared spectroscopy
ml	Milliliter
mln tons	Million tonnes
mm	Millimeter
MoFPI	Ministry of Food Processing Industry
MS	Mass spectrometry
MT	Metric ton
MT/ha	Metric ton per hectare
NIR	Near-infrared
NIRs	Near-infrared spectroscopy
nm	Nanometer
NMR	Nuclear magnetic resonance
NTP	Non-thermal processing
°C	Degree Celsius
PC	Phosphatidylcholine
PE	Phosphatidylethanolamine
PEF	Pulsed electric field
PG	Phosphatidylglycerol
PHL	Post-harvest losses
PHM	Post-harvest management
PHT	Post-harvest technologies
PI	Phosphatidylinositol
PLD	Phospholipase D
PRPs	Prerequisite programs
PZT	Lead zirconate titanate
QuEChERS	Quick Easy Cheap Effective Rugged and Safe
R^2	Coefficient of regression
RJC	Real juice content
RMSECV	Root mean square error of cross-validation
RMSEP	Root mean square error of prediction
RPD	Residual predictive deviation
R_{pred}	Correlation coefficient for prediction
RRLC-MS/MS	Rapid resolution liquid chromatography tandem mass spectrometry
RTS	Ready to serve
SAR	Systemic acquired resistance
SE	Solid-liquid extraction
SE	Soxhlet extraction

SFE	Supercritical fluid extraction
sp.	Species (Singular)
SPE	Solid phase extraction
spp.	Species (Plural)
sq Km	Square kilometer
SSC	Soluble solid content
TA	Titratable acidity
TAC	Total anthocyanin content
TPC	Total phenolic content
TSEC	Two-stage evaporative cooler
TSS	Total soluble solids
UAE	Ultrasound-assisted extraction
UAE	Ultrasonic-assisted extraction
UPLC-Q-TOF-MS	Ultra-high performance liquid chromatography with quadrupole time-of-flight mass spectrometry
UV	Ultraviolet
VALDS–ME	Vortex-assisted low density solvent–microextraction
var.	Variety
Vis/NIRs	Visible to near-infrared spectroscopy
viz.	Videlicet; namely
VOCs	Volatile organic compounds
vol.(s)	Volume(s)
w.r.t.	With respect to
WE	Water extraction
WHO	World Health Organization
ZECC	Zero energy cool chamber
Ψl	Leaf water potential

List of Figures

List of Tables

Part I
Introduction

Chapter 1
Advances in Postharvest and Analytical Technology of Horticulture Crops: A Review

Mohan Lal Jat, R. K. Jat, Jitendra Singh Shivran, Rajesh Mor, Rakesh Kumar, and Naleeni Ramawat

Abstract Globally, small-scale farmers control the majority of agricultural systems. However, these growers have limited access to preharvest and postharvest technologies because of their high upfront costs. Without these technologies, postharvest management of fruits and vegetables would lose between 25% and 50% of their potential. A simple postharvest infrastructure for product preservation would have prevented product rotting due to improper handling, shipping, packing, and storage, which is what caused these large losses. High postharvest losses of products pose a challenge to global food security, nutrition, and availability because the producer maintains less of his entire quality produce. As a result, there is less of these food commodities available net for consumption. As a result, this chapter provides a comprehensive overview of current concepts and techniques for postharvest management in horticultural crops worldwide. Because of postharvest research and technology, the horticultural sector can provide consumers with fresh, safe, and nutrient-dense goods that follow the supply chain from farm to mouth. Enhancing postharvest research, innovation, instruction, and training might significantly lower food waste and losses. To combat delays at various handling levels and reduce postharvest losses that occur during harvesting, packaging, and shipping in wholesale and retail markets, a wide range of cutting-edge technologies and tactics are cur-

M. L. Jat
Department of Horticulture, CCS Haryana Agricultural University, Hisar, Haryana, India

R. K. Jat
Department of Fruit Science, College of Horticulture, SDAU, Jagudan, Gujarat, India

J. S. Shivran
Department of Horticulture, College of Agriculture, GBPUAT, Pantnagar, Uttarakhand, India

R. Mor
Department of Fruit Science, College of Horticulture, MHU, Karnal, Haryana, India

R. Kumar · N. Ramawat (✉)
College of Agriculture, Agricultural University Jodhpur, Jodhpur, Rajasthan, India

M. Thakur, T. Belwal (eds.), *Advances in Postharvest and Analytical Technology of Horticulture Crops*, https://doi.org/10.1007/978-981-97-7247-6_1

rently being used. Further research should focus on combining various new technologies with postharvest techniques and integrating relevant enhanced practices into an existing value chain and marketing framework.

Keywords Postharvest losses · Quality · Commodity · Consumption · Harvesting and storage

1.1 Introduction

With horticultural crop production of approximately 342.33 million MT from a 28.08 million ha area and productivity of 12.08 MT/ha, China leads the world in fruit and vegetable production, while India comes in second. These two countries account for approximately 10.5% and 11.2% of global fruit and vegetable production, respectively (Anonymous 2021). The horticulture sector contributes over 30% of the agricultural GDP, indicating that demand for horticultural products has been rising over time in both the domestic and foreign markets. Both domestic and foreign markets are seeing a sharp increase in the demand for fruits and vegetables. Because of the varied climatic conditions in India, there is ample opportunity to cultivate a wide variety of fruits and vegetables throughout the year in one or more regions of the nation.

Nowadays, people encounter many obstacles, primarily in the form of locating fresh produce that is both nutrient-dense and of excellent quality. The global demand for fresh produce is increasing rapidly due to various factors such as urbanization, growing middle class, increased disposable income, and shifting consumer preferences. Fresh fruit and vegetable sales are expected to reach a valuation of over 200 billion USD by 2027, from an estimated 144 billion USD in the current worldwide market (Stanaway et al. 2022). Fruit and vegetables should always be included in a healthy, well-balanced diet. Chen et al. (2022) claim that they are rich in essential vitamins, minerals, antioxidants, and dietary fibers, all of which have the ability to heal a range of ailments and deficiencies.

Fresh produce is susceptible to spoiling due to microbial activity and different physiological changes that may occur after harvest, giving them a short shelf life. When a significant amount (between 25% and 50%) of fruits and vegetables are lost after harvest, it is referred to as "postharvest losses." These losses occur on a worldwide level. Roughly one-third of the total amount of food produced worldwide is lost due to this (Bancal and Ray 2022). A further basic issue facing agricultural researchers and policymakers is making sure that by 2050, over 9.1 billion people will have access to wholesome food. As a result, it is projected that by 2050, food production will increase by 50% in order to meet the demands of the world food supply (Parfitt et al. 2010).

1.2 Nature of Postharvest Losses

The losses that happen after harvesting are known as postharvest losses. It is defined as the losses resulting from total qualitative loss, total quantitative loss, and total food waste (by consumers) from the time of harvest to the point of consumption. Five processes can lead to postharvest losses in horticultural commodities: production/harvest, handling and storage following harvest, processing, distribution, and consumption.

A number of losses are caused by inadequate infrastructure, inexperience, bad management, broken markets, or plain carelessness on the part of farmers.

(a) **Extend of Postharvest Loss:** It is obvious that to increase the amount of food that can be produced at the current level, postharvest loss estimation is crucial. An estimated Rs. 23,000 crores are wasted annually. According to the Swaminathan Committee, postharvest handling is responsible for 20–30% of losses during the various phases of grading, packing, transporting, and selling fresh or processed produce. Between 40% and 50% of food losses globally occur from fruits and vegetables, resulting in an annual loss of US$750 billion. Of these losses, 54% happen during production, postharvest, handling, and storage, while 46% happen during processing, distribution, and consumption. According to Chadha (2011), approximate losses in the Indian context are about 35–45% of the harvested fruits and vegetables during handling, storage, transportation, and so on, resulting in a loss of Rs. 40,000 crores per year.

(b) **Important Sites of Postharvest Losses:** Important sites where postharvest losses are reported in India at farmer's field (15–20%), packaging (15–20%), transportation (30–40%), and marketing (30–40%).

1.3 Factors Affecting Postharvest Loss

Numerous factors affect fruit and vegetable postharvest losses, leading to mechanical, hygienic, physiological, and physical losses. Horticultural product has a short shelf life as it goes through multiple metabolic processes.

Mechanical, physiological, pathological, and environmental variables are the main causes of fruit and vegetable postharvest loss (Yahaya 2005).

1.3.1 Mechanical Loss

Fruits and vegetables that are handled carelessly during harvesting, packing, shipping, storing, and so on are lost mechanically. The mechanical harm to fruits and vegetables can also be attributed to some insects and birds (Alao 2000; Hayatu 2000). Fruits and vegetables that have been bruised or cracked by mechanical damage are frequently more vulnerable to organism attack, which accelerates water loss and gaseous exchange. The majority of mechanical damage to fruits and vegetables is produced by pressure thrust during transportation, which can rupture inner tissues and cells even though it is frequently unnoticeable. During the course of natural aging, such produce deteriorates more quickly (Alao 2000).

1.3.2 Microbial Loss

Vegetables and fruits can also be attacked by microbes. Microbiological degradation is most commonly caused by molds, bacteria, yeast, and fungi. On the contrary, a significant portion of fruit and vegetable losses during the postharvest period are attributed to bacterial and fungal infections. The most common pathogens causing rots in produce are fungi, including *Alternaria, Botrytis, Diplodia, Monilinia, Phomopsis, Rhizopus, Penicillium, Fusarium*, and so on. *Pseudomonas, Erwinia*, and other bacteria cause a great deal of damage. Microbes that cause postharvest deterioration thrive in environments with high temperatures and relative humidity.

1.3.3 Environmental Factors

There are several environmental factors that can contribute to postharvest losses in fruits and vegetables. Here are a few examples:

- **Temperature:** One of the main causes of postharvest losses is inadequate temperature regulation. Because they are perishable, fruits and vegetables that are exposed to extreme heat or cold will swiftly deteriorate. While low temperatures can cause chilling injury and have an adverse effect on the quality and shelf life of food, high temperatures can hasten the ripening and decaying processes. Fruits breathe more quickly at high temperatures. High temperatures and relative humidity also have an impact on horticulture crop productivity because they accelerate fruit deterioration. FAO (2011) vegetables, while a drop in temperature slows the pace of microbial attack on postharvest losses of fruits and vegetables on various crops, particularly when the primary cause is mechanical or physiological, and the temperature is below 5 °C. Low but pathogenic and environmental variables lead to chill damage. Tropical mechanical factors are primarily associated with temperatures below freezing, while signs of chilling injuries may not always be apparent. Gathering: Harvesting at peak ripening occurs when the fruits and vegetables are kept at a cooling temperature, which is the primary factor influencing the product's quality and temperature. However, the fruits may not be apparent until they have reached the end of their shelf life. However, some farmers are able to harvest immature crops because vegetables are brought to room temperature (37 °C).

- **Humidity:** Postharvest losses may also be caused by improper humidity conditions. Overly humid conditions can encourage the growth of bacteria and fungi, which can cause fruits and vegetables to rot and degrade. Conversely, low humidity can lead to develop shriveling and dehydration.
- **Light:** For some fruits and vegetables, exposure to light might be harmful. Overexposure to light, particularly sunshine, can cause vitamin breakdown, pigment deterioration, and color loss. Produce should be transported and stored in suitable packaging that shields it from excessive light exposure.
- **Ethylene:** Fruits and vegetables naturally release ethylene, which is a plant hormone. While ethylene aids in the ripening process, excessive ethylene levels can hasten produce ripening and senescence, which can result in spoiling. To avoid premature deterioration, it is critical to control the amount of time that ethylene-sensitive fruits and vegetables are exposed to ethylene-producing ones.
- **Oxygen and Carbon Dioxide Levels:** As they breathe, fruits and vegetables use up oxygen and release carbon dioxide. Inadequate control of carbon dioxide and oxygen concentrations can compromise the quality and durability of produce. For instance, off-odors can arise from anaerobic respiration brought on by low oxygen levels, and off-flavors can be caused by excessive carbon dioxide levels.

Appropriate handling, storage, and transportation procedures are crucial to reduce postharvest losses brought on by these environmental variables. This entails regulating ethylene levels, shielding products from intense light, preserving the proper temperature and humidity, and maximizing the oxygen and carbon dioxide atmospheres during transit and storage.

1.3.4 Physiological Factors

The main causes of horticulture crops' postharvest losses were reportedly inadequate packing materials, poor transportation, and a lack of suitable storage facilities. Researchers have found that perishable goods are sold soon after harvesting without proper packaging or primary processing as there are insufficient farm storage facilities and packing stations (Atanda et al. 2011). During harvesting, the product is collected from fields using a variety of containers, including wooden boxes, baskets, plastic materials, and sacks. Improper handling increases the degree of produce damage. Packages need to be sufficiently sturdy and ventilated to prevent any physical or structural harm. For the convenience of handling, fruit is best packed in waxed cartons, wooden crates, or solid plastic containers rather than bags or open baskets, as the latter offer less protection to the food when piled high. Postharvest problems include improper packaging, overfilling containers, and combining different vegetable varieties. Most sellers lose money because they combine contaminated and broken goods with regular inventory.

1.4 Postharvest Technologies

Various losses can be minimized by various postharvest technologies. The postharvest technologies are divided into two segments basically:

- Traditional postharvest management technologies.
- Advanced or recent postharvest management technologies.

1.4.1 Traditional Postharvest Management Technologies

A variety of methods and approaches are included in traditional postharvest management technologies, which are intended to increase the shelf life, preserve quality, and lower postharvest losses of produced horticultural goods. These conventional techniques have been used extensively for a long time, even if newer technologies are still developing. The following are some instances of conventional postharvest management technologies. Cooling is a common traditional postharvest technique that lowers the temperature of harvested products to impede microbiological and physiological processes that cause deterioration. Room cooling, forced-air cooling, refrigeration, and hydro-cooling—which involves submerging in cold water—are examples of cooling techniques. Harvested crops can be dried to eliminate moisture, which lowers their water content and stops microbial growth. Air drying, sun drying, and mechanical drying with hot air or direct sunlight are examples of traditional drying methods. Onions, garlic, and potatoes are some crops that are cured using a different method. To encourage drying, taste development, and stable storage, harvested produce is exposed to particular environmental conditions, such as regulated temperature and humidity. Harvested produce is better protected from microbiological infection, physical damage, and moisture loss when it is packaged properly. Natural fibers like jute and bamboo, as well as crates, baskets, and sacks, are examples of traditional packaging materials.

Various storage structures are used in traditional postharvest management to establish ideal conditions for maintaining the quality of produced products. Traditional granaries, tunnel pits, silos, cellar storage, and warehouses are a few examples. These buildings can reduce exposure to light, control humidity and temperature, and offer pest protection. Traditional preservation techniques have been established by various cultures in response to local requirements. Pickling, fermenting, smoking, salting, canning, and jam-making are a few examples. These methods increase the shelf life of harvested produce by changing the microbial environment or preventing microbial growth. For marketing and quality assurance objectives, sorting and grading are crucial. Using hand-sorting techniques and visual inspection, traditional methods sort gathered products according to size, shape, color, and quality factors. Grading ensures that output is comparable.

Even though there has been some success with these more conventional postharvest management solutions, it is important to remember that newer inventions and technologies are constantly being created to further enhance postharvest practices. These could include postharvest treatments, controlled atmosphere (CA) storage, modified atmosphere packing, sophisticated cooling systems, and other scientific and technological developments meant to minimize losses and preserve the quality of harvested crops.

1.4.2 Advanced or Recent Postharvest Management Technologies

1.4.2.1 Postharvest Chemical Treatments

Postharvest management using chemicals is an age-old practice that is used particularly for the management of fragile horticultural crops. Chemical treatments such as nitric oxide, salicylic acid, 1-MCP, and oxalic acid are used commercially in the fruit industry (Ding et al. 2007; Hong et al. 2014; Junmatong et al. 2015; Razzaq et al. 2016). These treatments are intended to prevent fruit from developing mold infections, which can lead to spoilage. Their use has shown to be effective in maintaining fruit quality and extending shelf-life.

1.4.2.1.1 1-MCP Technology

The 1-MCP technology has had a significant influence on both scientific understanding of the role that ethylene plays in the metabolic processes of plants and the methods used to store and manage crops. It has been discovered that 1-MCP is an effective ethylene antagonist at the receptor level, and it has shown good results in terms of increasing shelf life and maintaining the quality of fruits and vegetables. Because of the significance of the consumption of bioactive chemicals and the antioxidant activity of fruit to human health, more study is required to investigate the effects of 1-MCP on a variety of enzymatic and non-enzymatic antioxidant systems in order to gain a better understanding of how 1-MCP might increase TAA (Valero et al. 2016). 1-MCP has a powerful effect on delaying fruit softening, suppressing browning, preventing deterioration, and delaying peel yellowing. 1-MCP has been widely used to maintain the quality of fruits during storage (Watkins 2006; Cheng et al. 2012). The effects of 1-MCP are significantly influenced both by the temperature at which it was stored and the duration of time it was kept there. In most

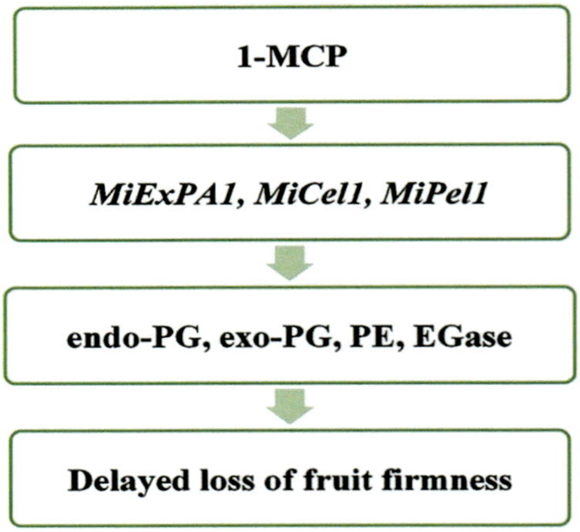

Fig. 1.1 Mechanism of 1-MCP in maintaining fruit firmness. (Source: Bambalele et al. 2021)

circumstances, the effects are negligible when exposed to low temperatures for a short period of time. Nevertheless, they become significant when the food is preserved for an extended period of time. To maximize its use in different circumstances, it is crucial to examine how 1-MCP affects the changes in fruit active chemicals during storage (Guan et al. 2015; Fig. 1.1).

1.4.2.1.2 Phospholipase D Inhibition Technology

Phospholipase D (PLD) is the essential enzyme that begins a chain reaction of catabolic cascades that ultimately results in the breakdown of the membrane. During the process of ripening, the membrane begins to break down, which is an important step in the development of the organoleptic quality of the fruit. However, if the trend of lipid breakdown is allowed to continue unchecked, it might shorten the shelf life of horticultural crops. This enzyme has been linked to a number of different aspects of cellular functions, most notably the responses to hormones as well as biotic and abiotic stressors. PLD is responsible for catalyzing the hydrolysis of membrane phospholipids such as phosphatidylcholine, phosphatidylethanolamine (PE), phosphatidylinositol, and phosphatidylglycerol (PG), among others, in order to produce phosphatidic acid and the appropriate head group (Kates 1955; Galliard 1980; Exton 1997). It was originally hypothesized that PLD plays a significant role in phospholipid catabolism in plants, starting up a lipolytic cascade in membrane degradation in response to aging and stress (Paliyath and Thompson 1987). Reactive oxygen species formed during stressful circumstances such as cold storage, postharvest handling, and so on also contribute to membrane degeneration (Paliyath and Droillard 1992).

The structure and mechanism of the enzyme PLD is complex. It is thought that a rise in intracellular calcium levels carried on by hormone stimulation is the mechanism that causes PLD to attach to the membrane. PLD cannot be able to catalyze the breakdown of membrane phospholipids if it only occurs in the cytosol.

The enzyme is considered to be activated by several cofactor-binding motifs and two active sites. As a result, PLD activity can be suppressed by various methods.

1.4.2.1.2.1 By N-Acylethanolamine

Some phospholipid head groups, like choline, have the ability to block PLD function at very high doses (>10 mM). Austin-Brown and Chapman (2002) reported that N-acyl ethanolamines (NAEs) inhibit the PLD (alpha isoform) precisely and effectively.

The ethanolamine molecule of PE can be acylated to generate N-acyl PE, which, although claimed to exist naturally, is not a common phospholipid in plants. The action of PLD on N-acyl PE would result in the formation of N-acylethanolamine as the head group. NAEs have been shown to be effective inhibitors of PLD activity. Shorter chain length NAEs (C12) are more effective inhibitors than longer chain length NAEs (e.g., C18). In addition, increasing the unsaturation of the acyl chain reduced inhibitory efficiency. Even though the catalytic domains of all PLDs have a high degree of sequence similarity, NAEs did not inhibit other PLD isoforms (beta and gamma).

1.4.2.1.2.2 By Hexanol and Hexanal

According to Paliyath et al. (2003), a primary alcohol such as hexanol and an aldehyde such as hexanal suppress PLD action. Both soluble and membrane-associated PLD are decreased by these components. The efficacy of primary alcohols in participating in the reaction tends to decrease as chain length increases, suggesting that the activity is impeded in the presence of water. PLD activity is suppressed by hexanal. Aldehydes lack the primary alcohol structure required for reaction participation and might block the reaction at the water-binding site. Hexanal is produced naturally during lipid peroxidation, which is mediated by lipoxygenase and hydroperoxide lyases and is responsible for the distinctive green flavor gained upon wounding in crops such as cucumber and beans. Also included in the GRAS category is hexanal. Hexanal's volatile nature is its key benefit when compared with PLD inhibitors. If necessary, the product can be subjected to a vapor, which is functionally equivalent to using 1-MCP. Hexanal, geraniol, ascorbic acid, vitamin E, and calcium chloride are all components of compositions employed by certain studies. Optimizing the mix for each individual fruit or vegetable is necessary.

Hexanal is only one active component in the hexanal formulation. Other components are included in the hexanal composition (Paliyath and Murr 2007), which is also known as an enhanced freshness formulation. Sprays are used before harvest, dips are used after harvest, and hexanal vapor treatments are also available.

1.4.2.2 Salicylic Acid

It is common practice to make use of salicylic acid and its derivatives in order to extend the postharvest life of fruits. This is accomplished by altering physiological processes and mitigating the adverse effects of a wide variety of biotic and abiotic stimulations. After harvest, fruits and vegetables are especially susceptible to attacks from several diseases. To protect themselves from the harmful effects of infections, plants employ self-defense mechanisms such as locally acquired resistance and systemic acquired resistance (Vlot et al. 2009). The synthesis of large amounts of salicylic acid is one of the mechanisms that defend the host from infections. Salicylic acid (SA) and its derivatives, acetylsalicylic acid (ASA) and methyl salicylate (MeSa), are hormones that play important roles in various phenological stages of plants, that is, from seed germination to flowering and fruit ripening. Although many studies also discussed their role and effects on inducing plant defense systems against different biotic and abiotic stresses (Tiwari et al. 2017; Koo et al. 2020), it has been reported that postharvest treatments with salicylates reduce decay and chilling injury in a wide range of fruits and improve other quality properties, such as appearance, texture, and nutritional compounds (Asghari and Aghdam 2010; Glowacz and Ree 2016).

Salicylic acid and its derivatives are also thought to be key disease-resistance regulators in plants (Klessig and Malamy 1994). Salicylic acid stimulates a number of defense genes, particularly those that encode pathogenesis-related (PR) proteins (Malamy et al. 1990; Bowles 1990), and several of these PR proteins have antimicrobial properties, such as chitinase or 1,3-glucanase activity (as acetylsalicylic acid promotes the expression of pathogenic genes and confers resistance); salicylic acid participates in the signal transduction; salicylates preserve postharvest quality by delaying fruit ripening, most likely by inhibiting ethylene production or activity (Srivastava and Dwivedi 2000). Table 1.1 provides a summary of some of the findings related to the impacts of SA on various harvested horticulture crops' qualitative attributes.

1.4.2.3 Edible Coatings

Edible coatings are of great importance in postharvest management because they preserve the quality of horticultural crops. Some physical changes can be prevented during storage by coating them with edible lipids, polysaccharides, and proteins. The primary purpose of the edible coating is to act as a barrier to respiration and transpiration. It enhances the quality and shelf life of fresh, minimally processed fruits and vegetables by acting as a number of functional compounds, such as antibacterial and antioxidant agents. Fruits and vegetables with edible coverings have their stomata covered, which reduces transpiration and, consequently, weight loss. It has been observed in numerous fruits, including apple, papaya, carrot, guava, plum, mango, apricot, banana, and orange (Prasad et al. 2018). It also prevents degradation by limiting dehydration, reducing respiration, enhancing texture quality,

Table 1.1 Concluded results of SA on some treated fruit crops

Fruit crops	Achievements	References
Punica granatum	Inhibition of PAL activity, retention of vitamin C content reduction of CI and EL	Sayyari et al. (2009), García-Pastor et al. (2020)
Actinidia delicious	Inhibition of ethylene production, ripening & decay control	Aghdam et al. (2009)
Prunus avium	Decay control, increase in CAT, GPX, b-1, 3-glucanase& chitinase	Xu and Tian (2008)
Malus domestica	Decreased ethylene, LOX activity, MDA content, DI, softening and respiration rates, and TSS and activated the SOD, POD, CAT, and APX enzymes	Mo et al. (2008)
Citrus reticulate	Acceleration of H_2O_2 accumulation, increase in SOD, GR, APX, and DHAR activity and ASA/DHASA and GSH/GSSG ratios, decrease in lipid peroxidation and MDA	Hung et al. (2007)
Fragraia annanasa	Marketability retention, decrease in ethylene production and fungal decay	Babalar et al. (2007)
Botriya jopanica	Inhibition of PAL, CAT, and POD, decrease in superoxide free radical production and CI	Cao et al. (2006)
Pyrus communis	Induction of resistance to diseases, increase in POD, PAL, GR, Chitinase and b-1. 3-glucanase, decrease in CAT and APX	Cai et al. (2006)
Malus domestica	Enhanced the biological efficacy of antagonist *C. laurentii*	Yu and Zheng (2006)
Prunus pyrifolia	Increase in firmness, APX, GR activity, AsA/DHAsA, and GSH/GSSG ratios, decrease in CI, DI induction of HSP101 and HSP73 genes	Wang et al. (2006)
Prunus avium	Inhibition of ACS. ACO, and LOX activity, suppression of ethylene and superoxide free radical production, increase in total SA content	Zhang et al. (2003)
Actinidia deliciosa	Decrease in fruit softening, pulp/peel ratio, reducing sugar content, invertase activity, and respiration rate, inhibition of cellulase, PG, xylanase, CAT, and POX activity	Srivastava and Dwivedi (2000)

assisting in the retention of volatile taste components, and lowering microbial contamination. Hydrophobic (lipid or wax-based coating), hydrophilic (polysaccharide-based coating), and hydrocolloids are the three types of edible coatings (Warriner et al. 2009).

1.4.3 Non-Chemical Treatments

Research has shown that applying non-chemical methods like heat treatment and UV irradiation can successfully preserve the postharvest quality of horticulture crops (Bambalele et al. 2021). These methods have been shown to be successful in lowering the frequency of illnesses contracted after harvest and extending the time that horticultural products can be kept in storage.

1.4.3.1 Hot Water Treatment Technology

Technology for postharvest heating has been used to improve agricultural shelf life and maintain sanitation in horticulture. Two popular and inexpensive heat treatment techniques are hot air treatment and hot water treatment (Bambalele et al. 2021). This method—in particular, immersion—involves submerging fruits or vegetables in hot water for a predetermined amount of time at a temperature that exceeds the maximum temperature allowed by the different stages of development for pests of concern. The most thermotolerant stage of development is usually killed by temperature and time. Hot water treatment is impacted by fruit type, shape, size, and physiological development (Ndlela et al. 2021). Before storing or retailing the fruits, they are submerged in hot water to prevent a number of postharvest illnesses. Fungicides and hot water treatment are most effective between 51 °C and 55 °C for 5–30 min depending upon the size of the fruit.

1.4.3.2 Ultraviolet-C Radiation

The ability of ultraviolet-C (UV-C) radiation technology to reduce microbial contamination and lengthen the shelf life of different agricultural goods has drawn attention as a possible tool for postharvest management in recent years. A particular range of ultraviolet light with wavelengths between 200 and 280 nm is referred to as UV-C radiation. Because ultrasonic radiation can change germs' DNA, it has been used as a germicide in surface disinfection. Furthermore, biological organs like fruits and vegetables can respond favorably to mild UV-C irradiation, which can have a variety of positive effects like extending their shelf life or increasing the amount of components that promote health (Ribeiro and Alvarenga 2012). With a wavelength of 190–280 nm, short-wave UV light is a non-thermal technology (Mohamed 2017). As a postharvest treatment, UV-C irradiation is used to improve fruit quality and increase the shelf life of agricultural commodities.

UV technology may be used for the postharvest treatment of fruits and vegetables for many different purposes (Ribeiro and Alvarenga 2012; Table 1.2):

- Selection of raw material
- Control of microorganisms that cause spoilage
- Modulation of ripening and senescence
- Control of microorganisms that cause disease in humans
- Enhancement of bioactive compound
- Induction of cross-stress resistance

Table 1.2 Experimental results related to postharvest UV treatment of fruit surface

Crop and its cultivar	UV rays	Achievements	References
Apple cv. Red Delicious	UV-C $\lambda = 254$ nm 7.5 kJ/m^2	It was shown that administering UV irradiation 96 h prior to *Penicillium expansum* inoculation offered the most effective resistance against infection.	De Capdeville et al. (2002)
Blueberry fruit cvs. Collins, Bluecrop	UV-C 0–4 kJ/m^2 Storage 7 days at 5 °C plus 2 days at 20 °C	There was no difference in stiffness or weight reduction with light. Response to 1–4 kJ/m^2 UV-C light was 10% reduced fruit mature rot (*Colletotrichum acutatum*, syn. *C. gloeosporioides*). Treatment intensity enhanced total anthocyanin, total phenolics, and ferric-reducing antioxidant power (FRAP).	Perkins-Veazie et al. (2008)
Mango cv. Tommy Atkins	UV-C, $\lambda = 254$ nm 4.9 and 9.9 kJ/m^2	They were kept at 5 °C for 14 days, while their quality and ability to withstand sickness were evaluated. They were then kept at 20 °C for 7 days. The fruit's texture and appearance were enhanced by the 4.9 kJ/m^2 treatment. The larger dose caused senescence to develop.	Aguilar et al. (2001)
Mango cv. Haden	UV-C, $\lambda = 254$ nm 2.46 and 4.93 J/m^2 Stored 18 days at 25 °C	Fruit exposed to UV-C remained fresher for longer.	Gonzales-Aguilar et al. (2007)
Oranges cv. Biondo Comune, Washington Navel, Tarocco, *Valencia*	UV-C, $\lambda = 254$ nm 0.5–3.0 kJ/m^2	A week at 20 °C and 4 weeks at 7 °C were used to evaluate disease resistance and quality. It affected *Valencia*	D'hallewin et al. (1999)
Peach cv. Elberta	UV-C, $\lambda = 254$ nm 0.4–40 kJ/m^2	Seasonal variations in higher-dose fruit damage were observed. UV light exposure of 0.5 kJ/m^2 reduced degradation. Early-harvested fruit only performed better below 1.5 kJ/m^2.	Stevens et al. (1998)

1.4.4 Storage Technologies

A number of commodities used in horticulture are seasonal and have a brief harvesting season. Fresh food storage will also help avoid market oversupply by giving customers access to a wide range of fruits, vegetables, and flowers for much of the year, especially in the off-season. The revenues of producers and farmers are increased by organized marketing and storage. Fresh produce is kept refrigerated to maintain its quality, minimally deteriorate, and increase its shelf life. One of the main reasons for the significant postharvest losses of horticulture goods is inadequate storage facilities. Slowing physiological processes including respiration, transpiration, maturation, and other biochemical transformations is the basic idea of storage. Proper storage aims to not only stop the spread of illness but also maintain the product's best quality for customers to prevent the onset of contamination.

1.4.4.1 Controlled Storage Technology

By adjusting the atmospheric conditions in which fruits and vegetables are stored, a technique known as "Controlled Atmosphere" storage can increase the fruits and vegetables' shelf life and maintain their quality. It entails establishing a space with regulated humidity, temperature, carbon dioxide, and oxygen levels. The best low temperatures in conjunction with CA have been used to maintain fruit quality while it is being stored. In California, storing fruits and vegetables involves keeping them in environments with low oxygen concentrations, high relative humidity, low temperature, and/or relatively high carbon dioxide concentrations (Valdez Fragoso and Mjica-Paz 2016). Jacques Etienne Berard investigated CA storage for the first time in France in the early 1800s (Dalrymple 1969). He discovered that fruits did not mature in an O_2-depleted environment. Other studies also investigated the effects of low levels of O_2 and high levels of CO_2 on ripening, but the basis for commercial application of CA storage was established by Kidd and West (1936) who studied the effects of O_2, CO_2, and C_2H_4 on respiration and ripening in pome fruits and berries. However, after the 1950s when CA technology improved, its usage became more common, and further developments and technical innovations during the 1990s expanded its use worldwide (Prange et al. 2005).

1.4.4.1.1 Benefits of CA Storage

- **Extended Shelf Life:** CA storage can significantly extend the shelf life of fruits and vegetables compared to conventional storage methods. By controlling the levels of oxygen and carbon dioxide, as well as temperature and humidity, the ripening process is slowed down, reducing the rate of decay and extending the storage life of the produce.
- **Preservation of Quality:** CA storage helps to preserve the quality attributes of fruits and vegetables. Maintaining optimal humidity levels prevents moisture loss and helps to retain texture, color, and flavor. Controlling the atmospheric conditions also helps to reduce the development of physiological disorders and decay, preserving the overall appearance and taste of the produce.
- **Reduced Losses:** CA storage can aid in lowering postharvest losses by reducing rot and delaying down the ripening process. This is especially advantageous for foods that spoil quickly, including perishable fruits and vegetables. It lowers the quantity of food that goes bad or is unsold by enabling growers, wholesalers, and merchants to store and transport produce for extended periods of time.
- **Extended Market Reach:** Fruits and vegetables can be transported over longer distances without experiencing substantial deterioration in quality because of CA storage. Producers now have the chance to reach markets that are far from

their places of production thanks to this. It increases company opportunities and revenue potential by enabling the export of fresh produce to foreign markets.

- **Flexibility in Harvesting and Marketing:** Timing flexibility for harvesting and commercialization is offered by CA storage. It enables farmers to gather their harvests at the height of maturity and preserve them until the right amount of demand or favorable prices arise. Better supply chain management and market planning may result from this.
- **Pest Control:** Certain insects and pests that damage fruits and vegetables that are kept in storage can be managed with the aid of CA storage. Lowering the oxygen content makes it harder for pests to survive, lowering the chance of infestation and the need for chemical treatments.

1.4.4.2 Modified Storage Technology

With MA storage, controlling the concentration of the gas surrounding the commodity in the atmosphere will be more difficult. The creation of polymeric films with a broad spectrum of gas permeability has altered the ambient atmosphere inside flexible film packaging. Concurrently, during MA storage, respiration and metabolism speed increase by a factor of two for every 10 °C increase in temperature. The temperature, O_2 concentration, and CO_2 level of the storage environment may be changed to reduce respiration. Both the amount of carbon dioxide and oxygen in the air have an impact on breathing rate. Overall, there may be a synergistic or additive effect. People breathe more slowly when the air's oxygen content falls below 10%. However, a process known as anaerobic respiration can start when oxygen levels fall below 2%, leading to the accumulation of ethanol and acetaldehyde. The favorable effect of MA on plant tissues is assumed to be caused by a drop in pH, which is brought on by the breakdown of CO_2 in plant tissues.

MAP is putting fruits in films or containers that are specifically made to control the gasses that surround the fruit. MAP produces an ideal storage environment that slows down respiration and prevents microbial development by changing the composition of the gas. This method can lower postharvest losses, maintain the texture and flavor of fruits, and increase their shelf life. MA packing works by changing the environment inside the package through the product's respiration and gas transfers through the packaging, resulting in an atmosphere that is higher in CO_2 (7–15%) and lower in O_2 (2–6%) (Zhang et al. 2015; Pandey and Goswami 2012). MA packaging success depends on different factors such as good quality of the initial product, good manufacturing practices, correct gas mixture, adequate refrigeration of the product, and appropriate packaging material (Brody et al. 1995; Martínez Ferrer et al. 2002). In fact, there are a wide variety of polymers and gas mixtures available for packaging fresh-cut produce that should be optimized for each commodity.

1.4.4.3 Hypobaric Storage Technology

Hypobaric storage works by postponing the ripening and senescence processes in order to extend the postharvest life of fruits and vegetables, much like modified or controlled environment storage does. During this process, the product is maintained in a refrigerated, vacuum-tight container that is continuously ventilated with humidified air at a relative humidity (RH) of 80–100%. Air pressure below 760 mmHg results in a drop in atmospheric oxygen content. Under continuously vented partial pressure, CO_2, ethylene, and other volatile metabolic byproducts swiftly diffuse out of the commodity and are flushed from the storage chamber. The delayed ripening and senescence brought on by the low partial pressure of oxygen and the low amounts of ethylene in fresh fruit and vegetables extend their storage life (Barkai-Golan 2001).

1.4.4.4 Emerging Postharvest Technologies

A variety of cutting-edge techniques are being created and put into practice to improve postharvest handling, increase shelf life, and preserve the quality of agricultural products. These techniques are collectively referred to as emerging postharvest technologies. These technologies take advantage of developments in data analytics, engineering, biology, and nanotechnology. These are some instances of newly developed postharvest technology:

- **Intelligent Packaging:** Sensors, indications, and data analysis are used by intelligent packaging systems to track and share information about the state of the product during transportation and storage. Real-time data on temperature, humidity, gas concentrations, and even the presence of pathogens can be obtained from these devices. Making decisions is aided by intelligent packaging, which also makes it possible to take prompt action to preserve product quality.
- **Nanotechnology Applications:** Potential advantages of nanotechnology for postharvest management exist. Fruits and vegetables can be protected from moisture loss, microbiological contamination, and gas exchange by applying nano-based coatings and films. Antimicrobial nanoparticles can be added to packaging materials to prevent the growth of pathogens. Additionally, monitoring and quality assessment can be done with nanosensors.
- **Cold Plasma Technology:** Applying a non-thermal plasma field to agricultural produce is known as cold plasma treatment. With its antimicrobial properties, this technology can prolong food safety, lower microbiological contamination, and increase shelf life. Research is being done on cold plasma treatment as a potential replacement for chemical sanitizers and as a means of managing spoilage microorganisms.
- **Precision Agriculture and Data Analytics:** The use of precision agriculture methods in postharvest management, including data collecting, analysis, and remote sensing, is growing. Hyperspectral imaging and near-infrared

spectroscopy are two examples of remote sensing technologies that can be used to evaluate product quality and spot flaws or damage. Predicting shelf life, maximizing storage conditions, and making decisions about distribution and inventory management are all aided by data analytics.

- **Biotechnology and Genetic Engineering:** A variety of biotechnological techniques, including genetic engineering, are being investigated to improve crop postharvest characteristics. Genetic alterations can be applied, for instance, to improve disease resistance, postpone ripening, and increase nutritional value. Enzymes or biocontrol agents for postharvest disease management and quality preservation can also be developed using biotechnology.

Many researchers have developed new technologies that have been applied to reduce the microbial load and preserve the freshness and quality characteristics of fruit and vegetables. These technologies include microwave heating, high hydrostatic pressure, pulsed electric fields, high hydrostatic pressure, ozone, and cold plasma. The potential of these technologies to maintain fruit quality needs more investigation (Fan and Wang 2021). These new postharvest technologies have enormous potential to address population growth demands while also increasing the food supply chain's sustainability and efficiency and lowering losses. To ensure their successful deployment and incorporation into current postharvest management methods, ongoing research, development, and commercialization initiatives are crucial.

1.5 Conclusion

This chapter gives an overview of the effects of biocontrol techniques and advanced postharvest technologies (such as microwave heating, high hydrostatic pressure, pulsed electric fields, controlled storage, modified storage, hypobaric storage, and postharvest chemical and nonchemical treatments applied to reduce the microbial load) on maintaining the high nutritional value of the harvested crop. Even though using this kind of postharvest treatment has shown excellent results, the commercial sector has not yet adopted it widely. Achieving a balance between the product's superior quality and the customers' peace of mind is crucial for the product's acceptance by the industry and customers alike. Further research is necessary to have a deeper understanding of and explore the potential uses of cutting-edge technologies on fresh food quality.

References

Aghdam MS, Mostofi Y, Motallebiazar A, Ghasemneghad M et al (2009) Effects of MeSA vapor treatment on the postharvest quality of Hayward kiwifruit. In: 6th International postharvest symposium. Antalya, Turkey

Aguilar GG, Wang CY, Buta JG, Krizek DT (2001) Use of UV-C irradiation to prevent decay & maintain postharvest quality of ripe 'Tommy Atkins' mangoes. Int J Food Sci Technol 36:767–773

Alao SEL (2000) The importance of post-harvest loss prevention. Paper presented at graduation ceremony of school of food storage technology. Nigerian Stored Products Research Institute, Kano, pp 1–10

Anonymous (2021) Indian Horticulture Database, National Horticulture Board, Ministry of Agriculture Government of India

Asghari M, Aghdam MS (2010) Impact of salicylic acid on post-harvest physiology of horticultural crops. Trends Food Sci Technol 21(10):502–509

Atanda SA, Pessu PO, Agoda S, Isong IU, Ikotun I (2011) The concepts and problems of post–harvest food losses in perishable crops. Afr J Food Sci 5(11):603–613

Austin-Brown SL, Chapman KD (2002) Inhibition of phospholipase D α by N-acylethanolamines. Plant Physiol 129:1–7

Babalar M, Asghari M, Talaei A et al (2007) Effect of pre- and postharvest salicylic acid treatment on ethylene production, fungal decay and overall quality of Selva strawberry fruit. Food Chem 105:449–453

Bambalele NL, Mditshwa A, Magwaza LS et al (2021) Recent advances on postharvest technologies of mango fruit: a review. Int J Fruit Sci 21(1):565–586

Bancal V, Ray RC (2022) Overview of food loss and waste in fruits and vegetables: from issue to resources. In: Fruits and vegetable wastes: valorization to bioproducts and platform chemicals. Springer Nature, Singapore, pp 3–29

Barkai-Golan R (2001) Postharvest diseases of fruits and vegetables: development and control. Elsevier

Bowles DJ (1990) Defense-related proteins in higher plants. Annu Rev Biochem 59:873–907

Brody LC, Abel KJ, Castilla LH, Couch FJ, McKinley DR, Yin G, Ho PP, Merajver S, Chandrasekharappa SC, Xu J, Cole JL (1995) Construction of a transcription map surrounding the BRCA1 locus of human chromosome 17. Genomics 25(1):238–247

Cai C, Li X, Chen KS (2006) Acetylsalicylic acid alleviates chilling injury of postharvest loquat (*Eriobotrya japonica* Lindl.) fruit. Eur Food Res Technol 223:533–539

Cao JK, Zeng KF, Jiang WB (2006) Enhancement of postharvest disease resistance in Ya Li pear (*Pyrus bretschneideri*) fruit by salicylic acid sprays on the trees during fruit growth. Eur J Plant Pathol 114:363–370

Chadha KL (2011) Opportunities of horticulture and horti-business in India. Indian J Hortic 68(1):1–5

Chen T, Xing J, El-Mogy M et al (2022) Editorial: nutritional quality formation and maintenance of horticultural crops. Front Plant Sci 13:1005338

Cheng Y, Dong YH, Ge W, Shen C, Guan J et al (2012) Effects of 1-MCP on chlorophyll degradation pathway-associated genes expression and chloroplast ultrastructure during the peel yellowing of Chinese pear fruits in storage. Food Chem 135(2):415–422

Dalrymple DG (1969) The development of an agricultural technology: controlled-atmosphere storage of fruit. Techno and Cult 10(1):35–48

De Capdeville G, Wilson CL, Beer S V Aist JR (2002). Alternative disease control agents induce resistance to blue mold in harvested 'red delicious' apple fruit. Phytopathology, 92(8):900–908

D'hallewin G, Schirra M, Manueddu E, Piga A, Ben-Yehoshua S (1999) Scoparone and scopoletin accumulation and ultraviolet-C induced resistance to postharvest decay in oranges as influenced by harvest date. J Am Soc Hortic Sci 124(6):702–707

Ding ZS, Tian SP, Zheng XL, Zhou ZW (2007) Responses of reactive oxygen metabolism and quality in mango fruit to exogenous oxalic acid or salicylic acid under chilling temperature stress. Physiol Plant 130:112–121

Exton JH (1997) Phospholipase D: enzymology, mechanisms of regulation, and function. Physiol Rev 77:303–320

Fan X, Wang W (2021) Quality of fresh and fresh-cut produce impacted by nonthermal physical technologies intended to enhance microbial safety. Crit Rev Food Sci Nutr 62(2):362–382
FAO (2011) Global food losses and waste: extent causes and prevention. FAO
Galliard T (1980) Degradation of acyl lipids: hydrolytic and oxidative enzymes. In: Stumpf PK, Conn EE (eds) The biochemistry of plants, a comprehensive treatise, vol 4. Academic Press, New York, p 3
García-Pastor ME, Zapata PJ, Castillo S, Martínez-Romero D, Guillén F, Valero D, Serrano M (2020) The effects of salicylic acid and its derivatives on increasing pomegranate fruit quality and bioactive compounds at harvest and during storage. Front Plant Sci 11 668
Glowacz M, Rees D (2016) Using jasmonates and salicylates to reduce losses within the fruit supply chain. Eur Food Res Technol 242:143–156
Gonzales-Aguilar J, Moreno M, Fulcheri L (2007) Carbon nanostructures production by gas-phase plasma processes at atmospheric pressure. Phys D Appl Phys 40:2361–2374
Guan J, Hu M, Shen C, Zhou S, Cheng Y et al (2015) Effects of 1-methylcyclopropene on active composition in fruits. In: Processing and impact on active components in food. Elsevier, pp 133–137
Hayatu M (2000) Post-harvest physiological studies of some selected members of family Solanaceae. Kano, pp 2–25
Hong K, Gong D, Xu HS, Wang S, Jia Z, Chen JL et al (2014) Effects of salicylic acid and nitric oxide pre-treatment on the expression of genes involved in the ethylene signalling pathway and the quality of postharvest mango fruit. N Z J Crop Hortic Sci 42(3):205–216
Hung RH, Liu JH, Lu YM et al (2007) Effect of salicylic acid on the antioxidant system in the pulp of Cara Cara navel orange (*Citrus sinensis* L. Osbeck) at different storage temperatures. Postharvest Biol Technol 47:168–175
Junmatong CB, Faiyue S, Rotarayanont J, Uthaibutra D et al (2015) Cold storage in salicylic acid increases enzymatic and non-enzymatic antioxidants of Nam Dok Mai No. 4 mango fruit. Sci Asia 41:12–21
Kates M (1955) Hydrolysis of lecithin by plant plastid enzymes. Can J Biochem Physiol 33:575–589
Kidd F, West C (1936) Gas-storage of fruit IV.: Cox's Orange pippin apples. J Pomol Hort Sci 14(3):276–294
Klessig DF, Malamy J (1994) The salicylic acid signal in plants. Plant Mol Biol 26:1439–1458
Koo YM, Heo AY, Choi HW (2020) Salicylic acid as a safe plant protector and growth regulator. Plant Pathol J 36:1–10. https://doi.org/10.5423/PPJ.RW.12.2019.0295
Malamy J, Carr JP, Klessig DF (1990) Salicylic acid: a likely endogenous signal in the resistance response of tobacco to viral infection. Science 250:1002–1004
Martínez-Ferrer M, Harper C, Pérez-Muntoz F, Chaparro M (2002) Modified atmosphere packaging of minimally processed mango and pineapple fruits. J Food Sci 67(9):3365–3371
Mo Y, Gong D, Liang G, Han R, Xie J, Li W (2008) Enhanced preservation effects of sugar apple fruits by salicylic acid treatment during postharvest storage. J Sci Food Agric 88:2693–2699
Mohamed IAI (2017) Effects of X-rays, gamma rays and ultra violet light on the quantity and quality of groundnut (*Arachis hypogaea* L.) seed oil. (Doctoral dissertation,. University of Gezira
Morris SW, Vernooij B, Titatarn S et al (1998) Induced resistance responses in maize. Mol Plant Micr Interact 11:643–658
Ndlela S, Mwando NL, Samira A et al (2021) Advances in postharvest disinfestation of fruits and vegetables using hot water treatment technology–updates from Africa. In: Postharvest technology–recent advances, new perspectives and applications. IntechOpen
Paliyath G, Droillard MJ (1992) The mechanisms of membrane deterioration and disassembly during senescence. Plant Physiol Biochem 30:789–812
Paliyath G, Murr DP (2007) Compositions for the preservation of fruits and vegetables. US patent #7,198,811
Paliyath G, Thompson JE (1987) Calcium and calmodulin regulated breakdown of phospholipids by microsomal membranes from bean cotyledons. Plant Physiol 83:63–68

Paliyath G, Pinhero RG, Yada RY et al (1999) Effect of processing conditions on phospholipase D activity of corn kernel subcellular fractions. J Agric Food Chem 47:2579–2588
Paliyath G, Yada RY, Murr DP, et al (2003) Inhibition of phospholipase D. US patent #6,514,914; Canadian patent #2,298,249 issued in 2007
Pandey SK, Goswami TK (2012) Modelling perforated mediated modified atmospheric packaging of capsicum. Int J Food Sci Technol 47(3):556–563
Parfitt J, Barthel M et al (2010) Food waste within food supply chains: quantification and potential for change to 2050. Philos Trans R Soc B Biol Sci 365:3065–3081
Perkins-Veazie P, Collins JK, Howard L (2008) Blueberry fruit response to postharvest application of ultraviolet radiation. Postharvest Biol Technol 47(3):280–285
Prange RK, DeLong JM, Daniels-Lake BJ, Harrison PA (2005) Innovation in controlled atmosphere technology. Stewart Postharvest Review 1(3):1–11
Prasad K, Jacob S, Siddiqui MW (2018) Fruit maturity, harvesting, and quality standards. In: Preharvest modulation of postharvest fruit and vegetable quality. Academic Press, pp 41–69
Razzaq K, Singh Z, Khan A, Khan SAKU, Ullah S (2016) Role of 1-MCP in regulating 'Kensington Pride' mango fruit softening and ripening. Plant Growth Regul 78(3):401–411
Ribeiro C, Alvarenga N (2012) Prospects of UV radiation for application in postharvest technology. Emir J Food Agric 24(6):586–597
Salehi F (2020) Edible coating of fruits and vegetables using natural gums: a review. Int J Fruit Sci 20:570–589
Sani MY, Alao SEL (2006) Assessment of post-harvest fungi of tomato and pepper in selected irrigation areas of Kano State. Int J Res Stud Biosci 1:53–56
Sayyari M, Babalar M, Kalantari S et al (2009) Effect of salicylic acid treatment on reducing chilling injury in stored pomegranates. Postharvest Biol Technol 53:152–154
Srivastava MK, Dwivedi UN (2000) Delayed ripening of banana fruit by salicylic acid. Plant Sci 158:87–96
Stanaway JD, Afshin A, Ashbaugh C, Bisignano C, Brauer M, Ferrara G et al (2022) Health effects associated with vegetable consumption: a burden of proof study. Nat Med 28:2066–2074
Stevens C, Khan VA, Lu JY, Wilson CL, Pusey PL, Kabwe MK, Igwegbe ECK, Chalutz E, Droby S (1998) The germicidal and hormetic effects of UV-C light on reducing brown rot disease and yeast microflora of peaches. Crop Prot 17(1):75–84
Tiwari S, Lata C, Chauhan PS, Prasad V, Prasad M (2017) A functional genomic perspective on drought signaling and its crosstalk with phytohormone-mediated signaling pathways in plants. Curr Genomics 18:469–482. https://doi.org/10.2174/1389202918666170605083319
Valero D, Guillén F, Valverde JM et al (2016) Recent developments of 1-methylcyclopropene (1-MCP) treatments on fruit quality attributes. In: Eco-friendly technology for postharvest produce quality. Academic Press, pp 185–201
Vlot AC, Dempsey DA, Klessig DF (2009) Salicylic acid, a multifaceted hormone to combat disease. Annu Rev Phytopathol 47:177–206
Wang L, Chen S, Kong W et al (2006) Salicylic acid pre-treatment alleviates chilling injury and affects the antioxidant system and heat shock proteins of peaches during cold storage. Postharvest Biol Technol 41:244–251
Warriner K, Huber A, Namvas A (2009) Recent advance in microbial safety of fresh fruits and vegetables. Adv Food Nutr Res 7:155–208
Watkins CB (2006) The use of 1-methylcyclopropene (1-MCP) on fruits and vegetables. Biotechnol Adv 24:389–409
Xu X, Tian S (2008) Salicylic acid alleviated pathogen-induced oxidative stress in harvested sweet cherry fruit. Postharvest Biol Technol 49:379–385
Yahaya SM (2005) Contribution of harvest to pathogenic and nonpathogenic losses of vegetables grown in Kano State-Nigeria. Bayero University, Kano
Yu T, Zheng XD (2006) Salicylic acid enhances biocontrol efficacy of the antagonist *Cryptococcus laurentii* in apple fruit. J Plant Growth Regul 25:166–174

Zhang Y, Chen K, Zhang S, Ferguson I (2003) The role of salicylic acid in postharvest ripening of kiwifruit. Postharvest Biol Technol 28:67–74

Zhang X, Wang H, Li N, Li M, Xu X (2015) High CO2-modified atmosphere packaging for extension of shelf-life of chilled yellow-feather broiler meat: a special breed in Asia. LWT-FOOD Sci Technol 64(2):1123–1129

Chapter 2
Market-Driven Innovative Horticultural Products

Alka Joshi and Bindvi Arora

Abstract The market is always open to receiving new and creative products. To be competitive in the rapidly expanding food market, producers must develop novel goods with distinctive flavors, textures, and shapes. The choice of ingredients is essential because it can either avert or encourage potential issues with the final product. Product development is an art in which it is crucial to comprehend the function of each component and the science underlying how process variables affect finished goods. One, a few, or numerous changes based on food science concepts could improve the product market in the end. Innovation is meaningless if it has no commercial application. The dynamics of a product can be greatly impacted by the addition, substitution, or removal of any element at a minute level. For this reason, the chain of product creation includes the selection of ingredients, additives, unit activities, process duration, endpoint determination, packaging, and storage. For innovation to be guaranteed buyers, it must be economically viable. This chapter discusses several viable, creative concepts and products based on horticultural produce. India has a large population and is a prospective market; thus, there is a great deal of room for creative items in the Indian market.

Keywords Food science · Horticultural products · Biocolorants · Product development

2.1 Introduction

India is renowned for its wide range of products, which includes tropical, subtropical, temperate, and arid fruits and vegetables. India is a fruit and vegetable basket full of wasted produce. According to the data from the Ministry of Agriculture and Farmers Welfare (2019–2020), there were 26.46 million hectares of horticultural

A. Joshi (✉) · B. Arora
Division of Food Science and Postharvest Technology, ICAR-Indian Agricultural Research Institute, New Delhi, Delhi, India

M. Thakur, T. Belwal (eds.), *Advances in Postharvest and Analytical Technology of Horticulture Crops*, https://doi.org/10.1007/978-981-97-7247-6_2

produce cultivated in India. This resulted in the production of 320.77 MT of horticultural produce, or roughly 15% of the world's fruit production and roughly 10% of its vegetable production. According to the data from ANGRAU University's General Agriculture Current Affairs–2023, horticulture produce was produced on 27–28 million hectares in 2021–2023. This yielded 342.33 MT of horticultural produce in 2022, with an estimated 350 MT of horticultural output predicted in 2023. Indians already eat a wide variety of underappreciated fruits and vegetables, including jujube, jamun, wood apple, and phalsa.

2.2 Challenges and Opportunities

Horticultural production has been higher than cereal grain production over the past 2–3 years. The nation produced 315.7 MT of food grains overall in 2021–2022, less than the horticultural yield indicated by the previously mentioned data. Because there are so many difficulties with fruits and vegetables, there are lots of chances for researchers to make goods, packages of methods, and so on for horticulture produce that is special to a particular type. All horticultural output falls under the category of perishable commodities in terms of perishability, with the exception of certain varieties of apples, potatoes, onions, garlic, and other vegetables, which are distinguished by their shorter shelf lives and wide range of uses.

With the exception of vegetable crops, horticulture takes longer to create diversity; therefore, handling, morphological, and sensory constraints must be taken into consideration. It is mostly eaten raw, with little to no processing, or with just slight structural changes. As a result, the general people may receive heat-labile substances including vitamins and antioxidants from horticulture matrices. As the majority of produce grown for horticulture lacks starch, it is less versatile and flexible in terms of product creation than cereal grains such as rice and wheat. However, so many tuber crops are able to overcome this obstacle. Due to the wax coating that covers the entire fruit or vegetable body—particularly those that grow above ground, such as tomatoes, apples, and brinjal—the natural structure of horticulture produce prevents any fortification from adhering to its skin. This presents a potential for the use of or invention of a method that would allow a nutrient, fortifier, or bioactive to integrate naturally into the carrier matrix without requiring genetic modification. By applying an edible covering to entire fresh produce and infusing it with a vacuum-pumped fruit and vegetable matrix, a lot of success in this field (Prajapati et al. 2022).

In this chapter, few concepts of innovative market-driven fruits and vegetable products available in the Asian markets are discussed as under:

2.3 Products and Concepts

2.3.1 Biocolourants

A survey conducted in 2007 by McCann et al. (2007) anticipated the size of the global food color business to be 2.97 billion US dollars by 2025. Certain fruits and vegetables contain concentrated natural colorants that are GRAS-approved and/or have a long history of use among the indigenous people. For instance, in the case of the red pigment betalains, Cacti (*Opuntia ficus-indica*) and beetroot (*Beta vulgaris* L.) have GRAS status if the colorant is removed physically, that is, without the need for chemical modification. However, betalains have a long history of use in Indian culture as food colorings. They are derived from the blooms of Rhododendron species in hilly regions of India and from cockscomb in the Kashmir valley. It is generally advised to employ GRAS ingredients in the food sector in order to thrive both domestically and internationally; in this regard, beetroot betalains present enormous prospects. Beetroot betalain may be extracted using a typical process and is stable throughout a broad pH range (Thippeswamy et al. 2022a). Still, the scientists tweak the method even more. There is evidence to support the use of beetroot betalains or enzyme-assisted extraction for faster and higher yields (Joshi et al. 2023). Black carrots have already seen the optimization of a similar approach. Nonetheless, the makeup of the cell wall and the location of pigment determine the kind of enzyme and the dosages. Generally speaking, the vacuolar portion of the cell contains pigment and antioxidants. The extracted colorant can be stored well in refrigerator conditions through the hot fill pasteurization technique. This colorant can also be converted to powder through spray drying for further use or it can also be utilized in liquid form.

Both these colorants give a red to purple color tinge depending on the pH of the products. They both are the GRAS substances. However, some basic differences occur in both the pigments. Under high salt concentrations, betalains are less stable; however, anthocyanins are more stable under higher salt concentrations and therefore can be used for pickle, kanji, and sauerkraut preparation. Betalains are stable at low pH, but anthocyanins are not. Betalains are compatible with ascorbic acid; however, anthocyanins lose their color stability in the presence of ascorbic acid. Unlike betalains, anthocyanins are thermo and photosensitive.

Besides underutilized produce, wastes can also be utilized for color and bioactive extraction, for example, pea pod pomace can be used to impart pista green color in bakery products (muffins) (Pooja et al. 2022). All biocolorants must be evaluated for a detailed stability study as was done for betalains by Thippeswamy et al. (2022b) to predict potential matrices in which colorant can be used. Thakur and Modi (2024) also did a comprehensive review on the sources, extractions, and applications of the biocolorants in various food products.

2.3.2 *Innovative Scientific Methodologies*

Innovation can come from value addition alone, as well as from the use of established scientific procedures. Thippeswamy et al. (2022c) successfully co-pigmented betalains pigment with black carrot anthocyanins for the first time because both substances are soluble in water. Because anthocyanins from black carrots are GRAS pigments with particularly high photo and thermal stability, their co-pigmentation improves the durability of betalains. They can both be utilized at the same time because scientific testing has shown that they work well in chips and papaya candies. The co-pigmented papaya candy can also be utilized in baked goods because of its thermal resilience. Extruded goods can also employ the co-pigmented colorant.

Two ELISA-based scientific methodologies for phytate and ascorbate estimation have been standardized on the potato matrix (Joshi et al. 2021) using standard methodologies of Haug and Lantzsch (2006) and Ranganna (2007), respectively, by which 30 samples in triplicates can be done at one go which reduces the utilization of chemical, manpower, and time. Similar to this, the following methodologies have also been standardized by the authors:

- Anthocyanins-rich papaya candy: innovative fusion (Joshi et al. 2020).
- Color quantification of food through RGB color cube concept (Joshi et al. 2016a, b).
- Methodology for quick fat estimation from potato chips (Joshi et al. 2015a, b).
- Methodology for preparation of cellulose nanowhiskers as reinforced material for potato starch-based packaging (Raigond et al. 2014).

2.3.3 *Innovation in Existing Technology*

Nitrogen flushing, microwave processing, and vacuum frying are methods for handling potato chips that are typically processed through deep fat frying, which results in a high calorific value and rancidity (Joshi et al. 2015a, b). The technology for developing ready-to-serve beverages currently exists, and small and marginal business owners can use it for very little money. Here, cheap and unused fruit can be stabilized during a surplus. Barren terrain also contains some low-value produce, such as bael pulp, mulberry, jamun, and aonla. By using this low-cost resource, significant profits can be made. It is possible to try vacuum impregnation and infusion to make horticultural produce more nutrient-dense and bioactively condensed. Detailed methodologies on potato chips rich in calcium and zinc and potato fries rich in iron have been standardized and published (Joshi et al. 2016a, b; Tiwari et al. 2022).

2.4 Innovative Concepts for Commercialization

Technology related to capsicum: This creative approach to salsa has already been commercialized by ICAR. Red capsicum is a rich source of lycopene and has an inherent astringency. It can be used in place of tomatoes. Some of the items that are considered distinctive and inventive are beetroot and aonla crackers, mushroom candy, re-constitutable (instant) green leafy vegetable powder, and anthocyanin-rich aonla and papaya candy. These products are not typically seen in stores. Recently, candied citrus peel flavonoid-rich muffins with the name of FiberPro-Antioxidant-rich muffins (Sethi et al. 2022), UV-C treatment for shelf life extension of bitter gourd (Prajapati et al. 2021), and potato flour-based gluten-free cookies (ICAR-CPRI) have been released for commercialization in ICAR Foundation day, 2023. For this FiberPro-Antioxidant-rich muffins, muffins were developed by incorporating soy okara and candied citrus peel at an 85:15 ratio (wheat flour: okara). The functional muffins had 12% higher protein and 146% higher flavonoid content. Optimized products can be stored well up to the ninth day of storage in polypropylene packaging material. These innovations are developed keeping the current industrial and consumer demand in view. Therefore, many vacuum-fried products are available in the market, for example, unexplored produce such as pumpkin crisp technology has been developed and is ready for commercialization by ICAR-IARI. Heat-stable anthocyanin-rich composition and the process of its preparation have already been patented (Patent No. 321722, dated 27.09.2019) and licensed with three industries. Processes for extraction of low methoxyl pectin from jackfruit waste and technology for whole tomato crush are few of the technologies that the industry can take up very easily. Low methoxyl pectin is good for making gel for diabetic people. Whole tomato crush can be developed by concentrating the whole tomato mass to 2.5–3-fold and preserved with 0.5% glacial acetic acid and 200 ppm of sodium benzoate. It retained acceptable quality at 25 ± 5 °C for 1 year and can be utilized when the price of tomatoes is on hike. Germplasm evaluation for product-specific traits and minimal processing for difficult-to-wash, cut, and peel vegetables such as lotus stem are still some underexploited areas where innovative outcomes can be expected.

Except above, some innovative products available already in the marketplace are described below:

2.4.1 *Jellied Fruit*

Fruit juice (without pulp) is cooked with sugar and pectin, a natural gelling agent, to create jellied fruit items. Fruits naturally contain pectin, which aids in thickening the jelly. After cooking, the mixture cools and is poured into molds. As it cools, the jelly solidifies. A clear, jelly-like substance that may be consumed on its own or added to other foods is the end result. The most widely consumed jellied fruit products are made from strawberries, grapes, raspberries, oranges, pineapples, and mangos.

The majority of grocery stores carry these jellies, which are frequently offered in jars or single cups. They can be consumed on their own, spread over toast, or added to pies, cakes, and trifles, among other delicacies. Products made from jellied fruit are a tasty and adaptable cuisine that may be consumed in many different ways. Products made with jammed fruit provide consumers with nutritional advantages because they are low in calories and fat, high in vitamins and minerals, and sometimes high in fiber and antioxidants.

The market for jellied fruit is expanding because of several causes, such as growing consumer knowledge of the health advantages of eating fruits, growing need for quick snacks, rising disposable income, and the growing appeal of Western cuisine in India. The market's leading companies are GD Foods, Orkla ASA, Unilever, JM Smucker Company, Kraft Heinz Company, and Ritter Alimentos SA.

2.4.2 Fruit Yogurt

Yogurt that has been flavored with fruit is called fruit yogurt. Fruit can be diced or pureed, and it can be fresh, frozen, or tinned. Fruit yogurt can be eaten as a snack or dessert in addition to being a popular breakfast meal. In addition to being high in calcium and protein, fruit yogurt has live bacteria called probiotics that are helpful for the gut. A few of the top brands of fruit yogurt are Amul, Mother Dairy, Nestle, and Epigamia. Another new company that offers frozen fruit yogurt is called Cocoberry. India's market for fruit yogurt is expanding quickly. IMARC Group projects that their market will expand at a compound annual growth rate (CAGR) of 23.56% from 2023 to 2028. Numerous factors are contributing to this growth, such as growing consumer awareness of the health benefits of yogurt, especially fruit yogurt, the popularity of flavored yogurts, rising Indian disposable incomes driving up demand for high-end goods like fruit yogurt, and India's increasing urbanization driving up demand for packaged foods like fruit yogurt.

2.4.3 Fruit Puree

Fruits that have been boiled, pounded, pressed, blended, or sieved to the consistency of a creamy paste or liquid are used to make fruit purees. Any fruit can be used to make fruit puree, but the most common ones are strawberries, bananas, mangoes, peaches, and apricots. Fruit puree is a multipurpose ingredient that works well in sauces, baked products, yogurt, baked goods, and desserts. In addition, it is a wonderful source of fiber, vitamins, and minerals. It is a flexible component that works well in many different recipes. India's market for fruit puree is expanding quickly. A report published by Market Research Future projects that the market will expand at a CAGR of 7.5% from 2022 to 2028. A few major companies, including Mother Dairy Fruit & Vegetable Products, Amul, and Dabur India, control the majority of the Indian fruit puree market. These companies are making significant investments in R&D in order to launch new goods and increase their market share. Fruit purees are also a very popular baby food product; Happa Foods, Happy Tot, and Slurrp Farms are just a few of the startups that provide a variety of pureed fruits. Nestle's popular baby food brand Gerber provides a range of pureed fruits and veggies. One organic baby food company, Earth's Best, has a range of pureed fruits, veggies, and cereal mixtures.

2.4.4 *Vegetable Sauce*

One kind of sauce prepared from vegetables is called vegetable sauce. It is a well-liked spice that enhances the taste and nutritional value of many different foods. A wide range of vegetables, such as tomatoes, onions, peppers, carrots, and mushrooms, can be used to make vegetable sauce.

There are several methods for making vegetable sauce. Several well-liked techniques are as follows:

- Sautéing: Using butter or oil, the veggies are cooked in a skillet until they become tender.
- Braise: Using this technique, vegetables are cooked in a liquid (broth or water) until they are soft.
- Roasting: Using this technique, vegetables are cooked in the oven until they are soft and browned.

Vegetables can be mixed until they have a smooth consistency after cooking. You can serve vegetable sauce either hot or cold. It can be added to rice or pasta dishes as a topping, used as a dipping sauce, or used to flavor soups and stews. A tasty and nutritious method to boost taste and nutrition in meals is using vegetable sauce. It is a flexible condiment that works well with many different recipes. It is anticipated that the tomato sauce business will continue to rule India. In addition to ketchup, the local palate favors tomato-flavored pasta, pizza, and other gourmet sauces. In India, there are a small number of very big companies, including Nestle India, Heinz India, and Hindustan Unilever.

In the upcoming years, there will probably be an increased need for vegetable sauces due to the rising appeal of foreign food and shifting nutritional preferences. In India, some common vegetable sauces are as follows:

- Pesto: It is an olive oil, basil, pine nuts, and parmesan cheese sauce. It is frequently used as a vegetable dip or pasta sauce.
- Tomato sauce: It is a mixture of herbs, garlic, onions, and tomatoes. It is a well-liked pasta sauce that doubles as a foundation for other sauces.

- Salsa: It is a mixture of herbs, chili peppers, tomatoes, and onions. This well-liked Mexican sauce is great as a topping for tacos and burritos or as a dip for chips.

2.4.5 *Vegetable Mayonnaise*

One kind of mayonnaise is called vegetable mayonnaise, and it is prepared by adding vegetables to regular mayonnaise. Cucumbers, carrots, and avocados are just a few of the vegetables that can be used to make vegetable mayo. A versatile condiment, vegetable mayo works well in a range of meals, such as salads, dips, wraps, and sandwiches. It can also be added to dishes like stir-fries and sauces as a culinary element. The market leaders in India for vegetable mayonnaise are Wingreens Farms, Veeba Foods, and Cremica Food Industries.

2.4.6 *Osmodried Fruits*

Osmotic dehydration is the method used to produce osmodried fruits. Fruits are soaked in a strong sugar solution for a certain amount of time in this procedure. Fruits get dehydrated during this period as their water content transfers into the sugar solution, prolonging the shelf life of fruits. Horticultural products become more concentrated in nutrients and flavor as a result of osmo-dehydration. Dried fruits make a handy and healthful snack. They are a wonderful source of antioxidants and fiber as well. Vitamin C, magnesium, potassium, and other vitamins and minerals are all present in good amounts in osmodried fruits. They work well with salads, yogurt, and trail mixes as well. The market is divided by type, with mango, apple, grapes, and other products included. The segment leader, mango, accounts for a significant share of the market. This is because there are many different mango kinds available and there is a large supply of mangoes in India. Sunsweet and Hero are a couple of the leading companies in the industry. To obtain a competitive edge in the market, the industry is currently concentrating on growing its distribution network, adding new products to its lineup, and initiating creative marketing initiatives.

2.5 Conclusions

Innovative items with outstanding industry acceptability and product appeal are always needed to be competitive in the market. Product diversification is a safe and sustainable alternative to making a total change. Bio-inspired products and chemicals, such as bio-colorants, bio-preservatives, and bio-antioxidants, have a vast potential to meet the needs of today's health-conscious consumers. Prioritize demand before meeting needs or creating new products. Postharvest practices for fruits and vegetables, such as processing, storing, or packing, can also be used to address gaps in the market.

References

Haug W, Lantzsch HJ (2006) Sensitive method for the determination of phytate in cereals and cereal products. J Sci Food Agric 34:1423

Joshi A, Raigond P, Singh B, Singh BP (2015a) Methodology for quick fat estimation from potato chips. ICAR-CPRI Newsletter. Jan–March 2015

Joshi A, Rudra SG, Sagar VR, Raigond P, Dutt S, Singh B, Singh BP (2015b) Development of low fat potato chips through microwave processing. J Food Sci Technol 53(8):3296–3303

Joshi A, Kar A, Rudra SG, Sagar VR, Varghese E, Lad M, Khan I, Singh B (2016a) Vacuum impregnation: a promising way for mineral fortification in potato porous matrix (potato chips). J Food Sci Technol 53(12):4348–4353

Joshi A, Sethi S, Arora B, Sharma RR, Sharma SK, Sagar VR (2016b) Colour quantification of food through RGB colour cube concept. ICAR News Lett 22(4):2–4

Joshi A, Kaur C, Prakash J (2020) Anthocyanins rich papaya candy: innovative fusion. ICAR News Lett 26(4):7–9

Joshi A, Kaundal B, Raigond P, Singh B, Sethi S, Bhowmik A, Kumar R (2021) Low-volume procedure to determine phytate and ascorbic acid in potatoes: standardization and analysis of Indian cultivars, vol 102. J Food Compos Anal, p 103998

Joshi A, Sethi S, Sindhu N, Tomar BS (2023) Enzyme-assisted extraction of beetroot betalains and its stabilization through copigmentation. In: Proceedings of national conference on underutilized food resources: nutrient composition, value addition and quality assurance organized by Mizoram University, p 28

McCann D, Barrett A, Cooper A et al (2007) Food additives and hyperactive behaviour in 3-year-old and 8/9-year-old children in the community: a randomised, double-blinded, placebo-controlled trial. Lancet 370:1560–1567

Pooja BK, Sethi S, Joshi A, Bhardwaj R, Bhowmik A, Grover M (2022) Investigation of physicochemical quality and textural attributes of muffins incorporated with pea pod powder as a source of dietary fiber and protein. J Food Process Preserv. https://doi.org/10.1111/jfpp.16884

Prajapati U, Asrey R, Varghese E, Singh AK, Singh MP (2021) Effects of postharvest ultraviolet-C treatment on shelf-life and quality of bitter gourd fruit during storage. Food Packag Shelf Life 28:100665

Prajapati U, Ksh V, Joshi A (2022) Extraction and use of bioactive components from underutilized horticultural crops. In: Bioactive components: a sustainable system for good health and well-being. Springer Nature, Singapore, pp 535–570

Raigond P, Raigond B, Kochar T, Singh B, Dutt S, Joshi A, Singh BP (2014) Methodology for preparation of cellulose nanowhiskers as reinforced material for potato starch based packaging published in ICAR-CPRI news letter. October–December 2014

Ranganna S (2007) Handbook of analysis and quality control for fruit and vegetable product, vitamins, 5th edn. New Delhi, Tata McGraw-Hill, India, pp 117–120

Sethi S, Joshi A, Seth K, Bhardwaj R, Yadav A, Grover M (2022) Phytonutrient content, antioxidant potential and acceptability of muffins functionalized with soy and citrus industry waste. J Food Process Preserv. https://doi.org/10.1111/jfpp.16606

Thakur M, Modi VK (2024) Biocolorants in food: sources, extraction, applications and future prospects. Crit Rev Food Sci Nutr 64(14):4674–4713

Thippeswamy B, Joshi A, Sethi S, Arora B, Tomar BS, Varghese E, Ajay Y (2022a) Extraction procedure of betalains pigments from hardy beetroot matrix and its stabilization. J Food Process Preserv. https://doi.org/10.1111/jfpp.16844

Thippeswamy B, Joshi A, Sethi S, Charanjit K, Tomar BS, Varghese E, Dahuja A (2022b) Stability enhancement of beetroot betalains through copigmentation. J Sci Food Agric. https://doi.org/10.1002/jsfa.11939

Thippeswamy B, Joshi A, Sethi S, Dahuja A, Charanjit K, Tomar BS, Varghese E (2022c) Chemical additives for preserving the Betalain pigment and antioxidant activity of red beetroot. Sugar Tech. https://doi.org/10.1007/s12355-021-01104-0

Tiwari P, Thakur M, Joshi A, Raigond P, Arora B (2022) Development of iron fortified potato fries through vacuum assisted processing strategies. J Food Sci Technol. https://doi.org/10.1007/s13197-022-05543-9

Chapter 3
Recent Trends in Postharvest Treatments for Fruits and Vegetables

Ram Asrey, B. R. Vinod, M. Menaka, Sajeel Ahamed, and Amit Kumar

Abstract The fruits and vegetables are called protective food, but their shelf life is very short due to higher water content and profound perishability. This warrants safe and sustainable approaches for treating fruits and vegetables after harvesting from the field. The postharvest treatment of fruits and vegetables has evolved immensely in recent years, driven by the need to enhance quality, extend shelf life, and reduce losses in the supply chain along with the safety of consumers. Green elicitors are being commercialized and involved in postharvest supply chain management, including physical treatments (ultraviolet-C, ozone, electrolyzed water, ultrasonication), plant hormones (salicylic acid, brassinosteroids, jasmonates, melatonin), novel signaling compounds (nitric oxide, polyamines, GABA, L-arginine), organic acids (ascorbic acid, oxalic acid), and edible coatings (chitosan, guar gum, gum arabica, starch, essential oil). The aim of green approaches is to maintain nutritional value, address food safety concerns, and minimize the effect of environmental stresses. The chapter also gives insight into the role of physical, biological, organic acids, and edible coatings in retaining health-benefiting compounds and preserving the freshness of fruits and vegetables. Emphatically, the content focused on various facets of postharvest treatments, their impact on the retention of secondary metabolites, and synergy with storage life and overall quality of the produce.

Keywords Fruits and vegetables · Postharvest treatment · Shelf life · Quality · Physical treatment · Edible coating

R. Asrey (✉) · B. R. Vinod (✉) · M. Menaka · S. Ahamed
Division of Food Science and Postharvest Technology, ICAR-Indian Agricultural Research Institute, New Delhi, Delhi, India

A. Kumar
Department of Fruit Science, Sardar Vallabhbhai Patel University of Agriculture and Technology, Modipuram, Meerut, Uttar Pradesh, India

M. Thakur, T. Belwal (eds.), *Advances in Postharvest and Analytical Technology of Horticulture Crops*, https://doi.org/10.1007/978-981-97-7247-6_3

3.1 Introduction

The postharvest phase of fruits and vegetables plays a pivotal role in determining their shelf life, quality, and nutritional value. With the global demand for fresh produce on the rise, there is an increasing need to enhance postharvest technologies and treatments to ensure that these essential commodities reach consumers in optimal condition. Recent years have witnessed significant advancements in postharvest treatments, ranging from innovative preservation techniques to the development of novel coatings. These breakthroughs have not only addressed the challenges posed by postharvest losses and food waste but also paved the way for a more sustainable and efficient supply chain in the horticulture sector. The importance of postharvest treatments cannot be overstated, as they directly influence the overall economic and nutritional aspects of the fresh produce industry. Fruits and vegetables are highly perishable commodities that are susceptible to various physiological, biochemical, and microbial changes after being harvested. Factors such as moisture loss, enzymatic activity, pathogen invasion, and ethylene production can accelerate deterioration and compromise the quality of these products. Therefore, effective postharvest interventions are critical in preserving the freshness, appearance, and nutritional content of fruits and vegetables, ultimately contributing to food security and minimizing losses. In recent years, researchers and scientists have been diligently exploring innovative methods to enhance the postharvest life of fresh produce and improve their marketability. One of the prominent areas of advancement is the development of physical treatments (ultraviolet-C (UV-C), ozone, electrolyzed water (EW), ultrasonication), plant hormones (salicylic acid (SA), brassinosteroids (BRs), jasmonates, melatonin (MT)), novel signaling compounds (nitric oxide, polyamines (PAs), GABA, L-arginine), organic acids (ascorbic acid (AsA), oxalic acid), and edible coatings (chitosan, guar gum, gum arabica, starch, essential oil). The adoption of physical treatments underscores the industry's commitment to food safety and environmental consciousness. These methods not only extend shelf life but also address concerns related to microbial contamination and pesticide residues, ensuring that consumers can enjoy fruits and vegetables with greater confidence (Meena et al. 2024). Plant hormones and novel signaling compounds have emerged as powerful tools to fine-tune postharvest physiology and enhance the market appeal of produce. By harnessing the regulatory roles of these substances, postharvest researchers are unlocking innovative ways to delay senescence, improve color retention, and enhance antioxidant defenses. These advancements not only impact the longevity of produce but also contribute to a more sustainable supply chain by minimizing losses due to premature spoilage. The integration of organic acids and their roles in preserving postharvest quality, particularly through combating enzymatic browning and oxidative stress, showcases the potential of natural compounds to replace or complement traditional chemical treatments. Overall, these techniques involve altering the metabolism they produce to regulate respiration rates, inhibit microbial growth, and delay ripening processes, leading to improved storage durations and quality retention of fruits and vegetables.

3.2 Melatonin

MT (*N*-acetyl-5-methoxytryptamine) is a versatile signaling chemical found in a wide range of plant species (Ze et al. 2021). It is an indoleamine compound with low molecular weight which was initially identified in *Hypericum monogynum* (Wu et al. 2021a). Phytomelatonin is biosynthesized from tryptophan in four sequential enzymatic processes involving five primary enzymes in the catalytic process. Plants use MT for a variety of physiological functions. MT is a safe and nontoxic molecule that has been investigated extensively in postharvest preservation for its function in extending the shelf life and quality of fruits and vegetables, as well as promoting crop output and maintaining food safety in an environmentally responsible manner. MT also acts as an ethylene antagonist. MT does not directly inhibit ethylene biosynthesis, but it may inhibit ethylene biosynthesis by increasing NO content. Pear fruit treated with MT (0.1 mM) resulted in reduced ethylene biosynthesis, maintained firmness, and downregulated relative expression of the PcACS1, PcACS2, PcACO1, and PcACO2 genes (Zhai et al. 2018). Similarly, postharvest MT application may regulate mango fruit ripening and softening by inhibiting ethylene and ABA biosynthesis (Liu et al. 2020). Exogenous MT has been shown in several trials to greatly delay the postharvest ripening of several fruits, including apples, sweet cherries, bananas, pears, peaches, and mangoes (Tijero et al. 2019). It can also improve quality (Sun et al. 2014), maintain nutritional quality with enhanced total phenols and anthocyanins accumulation (Aghdam and Fard 2017), reduce postharvest browning (Luo et al. 2021), induce chilling tolerance through activation of antioxidant system (Gao et al. 2016b), improve disease resistance, and reduce decay incidence of various fruits such as litchis, strawberries, and kiwifruits during postharvest storage. MT treatment lowers fruit decay, respiration, and ethylene production rates by inhibiting the expression of ethylene synthesis genes such as MdACO1, MdACS1, MdAP2.4, and MdERF109, thereby delaying fruit ripening and senescence (Onik et al. 2021). In addition, fruit softening is also reduced by suppressing the activities of cell wall hydrolyzing enzymes such as pectin methylesterase (PME), polygalacturonase (PG), cellulase, and β-galactosidase (Qu et al. 2022). Exogenous MT treatment lowers the build-up of malondialdehyde (MDA) and hydrogen peroxide (H_2O_2), which is the major product of membrane lipid peroxidation, and reduces oxidative stress (Dong et al. 2021). MT increased levels of total phenolics and flavonoid contents, resulting in higher antioxidant activity. This resulted in enhanced ROS scavenging superoxide dismutase (SOD), peroxidase (POD), and CAT enzyme activities (Rastegar et al. 2020). It can improve disease resistance and prevent deterioration in a variety of crops during the storage period, including bell pepper, litchis, strawberries, and kiwifruits (Ahamad et al. 2024a; Gao et al. 2016a, b; Aghdam and Fard 2017).

MT treatment (600 μmol L^{-1}, 2 h) successfully delayed fruit softening and reduced anthracnose incidence in guava fruit cv. Zhenzhu. It effectively increased antioxidant potential and lowers oxidative stress to the fruit by lowering superoxide anions, H_2O_2, and MDA levels, as well as improving overall antioxidant capacity and enhancing enzymatic and non-enzymatic antioxidants. CAT, SOD, ascorbate peroxidase (APX), and glutathione reductase (GSH) activities were all significantly increased by MT. The defense-related enzymatic activity of chitinase (CHI) and phenylpropanoid pathway enzymes, such as phenylalanine ammonia lyase and 4-coumaric acid-CoA-ligase, was also improved by this treatment. MT inhibited the activities of lipase, LOX, and phospholipase D (PLD), all of which are involved in lipid metabolism (Fan et al. 2022).

3.3 Brassinosteroids

BRs represent a novel group of plant hormones, characterized as natural polyhydroxylated steroidal compounds. The initial discovery of these hormones originated from *Brassica napus* pollen, leading to the naming of the first isolated compound as "brassinolide" (Ali 2017). In 1982, a similar growth substance, resembling brassinolide, was subsequently identified in insect galls of chestnut and was termed "castasterone." Collectively, these two compounds are collectively referred to as BRs.

BRs such as 24-epibrassinolide, brassinolide, and 28-homobrassinolide are frequently employed in various concentrations to modulate both preharvest and postharvest aspects (Aghdam et al. 2015; Nawaz et al. 2017). The exogenous application of BRs exerts multiple effects, including the inhibition of cell wall degradation and softening processes driven by enzymes like pectin esterase and PG (He et al. 2018). BRs also play roles in regulating sugar and energy metabolism (Lu et al. 2019), curbing lipid peroxidation, maintaining robust membrane integrity (Aghdam and Mohammadkhani 2013), overseeing ripening and color metabolism (Tavallali 2018), restraining peroxidase and polyphenol oxidase (PPO) coenzymes, ultimately leading to a delay in enzymatic browning (Gao et al. 2015), and reducing the production of reactive oxygen species (ROS) (O^-, H_2O_2) (Gao et al. 2016a, b), significantly impacting on a variety of physiological responses throughout the postharvest lifespan, thereby contributing to the maintenance of quality in diverse fruits and vegetables.

BRs have emerged as effective agents in mitigating chilling injury (CI) in fruits by increasing membrane integrity through the reduction of PLD and lipoxygenase (LOX) activities while simultaneously elevating the unsaturated fatty acid/saturated fatty acid ratio (Menaka et al. 2024a; Ahamad et al. 2024b). Furthermore, BRs contribute to disease prevention by fortifying defense mechanisms and inducing systemic acquired resistance. Their impact on reducing fruit decay is attributed to enhanced disease resistance and delayed senescence, rather than direct fungal toxicity, as noted by Champa et al. (2015). In the context of fruit ripening, BRs facilitate the softening process by augmenting the activity of softening enzymes such as PME and PG, along with modulating water-soluble pectin and protopectin content. Additionally, BRs elevate calcium content in the fruit pericarp, leading to a significant reduction in fruit cracking (Ekinci et al. 2018). Li et al. (2018) demonstrated that applying EBR at a concentration of 40 μM L^{-1} for 10 min at 25 °C effectively curbed CI in harvested bananas during a 12-day cold storage at 8 °C, leading to a reduced CI index. Treated fruits exhibited elevated chlorophyll fluorescence, total soluble solids (TSS), and TSS-to-titratable acidity ratio, along with diminished electrolyte leakage and MDA levels. In another context, during ambient storage, a concentration of 5 μM L^{-1} of EBR for 10 min prolonged the postharvest senescence of kiwifruit cv. Huayou, as noted by Lu et al. (2019). EBR's capacity to uphold membrane lipid integrity, suppress starch conversion, and disrupt the futile cycle of sucrose synthesis and degradation may underlie its role in regulating senescence. In treated kiwifruit, firmness loss was decelerated, while weight loss and TSS content increased.

3.4 γ-Aminobutyric Acid

γ-Aminobutyric acid (GABA) is a non-proteinaceous free amino acid that is prevalent across diverse organisms, spanning prokaryotes to eukaryotes (Bown et al. 2006). In plants, GABA serves as an endogenous component, participating in functions such as pH regulation, nitrogen storage, and plant development while affording protection against various abiotic stresses such as drought, oxidative stress, salinity, and cold conditions (Shelp et al. 2012). The application of exogenous GABA has shown promise in increasing disease resistance and preservation of postharvest fruits and vegetables. This includes maximized activity of enzymes such as POD, SOD, and CAT, leading to delayed pear browning (Li et al. 2019). Moreover, exogenously applied GABA has demonstrated efficacy in inducing resistance against *Alternaria alternata*-induced black spots in tomato fruit (Yang et al. 2017).

The augmented GABA activity exhibits a more pronounced effect in cultivars resilient to chilling, indicating that GABA catabolism revitalizes the tricarboxylic acid cycle, yielding reducing equivalents and energy that potentially counteracts oxidative damage. In grapes treated with 40 mM GABA, there was an elevation in phenolic compounds and antioxidant capacity, attributed to the suppression of PPO activity. Consequently, GABA emerges as a valuable option to uphold internal quality, mitigate fungal decay, and alleviate CI in grapes during the postharvest phase (Asgarian et al. 2022). GABA treatment effectively curbed the rate of weight loss and decay in postharvest Nanfeng tangerine fruit. Although the respiration rate of Nanfeng tangerine peaked at 80 days of storage, GABA treatment curtailed this peak, resulting in reduced nutrient consumption through respiratory metabolism (Chen et al. 2021). In "Valencia" sweet orange, GABA application led to elevated endogenous GABA levels, select amino acids, and various phytohormones, including auxins, salicylates, jasmonic acid (JA), and abscisic acid (Hijaz et al. 2018). On a molecular level, the GABA shunt functions as a signaling molecule, accumulating within the cytosol to serve as a crucial scavenger of ROS. This pivotal role in ROS scavenging contributes to the plant's defense against oxidative damage (Ansari et al. 2021). Notably, when applied as a preharvest treatment at a concentration of 50 mM, GABA exhibited its highest effectiveness in lemon fruit. This treatment not only extended shelf life but also improved overall quality by enhancing antioxidant activity, which effectively counteracted ROS activity (Badiche-El Hilali et al. 2023).

3.5 L-Arginine

L-arginine is an amino acid with versatile functions and is considered essential due to its role in synthesizing important signaling molecules such as NO and PAs. This unique property enables L-arginine to enhance stress resistance in plants (Khan et al. 2023). The accumulation of these signaling molecules, facilitated by L-arginine, proves beneficial in mitigating chilling damage and extending the shelf life of horticultural products (Sohail et al. 2021; Pakkish and Mohammadrezakhani 2022). In the context of postharvest applications, immersion of harvested produce in L-arginine solutions has shown promising results. This treatment has been found to effectively prevent browning and increase the storage duration of lettuce and oranges (Wills and Li 2016). Additionally, L-arginine has demonstrated its ability to impede disease development in tomatoes (Zhang et al. 2017), delay senescence in broccoli (Sohail et al. 2021), and counteract CI in pomegranates (Babalar et al. 2018). These findings highlight the potential of L-arginine as a valuable tool for enhancing the postharvest quality and longevity of various horticultural products. In mushrooms and persimmon fruit, L-arginine treatment has been shown to reduce chilling damage, resulting in decreased accumulation of MDA and reduced electrolyte leakage (Li et al. 2019; Khan et al. 2023). This indicates that pre-storage treatments with L-arginine could offer a safe and commercially viable strategy, given its GRAS status, to address postharvest decay, mitigate chilling stress, and enhance the nutraceutical qualities of horticultural items, ultimately extending their shelf life.

L-arginine has been the subject of extensive study due to its positive impact on enhancing cold tolerance in postharvest crops. For instance, Babalar et al. (2018) discovered that arginine treatment at a concentration of 1 mM led to heightened enzymatic and non-enzymatic antioxidant capacities in pomegranate fruit. This enhancement in antioxidant defenses contributed to improved cold tolerance and a reduction in the occurrence of CI. Similarly, arginine has demonstrated its ability to alleviate CI in papaya (Huang et al. 2021) and cucumber fruits (Hou et al. 2023). Zhang et al. (2017) conducted a study on tomato fruit to investigate the combined effect of L-arginine (1 mM) and methyl salicylate (0.05 mM). The results demonstrated that the synergistic treatment not only heightened the activities of important antioxidant enzymes such as SOD, CAT, and POD but also stimulated the expression of pathogenesis-related protein and the activities of defense-related enzymes such as phenylalanine ammonia lyase (PAL), PPO, β-1,3-glucanase, and CHI. These effects persisted over most of the storage period and were associated with a notable reduction in the incidence of gray mold in treated tomato fruits. Additionally, the combined treatment led to an increase in the accumulation of PAs, total phenolics, and NO, contributing to enhanced fruit quality and disease resistance. Shu et al. (2020) conducted research on postharvest strawberries and found that treatment with arginine (1 mM) triggered the NO synthesis pathway, leading to elevated production of NO. This increase in NO production served as a signaling molecule, contributing to improved fruit quality and the suppression of decay. In a similar vein, arginine treatment was shown to effectively mitigate fruit decay and enhance the overall quality of blueberries. This improvement was reflected in reduced weight loss, improved firmness, and higher soluble solid content. Mechanistically, arginine-mediated activation of the antioxidative defense system played a role in reducing oxidative damage caused by ROS.

3.6 Cold Plasma

Plasma is often described as the fourth state of matter, characterized by ionized gas containing free electrons, ions, reactive atoms, neutral molecules, and photons in a metastable state with a near-zero net electrical charge (Misra et al. 2011). Low-temperature plasma can be categorized into thermal and nonthermal (cold) plasma based on thermodynamic equilibrium. Various electrical discharges such as corona discharge, glow discharge, radiofrequency discharge, dielectric barrier discharge (DBD), pulsed corona discharge, microwave discharge, and plasma jet are used to generate nonthermal plasma or cold plasma. Cold plasma has demonstrated effectiveness in pathogen inactivation on the surface of fresh produce. While cold plasma generates free radicals, these reactive species can lead to rapid oxidation, potentially causing lipid deterioration and the breakdown of natural antioxidants, which may negatively impact the taste and aroma of food (Sonawane et al. 2020). The potential of cold plasma to extend the shelf life of fruit and vegetables while preserving their quality is promising. However, it is important to note that the reactive species present in cold plasma can interact with various components of fruits, leading to potential effects on both the physicochemical and sensory attributes of different fresh produce (Muhammad et al. 2018).

Cold plasma has demonstrated effective inactivation of aerobic bacteria and molds on blueberries, leading to a reduction in decay rate during storage. Furthermore, cold plasma treatment has been shown to increase antioxidant content and enhance antioxidant enzyme activity in blueberries, contributing to the maintenance of lower levels of ROS in the fruit. In terms of postharvest preservation, the optimal exposure time for blueberry sterilization was found to be 60 s when generating cold plasma at 12 kV and 5 kHz (Ji et al. 2020). For short-term cold plasma treatment (≤ 15 min), minor effects were observed on factors such as firmness, pH, oxidation-reduction potential, and anthocyanin concentration, while improvements were noted in color and a decrease in lipid peroxidation, thus positively impacting postharvest quality. However, a 20-min cold plasma treatment led to significant oxidative damage to blueberry peels, resulting in softening of the fruit and a decrease in anthocyanin content (Hu et al. 2021).

Zhou et al. (2019) demonstrated that cold plasma exposure effectively suppressed the growth of Botrytis cinerea, leading to improved preservation of blueberries. Another study explored the potential of DBD cold plasma treatment for decontaminating strawberries within sealed packages. A 5-min DBD cold plasma treatment reduced surface microflora on strawberries by up to 3.3 log cycles without significant impacts on color, firmness, or respiration rate (Misra et al. 2014). For fresh wolfberries, optimal DBD pre-treatment conditions were identified as 13.64 kV, 70 s, and 2.7 kHz. This treatment extended the shelf life of wolfberries from 2 to 5 days at room temperature, mitigating decay, weight loss, and firmness reduction by inactivating microorganisms and inhibiting respiration. Additionally, this approach preserved high levels of polysaccharides, titratable acid, and carotenoids while inhibiting the decline of phenols, flavonoids, AsA, and antioxidant activity during storage (Du et al. 2024).

3.7 Ultrasonication

Ultrasound (US) is a type of energy produced through sound waves at frequencies exceeding the audible range for humans (above 16 kHz). Frequencies of 20–50 kHz are commonly used, as they facilitate cavitation more effectively (Mason 1990). Its potential has been recognized in the food industry since the 1970s and recently, accepted as an effective treatment to destroy harmful microorganisms in fruits and vegetables owing to its cavitation effect. The formed cavity grows and collapses generating localized mechanical (generation of turbulence, compression, rarefaction, and shearing effect) and chemical energy (production of free radicals), and physical effects include extreme temperature and pressure in and around the produce (Yuting et al. 2013). Raising the intensity of US in a probe system has shown increasing antimicrobial efficacy, yet excessive intensities were linked to detrimental impacts on fruit quality during storage. Many previous works have suggested that US could enhance defense mechanisms, antioxidant activities, secondary metabolites biosynthesis, control postharvest quality, and induce pathogenic resistance in fruits and vegetables (Chen et al. 2008).

US-driven cavitation plays a crucial role in breaking down contaminant molecules, aided by the collapse of cavitation bubbles, resulting in the breakdown of both contaminant and water molecules. This process generates reactive OH radicals with potent oxidation capabilities against organic pollutants (Seymour et al. 2002). Utilizing ultrasonic treatment for reducing microbial load in strawberries has been found to maintain fruit firmness while enhancing antioxidant activity. Gani et al. (2016) reported noteworthy reductions of 2.0 and 1.2 logs in bacterial counts and yeast/mold, respectively. Notably, color, firmness, vitamin C content, and antioxidant activity were also well-preserved. However, it is worth noting that prolonged exposure times beyond 40 min at 33 kHz and 60 W did not extend shelf life further, potentially causing damage to strawberry samples. Another investigation demonstrated that ultrasonic treatment at 25 kHz and 26 W L^{-1} induced effects akin to those of abiotic stress, resembling the responses triggered by wounding, in Romaine lettuce (Yu et al. 2015). A plausible response mechanism was proposed to explain these effects. The treated lettuce exhibited notable improvements in antioxidant activity, increased firmness, delayed enzymatic browning, and the absence of deterioration during storage. The combination of mild heat with ultrasonic treatment has been investigated, revealing a more significant impact of ultrasonication (35 kHz, 120 W, and 21.4 W/L) at 65 °C on red bell pepper, watercress, and strawberry. This combined approach resulted in improved retention of quality and notable reductions of 7.43, 8.24, and 8.10 log CFU g^{-1} for *L. innocua*, total mesophiles, and total coliforms, respectively (Alexandre et al. 2011).

3.8 UV Radiation

UV treatment is widely used to control the decay of fresh produce and minimally processed commodities. UV radiation (100–400 nm) has been classified as UV-A (long waves, 400–315 nm), UV-B (medium waves, 315–280 nm), and UV-C (short waves, 280–100 nm). UV-C at 254 nm being more energetic and possessing the highest germicidal activity is used for surface disinfection and has low penetration into solids (Vinod et al. 2023). Even the application of low or sub-lethal doses of UV-C light induces disease resistance in the host (this phenomenon is also called "hormesis") (Gómez-López et al. 2013). Mechanisms include pyrimidine dimers formation (DNA damage), disruption of cytoplasmic membrane integrity and cellular enzyme activity, increase in the level of antioxidants (α-tocopherol, β-carotene, and AsA) and phytoalexins (scoparone and scopoletin), and stimulation of biosynthesis of phenols and defense-related enzymes such as phenylalanine ammonia-lyase (Gómez-López et al. 2013).

Razali et al. (2015) reported that firmness, TSS, titratable acidity, and antioxidant capacity (AsA, total phenolic content, and DPPH activity) were proven to increase significantly with longer exposure to UV-C treatment (60 min). Furthermore, the microbial count of guava fruit exposed to UV-C light was reduced as it resulted in the disruption of membrane depolarization and abnormal ionic flow of protein as well as introduced the set of mutations in the bacterial DNA. Treatment with UV-C for a duration of 40 min has demonstrated notable benefits in enhancing antioxidant activity during a 16-day storage period (Prajapati et al. 2021). This treatment effectively inhibited DPPH by 57%, resulting in a 61% increase in total phenols, a remarkable 134% increase in total carotenoid content, and a substantial elevation of total chlorophyll by 61.9%, when compared with the control group. Furthermore, the UV-C treatment maintained fruit firmness, minimized weight loss, and decreased the percentage of fruit decay. Notably, prolonged exposure to UV-C seemed to intensify abiotic stress in bitter gourd fruit, leading to the synthesis of defense-related compounds like MDA and proline. These compounds collaborated to mitigate stress effects and sustain higher levels of secondary metabolites (Prajapati et al. 2021). In a study by Zhang et al. (2021), exposing nectarine fruit to UV-C treatment at 3 kJ m^{-2} resulted in reduced lesion size and disease index caused by *Rhizopus stolonifer* infection. This treatment bolstered the activity of vital antioxidant enzymes such as SOD, CAT, and APX, along with elevating levels of AsA and GSH. Furthermore, UV-C treatment triggered the activation of the phenylpropane metabolic pathway and anthocyanin biosynthesis, leading to heightened concentrations of phenolic compounds, total flavonoids, anthocyanins, and lignin. Importantly, the treatment also enhanced the activities of CHI and GLU. These findings emphasize UV-C treatment's potential in fortifying the defense mechanisms and phytochemical composition of nectarine fruit against disease while enhancing its antioxidant profile (Menaka et al. 2024b).

3.9 Electrolyzed Water

EW is produced through the electrolysis of neutral salt solutions in an electrolytic cell, resulting in either acidic or alkaline solutions. This process involves a cathode and an anode within the electrolytic cell, often separated by a diaphragm. When using NaCl as the electrolyte, EW exhibits strong disinfection and cleaning properties, making it a commonly used form of EW (Rahman et al. 2016). The ability of EW to hinder the activity of enzymes that degrade cell walls may be linked to its capacity to reduce ethylene production. Ethylene is a significant regulator in the ripening and senescence processes of postharvest fruits, influencing various metabolic pathways associated with maturation and aging. Consequently, the activity of enzymes that break down cell walls is directly influenced by ethylene levels (Zhang and Jiang 2019).

EW has gained significant traction in the field of horticultural technology owing to its eco-friendliness, cost-effectiveness, and user-friendly application (Rahman et al. 2016). Initially utilized as a cleansing and sanitizing agent, EW has demonstrated remarkable potential in ensuring food safety, with the capability to swiftly eliminate various microorganisms such as viruses, bacteria, and fungi (Rahman et al. 2016). Recent research highlights the multifaceted benefits of EW treatment beyond mere microbial control in postharvest fruits and vegetables. Notably, it has been shown to enhance the overall quality of stored fruits and vegetables by retarding senescence, managing diseases, and mitigating CI (Aday 2016; Hussien et al. 2017; Shi et al. 2020). Additionally, studies have indicated that EW treatment can effectively eliminate pesticide residues from fresh F and V (Qi et al. 2018). Given these advancements, EW treatment holds great promise as a highly effective and promising method for the preservation of postharvest fruits and vegetables.Slightly acidic EW (SAEW) treatment has exhibited its effectiveness in suppressing the proliferation of *Escherichia coli* O157:H7 and *Listeria monocytogenes* on various fresh fruits such as apples, mandarins, and tomatoes when employed on an industrial scale (Chen et al. 2019). Additionally, the use of SAEW treatment has displayed a significant decrease in the population of Salmonella on lettuce. The duration of exposure was directly related to the improved efficacy of sterilization. Notably, a reduction of 2.73 Log CFU g^{-1} was evident following a 30-s treatment, while a reduction of 4.06 Log CFU g^{-1} was evident after 45 s of treatment (Cap et al. 2020). Numerous investigations have highlighted the potential of EW in mitigating the rate of respiration and the production of ethylene in postharvest fruits and vegetables. The application of acidic EW (AEW) treatment on longan fruit has been found to enhance the integrity and diminish permeability of cell membranes and consequently bolster resistance against fruit diseases.

This treatment approach also exhibited the capacity to impede the development of browning in stored longan fruit (Chen et al. 2020b). The observed inhibition of postharvest fruit respiration rate during storage through EW treatment is thought to be linked to its suppressive effect on microorganisms. In the case of climacteric fruits like peaches, the rapid aging process that follows the peak of respiration can be significantly slowed down with appropriate treatments (Zhang et al. 2019). The study conducted by Chen et al. (2019) examined the impact of AEW treatments on the storage quality and metabolism of ROS in blueberries during their storage period. The findings of this study indicated that AEW treatments contributed to the preservation of higher levels of SOD, CAT, and APX activities, thereby enhancing the antioxidant capacity of blueberries throughout the storage period. Similarly, the application of EW to satsuma orange fruit yielded beneficial outcomes. This treatment exhibited the ability to mitigate CI, regulate cell membrane permeability, and reduce MDA content, which is an indicator of oxidative stress. Notably, the results also demonstrated that EW treatment led to an increase in the activity of various antioxidant enzymes of satsuma orange fruits, including SOD, POD, and CAT (Shi et al. 2020).

3.10 Ozonation

Ozone (O_3) is naturally formed from oxygen as a result of lightning or UV light interaction, which undergoes quick auto-decomposition into molecular oxygen leaving no chemical residues. In 1997, ozone was designated as a GRAS disinfectant and was licensed as an antimicrobial agent in the treatment, storage, processing, and transportation of meats and horticulture goods by the US FDA in 2001 (Vinod et al. 2023). Ozonation is done in either a liquid or gaseous form which inactivates enzymes and microorganisms, oxidative destruction of hazardous pollutants, eradicates mycotoxins (Ong et al. 1996), enhances antioxidant capacity (Ali et al. 2014), and delays fruit softening, thereby extending the shelf life of the produce (Salvador et al. 2007). Ozone that is a strong oxidant acts upon the cell surface and oxidizes numerous cellular components, such as sulfhydryl groups of amino acids, unsaturated lipids and respiratory enzymes in the cell membrane, peptidoglycans in the cell envelope, enzymes and nucleic acids in the cytoplasm, as well as protein and peptidoglycans in spore mantles and virus capsids (Priyanka and Rastogi 2014; Sukarminah et al. 2017). Ozone treatment has been shown to effectively suppress ethylene production in fruits. Ethylene biosynthesis enzymes, such as ACC synthase and ACC oxidase, are downregulated by ozone treatment, resulting in reduced ethylene production (Vinod et al. 2024). This is mostly due to the ozone's strong oxidative capacity (Arrebola 2015).

Ozonated water has proven to be effective in reducing pesticide residues, as well as in suppressing respiration rates and decreasing weight losses in bell peppers. Moreover, it has been observed that ozonated water helps to maintain the green color of bell peppers and extends their overall storage life (Özen et al. 2020). In the case of peaches, treatment with ozone at a concentration of 200 ppb led to a substantial reduction in decay incidence caused by *Monilinia* sp. during cold storage. The effectiveness of this ozone treatment ranged from a reduction of 10% to as much as 70% in decay incidence (D'Ortenzio et al. 2023). Exposure to ozone at a concentration of 0.5 ppm for 30 min effectively suppressed the respiration rate and delayed sensory quality losses in pomegranates stored at 6 °C. This suppression of metabolic activities contributed to the preservation of pomegranate quality during storage (Buluc and Koyuncu 2020). Similarly, a study by Chen et al. (2020a, b) found that cantaloupe subjected to ozone treatment at a concentration of 15 mg m^{-3} exhibited several favorable outcomes. These included lower respiration rates, reduced ethylene production rates, and decreased decay incidence. Additionally, the ozone-treated cantaloupe demonstrated the retention of higher levels of firmness, pectin content, reducing sugars, and titratable acidity.

3.11 Salicylic Acid

Salicylic acid (SA) (2-hydroxybenzoic acid, $C_7H_6O_3$), a ubiquitous phenylpropanoid compound, holds a pivotal role as an endogenous signaling molecule with robust metabolic and physiological influence on horticultural crop growth and development. SA biosynthesis in plants occurs through discrete enzymatic pathways: either from chorismate via isochorismate (chloroplast-localized) or from phenylalanine (cytoplasm-localized) (Wildermuth et al. 2001; Leon et al. 1993). SA has recently gained traction in postharvest technologies for various horticultural crops, earning the status of generally recognized as safe (GRAS). SA and its derivative, acetylsalicylic acid, have demonstrated their ability to curb ethylene biosynthesis in crops such as pear, apple, and others, suggesting SA's role as an ethylene antagonist. Inhibition of ethylene biosynthesis and delay in fruit ripening is attributed to the suppression of both amino cyclopropane-1-carboxylic acid synthase (ACS) and oxidase (ACO) production and activity during the early stage of fruit ripening (Loay and Taher 2018). Additionally, SA has been linked to diminished lipid oxidation, membrane senescence, reduced CI, and browning reactions, enhanced firmness, anthocyanin synthesis, and antioxidant capacity. Moreover, SA application preharvest or postharvest has exhibited efficacy against fungal decay by invoking defense responses and enhancing antioxidant enzymes (Kazemi et al. 2011; Sayyari et al.

2009; Wei et al. 2011; Babalar et al. 2018; Yao and Tian 2005; Chan and Tian 2006; Xu and Tian 2008).

Postharvest application of SA or 5-sulfosalicylic acid at 2 mM concentrations for 5 min on guava fruits cv. Allahabad Safeda and stored at 10 °C for 12 days demonstrated minimal weight loss and color change while maintaining maximum firmness, total phenolics, and AsA content. In comparison to untreated samples, SA-treated fruits showed a delay in respiration and ethylene production, as well as a minimal rise in soluble solids content (Madhav et al. 2016). Postharvest treatment with 0.3% SA was found to effectively inhibit the activity of cell wall degrading enzymes and their corresponding gene expression. This treatment led to the preservation of higher levels of protopectin, cellulose, and hemicelluloses while reducing water-soluble protein content. As a result, the structural integrity of the cell wall was maintained, contributing to the alleviation of fruit senescence and the extension of storage life for "Jiashayou" pummelo fruit (Chen et al. 2023). Zhang et al. (2022) studied the combined effect of blue light (BL) and SA (2 mM) on strawberry postharvest quality during cold storage. The use of SA led to a delay in weight loss, prevention of decay, enhancement in the brightness of the fruit skin, and an increase in the amount of soluble proteins present. Additionally, the levels of total flavonoids, phenolics, anthocyanins, and proanthocyanidins, as well as the activities of SOD and APX, remained stable in strawberries treated with both BL and SA throughout the storage period.

Another investigation by Jiang et al. (2022) explored the impact of exogenous SA (1.0 mmol L^{-1} solution for 5 min) on the organoleptic quality and organic acid metabolism of rabbiteye blueberries during cold storage (4 °C). The application of SA led to reduced fruit decay and weight loss, delayed fruit softening, and a slower decline in TSS. The levels of titratable acidity and total organic acids remained consistent during the late storage period in SA-treated blueberries.

3.12 Nitric Oxide

Nitric oxide (NO) is a gaseous free radical serving as a pivotal plant intrinsic signal regulator. Its high reactivity leads to the formation of other oxides (NO_2, N_2O_3, and N_2O_4) when exposed to atmospheric oxygen. The application of a low concentration of NO gas in postharvest treatments for fruits and vegetables is associated with the inhibition of ethylene biosynthesis, effectively delaying fruit ripening and senescence (Mukherjee 2019). Notably, NO's direct inhibition of ethylene is not its primary mechanism; it's suggested that an ACO-NO complex formation occurs, which then interacts with ACC, yielding a stable ACC-ACO-NO complex that directly curbs ethylene biosynthesis (Manjunatha et al. 2010). Furthermore, NO can directly react with ACO through S-nitrosylation, leading to reduced ACO activity and subsequently lowering overall ethylene production (Manjunatha et al. 2010).

Prior investigations have asserted that NO has demonstrated utility in decreasing respiration rate, inhibiting ethylene production, reducing enzyme activity, mitigating disease occurrence, and altering skin coloration in fruits (Palma et al. 2019). NO offers a promising avenue for enhancing texture retention during storage, as evidenced by Zhu et al. (2010). It facilitates improved preservation of cellular constituents, including pigments, titratable acidity, and antioxidants, by mitigating cellular membrane disintegration and reducing electrolyte leakage (Palma et al. 2019). Cold storage-related issues like CI, fresh browning, and translucency are also alleviated, as observed by Singh et al. (2009). Furthermore, NO displays high efficacy in inducing resistance against systemic infections and safeguarding against microbial pathogens like *Penicillium italicum, Rhizopus nigricans, Aspergillus niger*, and *Monilinia fructicola*. Consequently, treated fruits experience extended postharvest shelf life. By preserving cellular compartmentation, NO prevents direct enzyme-substrate interaction, thereby curbing enzyme activity, such as PPO. This novel application of NO underscores its potential for eco-friendly enhancement of fruit shelf life, marking a progressive step in postharvest research.

Sahu et al. (2020) observed sodium nitroprusside (SNP) treatment (1.0 mM for 5 min) on guava fruit delayed unwanted quality changes and extended the storage life up to 12 days at ambient storage conditions (20 °C). This treatment slowed the degradation of AsA, chlorophyll, phenolics, and flavonoids while minimizing physical quality deterioration (weight loss, decay). The presence of high antioxidant properties in treated fruits apparently helped to delay the onset of senescence. The application of the exogenous SNP has been investigated as a potential method to enhance disease resistance and extend the shelf life of postharvest muskmelon (Wang et al. 2021). In this study, SNP treatment was found to induce lignin production, promote wound healing, and enhance the activity of PAL and POD enzymes. The most effective concentration of NO was observed at 500 $\mu L\ L^{-1}$, which significantly inhibited rachis browning in postharvest table grapes (Wu et al. 2021a, b). This treatment delayed water loss, reduced phenol accumulation, suppressed PPO activity, and stimulated POD activity. As a result, it improved rachis freshness and extended the shelf life of the grapes. Additionally, the application of exogenous SNP at concentrations of 0.025, 0.05, and 0.1 mM, was shown to effectively reduce postharvest browning in mushrooms (Gao et al. 2023). This treatment helped maintain the antioxidant capacity of the mushrooms, preserved their hardness, minimized weight loss, and prevented enzymatic browning. These effects were achieved by downregulating the gene expression of browning-related enzymes, leading to reduced enzyme activity and delayed browning. Overall, the use of exogenous NO holds promise as a strategy to enhance the quality and extend the shelf life of various postharvest fruits and vegetables while addressing issues related to enzymatic browning and deterioration.

3.13 Polyamines

PAs represent a group of small aliphatic amines with low molecular weight, widely present in all plants. They play a pivotal role in various biological processes, encompassing plant growth, development, and stress reaction, as noted by Smith (1985).

Within the realm of plants, these molecules are intricately involved in growth, development, and responses to abiotic stressors. Among the prevalent PAs are Putrescine, Spermidine, and Spermine, ubiquitously present in plant cells at varying concentrations, from micromolar to millimolar. Their metabolism is closely governed by enzymatic regulations, heavily influenced by environmental factors, particularly stress conditions.

PAs have emerged as a novel class of plant growth regulators, contributing significantly to a wide array of physiological and biological processes within fruits and vegetables. Notably, free-form PAs have demonstrated anti-senescent properties, as evidenced by studies involving both internal and external applications (Valero et al. 1998). In postharvest fruit management, the impact of PAs is substantial, encompassing the inhibition of ethylene biosynthesis, enhancement of fruit firmness, reduction in respiration rate, mitigation of CI, attenuation of color alterations, and mitigation of mechanical damage. Both preharvest and postharvest treatments with PAs have been demonstrated to suppress ethylene biosynthesis and associated responses across various systems. This regulation contributes to enhanced fruit firmness through the inhibition of cell wall enzymes. Sobolev et al. (2014) and Pandey et al. (2014) have highlighted the reciprocal modulation of PAs and ethylene biosynthesis, potentially through direct or indirect competition for S-adenosylmethionine (SAM), a shared precursor for their synthesis. The application of PAs has been observed to downregulate the expression of key enzymes such as SAM decarboxylase, ACS, and ACO, thereby leading to a delay in fruit ripening (Khan et al. 2007).

Numerous research studies have highlighted various positive effects of interventions, including downregulation of ethylene biosynthetic enzymes, enhanced stability of cell membranes, increased antioxidant capacity, reduced occurrence of CI and physiological disorders, as well as delayed senescence processes. To extend the postharvest shelf-life of green bell pepper (*Capsicum annuum* L.), an exogenous application of two PAs, SPD and PUT, was conducted by Patel et al. (2019) at a concentration of 20 μM SPD + 20 μM PUT under storage at 4 ± 1 °C for 40 days. Over the storage period, both treated and untreated fruits exhibited gradual decreases in titratable acidity, protein content, CAT and POD activities, chlorophyll, and capsaicin content. Conversely, the proline content and antioxidant activity of the peppers showed a continuous increase as the storage duration extended. In a study by Kowaleguet et al. (2022), "Eureka" lemon fruits treated with a combination of PAs (0.5 mM/L PUT +0.5 mM L^{-1} SPD) demonstrated increased levels of TSS and titratable acidity, along with elevated activities of POD and PPO. Additionally, PAs led to a noticeable reduction in electrolyte leakage and showed less evidence of CI during cold storage at −2 °C for 45 days. Li et al. (2021) observed that the application of exogenous PAs effectively mitigated the initiation and progression of pericarp browning in pear fruit. This treatment resulted in elevated PA levels and improved the expression and activity of enzymes associated with PAs synthesis while concurrently reducing ROS levels. Furthermore, the exogenous PA treatment helped maintain the structural integrity and energy levels of pericarp mitochondria.

3.14 Jasmonates

Jasmonates, a class of intrinsic plant growth regulators, offer distinctive and potentially valuable attributes that impact plant growth and development under environmental stresses. The compounds methyl jasmonates (MeJA) and JA, which include its free acid form, are collectively referred to as jasmonates. Plants synthesize JA through the octadecanoid pathway, involving the activation of membrane lipids by phospholipase to release α-linolenic acid (Seo et al. 2001). MeJA, initially identified in *Jasminium grandiflorum* flower extracts for its pleasant scent (Demole et al. 1962), has emerged with significant effects on postharvest fruit and vegetable management. US FDA has approved this compound as a GRAS substance. Thus, its use may have full potential for enhancing the synthesis of bioactive compounds, inducing resistance against CI, and prolonging the shelf-life (Per et al. 2018). In addition, it induces oxidative stress and accumulation of secondary metabolites in plant cells (Ho et al. 2020). MeJA activates the defense system of fruit as related to increasing PAL and LOX activity and accumulation of total phenols (Gonzalez-Aguilar et al. 2004). It also increases the antioxidant capacity by stimulating ROS scavenging APX, SOD, glutathione peroxidase, ascorbate dehydroreductase, guaiacol peroxidase, GSH, and monodehydro-ascorbate reductase enzyme activities (Chanjirakul et al. 2006). Exogenous application of MeJA has been shown to trigger the release of volatiles and alter lipid and carbohydrate levels in plants (Zalewski et al. 2019). Studies have also indicated that exogenous MeJA can bolster plants' natural resistance to disease and cope with abiotic stressors, including cold (Singh et al. 2019). The postharvest application of MeJA introduces shifts in volatile profiles, enhances natural resistance against pathogenic fungi in harvested crops, and boosts the expression of heat shock proteins and C-repeat binding factors (Aghdam and Bodbodak 2013).

In a study by Cai et al. (2020), MeJA was observed to stimulate the release of peach aroma-related lactones and carotenoids, thereby enriching sensory attributes like taste and odor in postharvest fruit. Similarly, in ripe pears, MeJA induced the biosynthesis of aromatic volatiles, as reported by Qin et al. (2017). Treating olives with MeJA after harvest led to increased phenolic acids concentration, reduced levels of saturated fatty acids, and elevated content of beneficial unsaturated fatty acids such as oleic, linoleic, and linolenic acids (Flores et al. 2016). In the context of postharvest storage, jasmonates, including MeJA, play a vital role in mitigating CI in fruits and vegetables through diverse mechanisms. These mechanisms include enhancing membrane integrity, enhancing antioxidant systems, and influencing PAL and PPO activities, as noted by Dhami et al. (2023). Treating mandarin with MeJA at concentrations of 1, 2, and 3 nmol/L (via dipping) and storing them at 5–7 °C led to reduced weight loss, spoilage, and minimized firmness and juice content losses (Baswal et al. 2020). In tomatoes, treating with MeJA at 50 μg L^{-1} and storing at 2 °C resulted in elevated proline content, increased lycopene levels, and enhanced antioxidant capacity (Min et al. 2018).

3.15 Ascorbic Acid

AsA protects the cell membrane from lipid peroxidation and in conjunction with calcium salts and organic acids limits the activity of PPO and prevents browning and membrane disintegration (Oms-Oliu et al. 2010). AsA-based derivatives have mostly been employed as anti-oxidants, anti-browning agents, and anti-bacterial agents (Sogvar et al. 2016). Many studies have demonstrated that using the right amount of AsA to treat horticulture produce can extend their shelf life. Antibacterial properties have also been observed in freshly sliced banana (Yurdugül 2016), apple (Qi et al. 2011), and papaya (Tapia et al. 2008).

Fruit deterioration percentage, PWL, pH, and sugar/acid ratio were significantly lowered in AsA (200 ppm) treated Guava fruit cv. Golla (25 °C for 12 days) compared with control fruit. Furthermore, the soluble solid content, total acidity, total sugar, vitamin C, and overall phenolic content displayed elevated levels, suggesting an enhanced fruit quality. Moreover, an augmentation in the activities of SOD, POD, and CAT was observed (Azam et al. 2021). The impact of postharvest dipping in 2% AsA and 1% procyanidin (PN), either individually or combined, was investigated on the quality of "Sensation" mangoes during ripening under ambient conditions. The treated fruit exhibited better retention of green color, firmness, titratable acidity, and membrane stability index while showing lower levels of TSS, TSS/TA ratio, and weight loss compared with the control throughout the ripening process (Al-Qurashi et al. 2023). Furthermore, Wang et al. (2023) assessed the effect of a chitosan-based edible coating containing 1% AsA combined with 1% chitosan (CH) on *Annona squamosa* fruit stored at 15 ± 1 °C. The combined CH + AA coating effectively reduced firmness loss, visual deterioration, weight loss, and the accumulation of MDA content over the storage period. This composite coating also played a role in maintaining higher titratable acidity, moisture content, and antioxidant enzyme activities, which contributed to the preservation of quality. Additionally, the CH + AA coating mitigated fruit softening by regulating the activities of enzymes responsible for cell wall degradation and influencing their gene expression levels.

3.16 Oxalic Acid

It is a natural organic acid that plays a key role in systemic defense and environmental restraint and main regulator of calcium concentrations in plants (Zheng et al. 2012; Jin et al. 2014). The application of oxalic acid is a safe and promising strategy for preserving fruit quality and extending storage life through antioxidant and anti-browning activities (Zheng and Tian 2006). It plays a significant role in systemic strength, programmed cell death, homeostasis in plants, and anti-senescence in harvested fruits (Wu et al. 2011). Oxalic acid reduces fruit softening and exo-PG enzyme activity while increasing antioxidative enzyme activity (SOD, CAT, and POD). On the contrary, it also inhibits PPO activity, which minimizes non-enzymatic browning (Zheng and Tian 2006). Antimicrobial activity is linked to upregulation of several stress- and defense-related proteins (Wang et al. 2008), enhanced defense-related enzyme activity, phenolic compound buildup, and a delay in fruit ripening (Zheng et al. 2012).

Oxalic acid treatment has been demonstrated to enhance the antioxidant compounds (such as phenolics, anthocyanins, and carotenoids) as well as antioxidant enzymes, delay ethylene production, and inhibit CI in plums compared with control fruit during a 50-day period of postharvest cold storage (Martínez-Esplá et al. 2018). Compared with the control, postharvest oxalic acid (6 mM/L) treatments to "Maamoura" guava fruits dramatically reduced weight loss and decay percentage throughout cold storage (7 °C and 90% RH) period and shelf life. Furthermore, enhanced fruit firmness, soluble solid content, SSC/acid ratio, AsA concentration, and decreased titratable acidity were also observed (El-Gawad 2021). In the case of lotus roots, slices treated with oxalic acid (10 mmol L^{-1}) and then stored at 20 ± 1 °C for 5 days exhibited reduced browning, lower levels of superoxide anion, H_2O_2, electrolyte leakage, and MDA content. These treated slices also displayed improved visual quality, higher AsA, and total phenolic contents, in contrast to the untreated slices (Ali et al. 2020). Furthermore, Ali et al. (2021) investigated the integrated effect of AsA (40 mmol L^{-1}) and oxalic acid (2 mmol L^{-1}) on the discoloration of litchi fruit under both controlled atmosphere conditions of 5% CO_2 + 1% O_2 and at cold storage of 5 ± 1 °C for 28 days. The combined use of AsA and oxalic acid effectively inhibited the soluble quinones, the browning index, and the activity of PPO. Additionally, this treatment reduced weight loss and maintained higher levels of anthocyanins, total phenolics, membrane integrity, and the activities of antioxidant enzymes such as APX, CAT, GSH, and SOD.

3.17 Surface Coatings

Edible coating, a thin and uniform layer applied through spraying or dipping in accordance with good manufacturing practices (de Vasconcellos Santos Batista et al. 2019), is both colorless and odorless. Comprising hydrocolloids, polysaccharides (like chitosan, starch, cellulose, alginate, gums, and gelatin), proteins, lipids (essential oils and waxes), as well as nanoparticles and their composites (Sanchís et al. 2016), it serves as a versatile tool. Notably, edible coatings significantly contribute to preserving postharvest quality and improve the storage life of fresh fruits. Beyond this, they fulfill roles as antimicrobial agents, antioxidants, texture enhancers, dyes, flavor enhancers, nutritional supplements, spices, surfactants, emulsifiers, and plasticizers. Moreover, their strong adhesion to product surfaces, especially when combined with other materials, further expands their applications (Avramescu et al. 2020). Edible coatings commonly used in fruits and vegetables are listed in Table 3.1.

The establishment of an additional layer around the fruit effectively seals the stomata, resulting in decreased transpiration and respiration rates. This process

Table 3.1 Specific coating applications for various fruits and vegetables

Coating material	Composition	Fruit/vegetable	Effect of coating
Semperfresh	High concentrations of short-chain unsaturated fatty acid esters in sucrose esters, sodium compounds of CMC, and mixed mono- and diglycerides	Apple, tomato	Decreased color changes, preserved acid, lengthened shelf life, and preserved quality
		Banana	Delayed chlorophyll loss and reduced ethylene synthesis
		Zucchini	Less water loss and internal CO_2
Prolong	Mixture of sodium CMC, mono and diglycerides, and sugar fatty acid esters	Mango	Slows down ripening, weight loss, chlorophyll loss
		Pear	Retaining firmness, maintaining green epidermis, and maintaining titratable acidity
Tal prolong	Mixture of sucrose fatty acid esters, sodium CMC, and mono and diglycerides	Mango	Delayed ripening with extended shelf life
Nature-seal	Cellulose-based edible coating	Mango and tomato	Delayed ripening
		Carrot, pome fruit	Retard discoloration and carotene loss, and is a barrier to oxygen diffusion
Chitosan	Chitosan and tween 80	Strawberry, cucumber, and bell pepper	Antimicrobial coating
		Apple, pear, plum, and peach	Act as a gas barrier
		Litchi	Reduced weight loss and browning, improve storability
		Carrot	Reduced decay and improve appearance
		Tomato	Reduced respiration rate, ethylene production, and increased tithable acidity
		Pear	Low ethylene production and delayed ripening are both suggested by a high titratable acidity content

(continued)

Table 3.1 (continued)

Coating material	Composition	Fruit/vegetable	Effect of coating
Zein	Corn zein protein	Tomato	Extended shelf life, delayed color change, loss of firmness and weight
Nutri-save	NO carboxyl methyl and chitosan	Apple, pear, pomegranate	Increased shelf life
Brilloshine	Sucrose esters and wax (shine	Apple, avocado, melons and citrus	Protected, increased shine, and extended shelf life
Nu-coatFlo, Ban-seel	Sucrose esters of fatty acids and sodium salt of CMC	Apple, banana, cucumber, guava, melon, pear, and plum	Extended shelf life
Casein	Calcium caseinate	Bell pepper	Gas barrier that effectively prevents interior O_2 and CO_2, inhibits color change, and slows decay
FreshSeal	Polyvinyl alcohol, starch, and surfactant	Fruits	Extended shelf life

generates a modified atmosphere that serves as a semi-permeable barrier, hindering the movement of gases (O_2, CO_2) (Parven et al. 2020), moisture (Alali et al. 2018), and volatile compounds (Kumari et al. 2017). Consequently, this mechanism acts to impede both ripening and senescence. Moreover, this technique proves beneficial in mitigating adverse chemical reactions, reducing deterioration, preventing microbial contamination, and controlling a range of postharvest diseases (Hashemi and Khaneghah 2017; Thakur et al. 2018).

Loay and Taher (2018) studied the efficacy of chitosan/polyvinylpyrrolidone (CS/PVP) supplemented with SA at varied concentrations of 0, 1, and 2 mM on different maturity stages of guava fruits. Fruit coated with CS/PVP-SA 2 mM reduced water loss and browning incidence, which retained the color hue angle and firmness of the fruit at all stages of maturation. Furthermore, the presence of SA reduced cell wall degradation enzyme activities, resulting in a decrease in cell membrane permeability. By suppressing both enzymes PPO and PAL and preserving total phenol concentration in fruit tissue, browning symptoms are minimized in guava fruits. The chitosan 2% (w/v) and Rue essential emulsion induced a reduction of 97% of mycelial growth of *C. gloeosporioides* on "Maradol" papaya during storage for 12 days at 20 °C and 88% RH (Peralta-Ruiz et al. 2020). An investigation was carried out by Monzón-Ortega et al. (2018) on papaya involving composite films with *Aloe vera*, chitosan (15 g L^{-1}) along with essential oils (each 10 mL L^{-1}) of cinnamon and thyme. The study recorded greater firmness, lower respiration rate, and minimum internal color change during storage at room temperature (30 ± 2 °C and 80% RH). Furthermore, these combined coatings reduced the incidence and severity of fungal diseases by 40–50% as compared with control. In a study conducted by Bhan et al. (2022), it was demonstrated that a composite coating consisting of Guar gum and tamarind had a positive impact on the postharvest shelf life of Kinnow mandarin fruit at ambient conditions for 25 days. Coating led to a significant reduction in mass loss, a decrease in decay occurrence, and a slowed respiration rate. Additionally, the activity of enzymes responsible for fruit softening, such as PME and LOX, was noticeably suppressed by the composite coating. Notably, treated fruits exhibited lower TSS and preserved fruit firmness, acidity, AsA content, total phenolic compounds, and antioxidant activity, resulting in an enhanced sensory experience.

3.18 Conclusion

Recent advances in postharvest treatments for fruits and vegetables encompass a diverse array of strategies, each contributing to prolonged shelf life, enhanced quality, and reduced postharvest losses. In the face of global shifts toward sustainability and heightened consumer expectations, the insights gained from recent advancements will continue to guide the future of postharvest research and practice. By embracing these innovations and fostering interdisciplinary collaborations, we have the opportunity to shape a more resilient and efficient fresh produce supply chain, ultimately benefiting producers, distributors, and consumers alike. As we move forward, it is imperative that researchers, policymakers, and industry stakeholders continue to invest in and support these advancements, ensuring that the journey from farm to table is marked by freshness, quality, and a commitment to a more sustainable food future.

References

Aday MS (2016) Application of electrolyzed water for improving postharvest quality of mushroom. LWT Food Sci Technol 68:44–51. https://doi.org/10.1016/j.lwt.2015.12.014

Aghdam MS, Bodbodak S (2013) Physiological and biochemical mechanisms regulating chilling tolerance in fruits and vegetables under postharvest salicylates and jasmonates treatments. Sci Hortic 156:73–85. https://doi.org/10.1016/j.scienta.2013.03.028

Aghdam MS, Fard JR (2017) Melatonin treatment attenuates postharvest decay and maintains nutritional quality of strawberry fruits (*Fragaria×anannasa* cv. Selva) by enhancing GABA shunt activity. Food Chem 221:1650–1657. https://doi.org/10.1016/j.foodchem.2016.10.123

Aghdam MS, Mohammadkhani N (2013) Enhancement of chilling stress tolerance of tomato fruit by postharvest brassinolide treatment. Food Bioprocess Technol 7:909–914. https://doi.org/10.1007/s11947-013-1165-x

Aghdam MS, Naderi R, Sarcheshmeh MAA, Babalar M (2015) Amelioration of postharvest chilling injury in anthurium cut flowers by γ-aminobutyric acid (GABA) treatments. Postharvest Biol Technol 110:70–76. https://doi.org/10.1016/j.postharvbio.2015.06.020

Ahamad S, Asrey R, Singh AK et al (2024a) Melatonin treatment enhances bioactive compound retention, antioxidant activity and shelf-life of bell pepper (*Capsicum annuum* L.) during cold storage. Int J Food Sci Technol. https://doi.org/10.1111/ijfs.17154

Ahamad S, Asrey R, Vinod BR et al (2024b) Maintaining postharvest quality and enhancing shelf-life of bell pepper (*Capsicum annuum* L.) using brassinosteroids: A novel approach. S Afr J Bot 169:402–412. https://doi.org/10.1016/j.sajb.2024.04.049

Alali AA, Awad MA, Al-Qurashi AD, Mohamed SA (2018) Postharvest gum Arabic and salicylic acid dipping affect quality and biochemical changes of 'Grand Nain' bananas during shelf life. Sci Hortic 237:51–58. https://doi.org/10.1016/j.scienta.2018.03.061

Alexandre EMC, Santos-Pedro DM, Brandao TRS, Silva CLM (2011) Study on thermosonication and ultraviolet radiation processes as an alternative to blanching for some fruits and vegetables. Food Bioprocess Technol 4:1012–1019. https://doi.org/10.1007/s11947-011-0540-8

Ali B (2017) Practical applications of brassinosteroids in horticulture—some field perspectives. Sci Hortic 225:15–21. https://doi.org/10.1016/j.scienta.2017.06.051

Ali A, Ong MK, Forney CF (2014) Effect of ozone pre-conditioning on quality and antioxidant capacity of papaya fruit during ambient storage. Food Chem 142:19–26. https://doi.org/10.1016/j.foodchem.2013.07.039

Ali S, Khan AS, Anjum MA et al (2020) Effect of postharvest oxalic acid application on enzymatic browning and quality of lotus (*Nelumbo nucifera* Gaertn.) root slices. Food Chem 312:126051. https://doi.org/10.1016/j.foodchem.2019.126051

Ali S, Khan AS, Malik AU et al (2021) Combined application of ascorbic and oxalic acids delays postharvest browning of litchi fruits under controlled atmosphere conditions. Food Chem 350:129277. https://doi.org/10.1016/j.foodchem.2021.129277

Al-Qurashi A, Awad M, Ali M (2023) Exogenous ascorbic acid and procyanidin application affect quality of sensation mangoes during ripening. JAPS 33. https://doi.org/10.36899/japs.2023.3.0651

Ansari MI, Jalil SU, Ansari SA, Hasanuzzaman M (2021) GABA shunt: a key-player in mitigation of ROS during stress. Plant Growth Reg 94:131–149. https://doi.org/10.1007/s10725-021-00710-y

Arrebola EVA (2015) Advances in postharvest diseases management in fruits. In: Siddiqui MW (ed) postharvest biology and technology of horticultural crops, principles and practices for quality maintenance. Apple Academic Press, p 268

Asgarian ZS, Karimi R, Ghabooli M, Maleki M (2022) Biochemical changes and quality characterization of cold-stored 'Sahebi' grape in response to postharvest application of GABA. Food Chem 373:131401. https://doi.org/10.1016/j.foodchem.2021.131401

Avramescu SM, Butean C, Popa CV et al (2020) Edible and Functionalized Films/Coatings—Performances and Perspectives. Coatings 10:687. https://doi.org/10.3390/coatings10070687

Azam M, Hameed L, Qadri R et al (2021) Postharvest ascorbic acid application maintained physiological and antioxidant responses of Guava (*Psidium guajava* L.) at ambient storage. Food Sci Technol 41:748–754. https://doi.org/10.1590/fst.19820

Babalar M, Pirzad F, Sarcheshmeh MAA et al (2018) Arginine treatment attenuates chilling injury of pomegranate fruit during cold storage by enhancing antioxidant system activity. Postharvest Biol Technol 137:31–37. https://doi.org/10.1016/j.postharvbio.2017.11.012

Badiche-El Hilali F, Valverde JM, Diaz-Mula H et al (2023) Potential preharvest application of γ-aminobutyric acid (GABA) on improving quality of 'Verna' lemon at harvest and during storage. Agriculture 13:1397. https://doi.org/10.3390/agriculture13071397

Baswal AK, Dhaliwal HS, Singh Z et al (2020) Postharvest application of methyl jasmonate, 1-methylcyclopropene and salicylic acid extends the cold storage life and maintain the quality of 'Kinnow' mandarin (*Citrus nobilis* L. X C. *deliciosa* L.) fruit. Postharvest Biol Technol 161:111064. https://doi.org/10.1016/j.postharvbio.2019.111064

Bhan C, Asrey R, Meena NK et al (2022) Guar gum and chitosan-based composite edible coating extends the shelf life and preserves the bioactive compounds in stored Kinnow fruits. Int J Biol Macromol 222:2922–2935. https://doi.org/10.1016/j.ijbiomac.2022.10.068

Bown AW, MacGregor KB, Shelp BJ (2006) Gamma-aminobutyrate: defense against invertebrate pests? Trends Plant Sci 11:424–427. https://doi.org/10.1016/j.tplants.2006.07.002

Buluc O, Koyuncu MA (2020) Effects of intermittent ozone treatment on postharvest quality and storage life of pomegranate Ozone Sci Eng 43:427–435. https://doi.org/10.1080/01919512.2020.1816449

Cai H, Han S, Yu M et al (2020) The alleviation of methyl jasmonate on loss of aroma lactones correlated with ethylene biosynthesis in peaches. J Food Sci 85:2389–2397. https://doi.org/10.1111/1750-3841.15339

Cap M, Rojas D, Fernandez M et al (2020) Effectiveness of short exposure times to electrolyzed water in reducing Salmonella spp and Imidacloprid in lettuce. LWT 128:109496. https://doi.org/10.1016/j.lwt.2020.109496

Champa WAH, Gill MIS, Mahajan BVC et al (2015) Brassinosteroids improve quality of table grapes (Vitis vinifera L.) cv. flame seedless. Trop Agric Res 26:368. https://doi.org/10.4038/tar.v26i2.8099

Chan Z, Tian S (2006) Induction of H_2O_2-metabolizing enzymes and total protein synthesis by antagonistic yeast and salicylic acid in harvested sweet cherry fruit. Postharvest Biol Technol 39:314–320. https://doi.org/10.1016/j.postharvbio.2005.10.009

Chanjirakul K, Wang SY, Wang CY, Siriphanich J (2006) Effect of natural volatile compounds on antioxidant capacity and antioxidant enzymes in raspberries. Postharvest Biol Technol 40:106–115. https://doi.org/10.1016/j.postharvbio.2006.01.004

Chen B, Huang J, Wang J, Huang L (2008) Ultrasound effects on the antioxidative defense systems of Porphyridium cruentum. Colloids Surf B Biointerfaces 61:88–92. https://doi.org/10.1016/j.colsurfb.2007.07.009

Chen X, Tango D et al (2019) Disinfection efficacy of slightly acidic electrolyzed water combined with chemical treatments on fresh fruits at the industrial scale. FoodReview 8:497. https://doi.org/10.3390/foods8100497

Chen C, Zhang H, Zhang X et al (2020a) The effect of different doses of ozone treatments on the postharvest quality and biodiversity of cantaloupes. Postharvest Biol Technol 163:111124. https://doi.org/10.1016/j.postharvbio.2020.111124

Chen Y, Xie H, Tang J et al (2020b) Effects of acidic electrolyzed water treatment on storability, quality attributes and nutritive properties of longan fruit during storage. Food Chem 320:126641. https://doi.org/10.1016/j.foodchem.2020.126641

Chen X, Zhang Y, Luo Z, Yang X et al (2021) Effect of GABA treatment on fruit quality and preservation of nanfeng tangerine. In: The 8th international horticulture research conference, Nanjing

Chen C, Huang Q, Peng X et al (2023) Alleviatory effects of salicylic acid on postharvest softening and cell wall degradation of 'Jinshayou' pummelo (*Citrus maxima* Merr.): a comparative physiological and transcriptomic analysis. Food Chem 424:136428. https://doi.org/10.1016/j.foodchem.2023.136428

De Vasconcellos Santos Batista D, Reis RC, Almeida JM et al (2019) Edible coatings in postharvest papaya: impact on physical–chemical and sensory characteristics. J Food Sci Technol 57:274–281. https://doi.org/10.1007/s13197-019-04057-1

D'Ortenzio AL, Makau SM, Mancini V et al (2023) Effects of ozone treatment on postharvest decay of peach fruits in cold storage. Acta Hortic 1363:207–212. https://doi.org/10.17660/actahortic.2023.1363.31

Demole E, Lederer E, Mercier D (1962) Isolement et determination de la structure du jasmonate de méthyle, constituant odorant caractéristique de l'essence de jasmin. Helv Chim Acta 45:675–685. https://doi.org/10.1002/hlca.19620450233

Dhami KS, Asrey R, Vinod BR, Meena NK (2023) Postharvest methyl jasmonate alleviates chilling injury and maintains quality of 'kinnow' (*Citrus nobilis* Lour x C. *deliciosa* Tenora) fruits under differential storage temperature. Erwerbs-Obstbau 65(5):1–8. https://doi.org/10.1007/s10341-023-00880-1

Dong J, Kebbeh M, Yan R et al (2021) Melatonin treatment delays ripening in mangoes associated with maintaining the membrane integrity of fruit exocarp during postharvest. Plant Physiol Biochem 169:22–28. https://doi.org/10.1016/j.plaphy.2021.10.038

Du Y, Mi S, Wang H et al (2024) Intervention mechanisms of cold plasma pretreatment on the quality, antioxidants and reactive oxygen metabolism of fresh wolfberries during storage. Food Chem 431:137106. https://doi.org/10.1016/j.foodchem.2023.137106

Ekinci N, Gokbayrak Z, Cavuşoglu S, Akcay ME (2018) Influence of postharvest application of 28-homobrassinolide on storage quality of medlar fruit. Erwerbs-obstbau 61:113–118. https://doi.org/10.1007/s10341-018-0402-x

El-Gawad MGA (2021) Influence of propolis extract and oxalic acid on preserving quality of guava fruits during postharvest cold storage. Plant Arch 21:127–138. https://doi.org/10.51470/plantarchives.2021.v21.s1.024

Fan S, Xiong T, Lei Q et al (2022) Melatonin treatment improves postharvest preservation and resistance of guava fruit (*Psidium guajava* L.). Food 11:262. https://doi.org/10.3390/foods11030262

Flores G, Blanch GP, del Castillo MLR (2016) Effect of postharvest methyl jasmonate treatment on fatty acid composition and phenolic acid content in olive fruits during storage. J Sci Food Agric 97:2767–2772. https://doi.org/10.1002/jsfa.8104

Gani A, Baba WN, Ahmad M et al (2016) Effect of ultrasound treatment on physico-chemical, nutraceutical and microbial quality of strawberry. LWT 66:496–502. https://doi.org/10.1016/j.lwt.2015.10.067

Gao H, Kang L, Liu Q et al (2015) Effect of 24-epibrassinolide treatment on the metabolism of eggplant fruits in relation to development of pulp browning under chilling stress. J Food Sci Technol 52(6):3394–3401. https://doi.org/10.1007/s13197-014-1402-y

Gao H, Zhang Z, Lv X et al (2016a) Effect of 24-epibrassinolide on chilling injury of peach fruit in relation to phenolic and proline metabolisms. Postharvest Biol Technol 111:390–397. https://doi.org/10.1016/j.postharvbio.2015.07.031

Gao H, Zhang ZK, Chai HK et al (2016b) Melatonin treatment delays postharvest senescence and regulates reactive oxygen species metabolism in peach fruit. Postharvest Biol Technol 118:103–110. https://doi.org/10.1016/j.postharvbio.2016.03.006

Gao X, Wu W, Chen H et al (2023) Nitric oxide treatment delays quality deterioration and enzymatic browning of Agaricus bisporus via reactive oxygen metabolism regulation. Food Front 4:447–458. https://doi.org/10.1002/fft2.212

Gómez-López VM, Marín A, Medina-Martínez MS et al (2013) Generation of trihalomethanes with chlorine-based sanitizers and impact on microbial, nutritional and sensory quality of baby spinach. Postharvest Biol Technol 85:210–217. https://doi.org/10.1016/j.postharvbio.2013.05.012

Gonzalez-Aguilar GA, Tiznado-Hernandez ME, Zavaleta-Gatica R, Martnez-Tellez MA (2004) Methyl jasmonate treatments reduce chilling injury and activate the defense response of guava fruits. Biochem Biophys Res Commun 313:694–701. https://doi.org/10.1016/j.bbrc.2003.11.165

Hashemi SMB, Khaneghah AM (2017) Characterization of novel basil-seed gum active edible films and coatings containing oregano essential oil. Prog Org Coat 110:35–41. https://doi.org/10.1016/j.porgcoat.2017.04.041

He Y, Li J, Ban Q et al (2018) Role of Brassinosteroids in persimmon (*Diospyros kaki* L.) fruit ripening. J Agric Food Chem 66:2637–2644. https://doi.org/10.1021/acs.jafc.7b06117

Hijaz F, Nehela Y, Killiny N (2018) Application of gamma-aminobutyric acid increased the level of phytohormones in *Citrus sinensis*. Planta 248:909–918. https://doi.org/10.1007/s00425-018-2947-1

Ho T-T, Murthy HN, Park S-Y (2020) Methyl jasmonate induced oxidative stress and accumulation of secondary metabolites in plant cell and organ cultures. Int J Mol Sci 21:716. https://doi.org/10.3390/ijms21030716

Hou Y, Liu L, Liu Y et al (2023) N--Lauroyl-L-arginine ethyl ester hydrochloride combined with hot water treatment alleviates chilling injury of postharvest cucumber fruit. Sci Hortic 315:111986. https://doi.org/10.1016/j.scienta.2023.111986

Hu X, Sun H, Yang X et al (2021) Potential use of atmospheric cold plasma for postharvest preservation of blueberries. Postharvest Biol Technol 179:111564. https://doi.org/10.1016/j.postharvbio.2021.111564

Huang Q, Song H, Pan Y, Zhang Z (2021) Exogenous arginine enhances the chilling tolerance in postharvest papaya fruit by regulating arginine and proline metabolism. J Food Process Preserv 45:e15821. https://doi.org/10.1111/jfpp.15821

Hussien A, Ahmed Y, Al-Essawy A-H, Youssef K (2017) Evaluation of different salt-amended electrolysed water to control postharvest moulds of citrus. Trop Plant Pathol 43:10–20. https://doi.org/10.1007/s40858-017-0179-8

Ji Y, Hu W, Liao J et al (2020) Effect of atmospheric cold plasma treatment on antioxidant activities and reactive oxygen species production in postharvest blueberries during storage. J Sci Food Agric 100:5586–5595. https://doi.org/10.1002/jsfa.10611

Jiang B, Fang X, Fu D et al (2022) Exogenous salicylic acid regulates organic acids metabolism in postharvest blueberry fruit. Front Plant Sci 13:1024909. https://doi.org/10.3389/fpls.2022.1024909

Jin P, Zhu H, Wang L et al (2014) Oxalic acid alleviates chilling injury in peach fruit by regulating energy metabolism and fatty acid contents. Food Chem 161:87–93. https://doi.org/10.1016/j.foodchem.2014.03.103

Kazemi M, Hadavi E, Hekmati J (2011) Role of salicylic acid in decreases of membrane senescence in cut carnation flowers. Am J Plant Physiol 6:106–112. https://doi.org/10.3923/ajpp.2011.106.112

Khan AS, Singh Z, Abbasi NA (2007) Pre-storage putrescine application suppresses ethylene biosynthesis and retards fruit softening during low temperature storage in 'Angelino' plum. Postharvest Biol Technol 46:36–46. https://doi.org/10.1016/j.postharvbio.2007.03.018

Khan MA, Azam M, Ahmad S, Atiq M (2023) Improvement of physicochemicals, antioxidant system and softening enzymes by postharvest L-arginine application leads to maintain persimmon fruit quality under low temperature storage. J Food Meas Charact 17:2964–2977. https://doi.org/10.1007/s11694-023-01835-6

Kowaleguet MGGM, Chen F, Shi W et al (2022) Exogenous polyamines alleviate chilling injury of citrus limon fruit. Int Food Res J 29:698–706. https://doi.org/10.47836/ifrj.29.3.21

Kumari M, Mahajan H, Joshi R, Gupta M (2017) Development and structural characterization of edible films for improving fruit quality. Food Packag Shelf Life 12:42–50. https://doi.org/10.1016/j.fpsl.2017.02.003

Leon J, Yalpani N, Raskin I, Lawton MA (1993) induction of benzoic acid 2-hydroxylase in virus-inoculated tobacco. Plant Physiol 103:323–328. https://doi.org/10.1104/pp.103.2.323

Li T, Yun Z, Wu Q et al (2018) Proteomic profiling of 24-epibrassinolide-induced chilling tolerance in harvested banana fruit. J Proteome 187:1–12. https://doi.org/10.1016/j.jprot.2018.05.011

Li B, Ding Y, Tang X et al (2019) Effect of L-arginine on maintaining storage quality of the white button mushroom (*Agaricus bisporus*). Food Bioprocess Technol 12:563–574. https://doi.org/10.1007/s11947-018-2232-0

Li J, Luo M, Zhou X et al (2021) Polyamine treatment ameliorates pericarp browning in cold-stored 'Nanguo' pears by protecting mitochondrial structure and function. Postharvest Biol Technol 178:111553. https://doi.org/10.1016/j.postharvbio.2021.111553

Liu S, Huang H, Huber DJ et al (2020) Delay of ripening and softening in 'Guifei' mango fruit by postharvest application of melatonin. Postharvest Biol Technol 163:111136. https://doi.org/10.1016/j.postharvbio.2020.111136

Loay AA, Taher MA (2018) Influence of edible coatings chitosan/PVP blending with salicylic acid on biochemical fruit skin browning incidence and shelf life of guava fruits cv. 'Banati'. Sci Hortic 235:424–436. https://doi.org/10.1016/j.scienta.2018.03.008

Lu Z, Wang X, Cao M et al (2019) Effect of 24-epibrassinolide on sugar metabolism and delaying postharvest senescence of kiwifruit during ambient storage. Sci Hortic 253:1–7. https://doi.org/10.1016/j.scienta.2019.04.028

Luo T, Yin F, Liao L et al (2021) Postharvest melatonin treatment inhibited longan (Dimocarpus longan Lour.) pericarp browning by increasing ROS scavenging ability and protecting cytomembrane integrity. Food Sci Nutr 9:4963–4973. https://doi.org/10.1002/fsn3.2448

Madhav JV, Sethi S, Sharma RR, Rudra SG (2016) Impact of salicylic acid on the physiological and quality attributes of guava (*Psidium guajava*) fruits during storage at low temperature. Indian J Agric Sci 86:1172–1174. https://doi.org/10.56093/ijas.v86i9.61512

Manjunatha G, Lokesh V, Neelwarne B (2010) Nitric oxide in fruit ripening: trends and opportunities. Biotechnol Adv 28:489–499. https://doi.org/10.1016/j.biotechadv.2010.03.001

Martínez-Esplá A, Serrano M, Martínez-Romero D et al (2018) Oxalic acid preharvest treatment increases antioxidant systems and improves plum quality at harvest and during postharvest storage. J Sci Food Agric 99:235–243. https://doi.org/10.1002/jsfa.9165

Mason TJ (1990) Chemistry with ultrasound. Elsevier, London

Meena NK, Vinod BR, Menaka M et al (2024) Sustainable technologies prospects and potential applications in postharvest storage of fruits and vegetables. CRC Press eBooks, In, pp 42–62

Menaka M, Asrey R, Singh D et al (2024a) Preserving functional properties and inhibiting postharvest peel browning in guava during cold storage via 24-epibrassinolide application. Postharvest Biol Technol 216:113033. https://doi.org/10.1016/j.postharvbio.2024.113033

Menaka M, Asrey R, Vinod BR et al (2024b) UV-C irradiation enhances the quality and shelf-life of stored guava fruit via boosting the antioxidant systems and defense responses. Food Bioprocess Technol. https://doi.org/10.1007/s11947-024-03338-8

Min D, Li F, Zhang X et al (2018) SlMYC2 involved in methyl jasmonate-induced tomato fruit chilling tolerance. J Agric Food Chem 66:3110–3117. https://doi.org/10.1021/acs.jafc.8b00299

Misra NN, Tiwari BK, Raghavarao KSMS, Cullen PJ (2011) Nonthermal plasma inactivation of food-borne pathogens. Food Eng Rev 3:159–170. https://doi.org/10.1007/s12393-011-9041-9

Misra NN, Patil S, Moiseev T et al (2014) In-package atmospheric pressure cold plasma treatment of strawberries. J Food Eng 125:131–138. https://doi.org/10.1016/j.jfoodeng.2013.10.023

Monzón-Ortega K, Salvador-Figueroa M, Gálvez-López D et al (2018) Characterization of Aloe vera-chitosan composite films and their use for reducing the disease caused by fungi in papaya Maradol. J Food Sci Technol 55:4747–4757. https://doi.org/10.1007/s13197-018-3397-2

Muhammad AI, Xiang Q, Liao X et al (2018) Understanding the impact of nonthermal plasma on food constituents and microstructure—a review. Food Bioprocess Technol 11:463–486. https://doi.org/10.1007/s11947-017-2042-9

Mukherjee S (2019) Recent advancements in the mechanism of nitric oxide signaling associated with hydrogen sulfide and melatonin crosstalk during ethylene-induced fruit ripening in plants. Nitric Oxide 82:25–34. https://doi.org/10.1016/j.niox.2018.11.003

Nawaz F, Naeem M, Zulfiqar B et al (2017) Understanding brassinosteroid-regulated mechanisms to improve stress tolerance in plants: a critical review. Environ Sci Pollut Res 24:15959–15975. https://doi.org/10.1007/s11356-017-9163-6

Oms-Oliu G, Rojas-Graü MA, González LA et al (2010) Recent approaches using chemical treatments to preserve quality of fresh-cut fruit: A review. Postharvest Biol Technol 57:139–148. https://doi.org/10.1016/j.postharvbio.2010.04.001

Ong KC, Cash JN, Zabik MJ et al (1996) Chlorine and ozone washes for pesticide removal from apples and processed apple sauce. Food Chem 55:153–160. https://doi.org/10.1016/0308-8146(95)00097-6

Onik JC, Wai SC, Li A et al (2021) Melatonin treatment reduces ethylene production and maintains fruit quality in apple during postharvest storage. Food Chem 337:127753. https://doi.org/10.1016/j.foodchem.2020.127753

Özen T, Koyuncu MA, Erbaş D (2020) Effect of ozone treatments on the removal of pesticide residues and postharvest quality in green pepper. J Food Sci Technol 58:2186–2196. https://doi.org/10.1007/s13197-020-04729-3

Pakkish Z, Mohammadrezakhani S (2022) The effect of preharvest application of arginine on the postharvest quality of sweet cherry fruits during storage. Int J Fruit Sci 22:837–851. https://doi.org/10.1080/15538362.2022.2134272

Palma JM, Freschi L, Rodríguez-Ruiz M et al (2019) Nitric oxide in the physiology and quality of fleshy fruits. J Exp Bot 70:4405–4417. https://doi.org/10.1093/jxb/erz350

Pandey R, Gupta A, Chowdhary A et al (2014) Over-expression of mouse ornithine decarboxylase gene under the control of fruit-specific promoter enhances fruit quality in tomato. Plant Mol Biol 87:249–260. https://doi.org/10.1007/s11103-014-0273-y

Parven A, Sarker MR, Megharaj M, Meftaul IM (2020) Prolonging the shelf life of Papaya (*Carica papaya* L.) using Aloe vera gel at ambient temperature. Sci Hortic 265:109228. https://doi.org/10.1016/j.scienta.2020.109228

Patel N, Gantait S, Panigrahi J (2019) Extension of postharvest shelf-life in green bell pepper (*Capsicum annuum* L.) using exogenous application of polyamines (spermidine and putrescine). Food Chem 275:681–687. https://doi.org/10.1016/j.foodchem.2018.09.154

Per TS, Khan MIR, Anjum NA et al (2018) Jasmonates in plants under abiotic stresses: crosstalk with other phytohormones matters. Environ Exp Bot 145:104–120. https://doi.org/10.1016/j.envexpbot.2017.11.004

Peralta-Ruiz Y, Tovar CG, Sinning-Mangonez A et al (2020) *Colletotrichum gloesporioides* inhibition using chitosan-*Ruta graveolens* L essential oil coatings: Studies in vitro and in situ on Carica papaya fruit. Int J Food Microbiol 326:108649. https://doi.org/10.1016/j.ijfoodmicro.2020.108649

Prajapati U, Asrey R, Varghese E et al (2021) Effects of postharvest ultraviolet-C treatment on shelf-life and quality of bitter gourd fruit during storage. Food Pack Shelf Life 28:100665. https://doi.org/10.1016/j.fpsl.2021.100665

Priyanka BS, Rastogi NK (2014) Opportunities and challenges in application of forward osmosis in food processing. Crit Rev Food Sci Nutr 56:266–291. https://doi.org/10.1080/10408398.2012.724734

Qi H, Hu W, Jiang A et al (2011) Extending shelf-life of Fresh-cut 'Fuji' apples with chitosan-coatings. Inn Food Sci Emerging Technol 12:62–66. https://doi.org/10.1016/j.ifset.2010.11.001

Qi H, Huang Q, Hung YC (2018) Effectiveness of electrolyzed oxidizing water treatment in removing pesticide residues and its effect on produce quality. Food Chem 239:561–568. https://doi.org/10.1016/j.foodchem.2017.06.144

Qin GH, Wei SW, Tao ST et al (2017) Effects of postharvest methyl jasmonate treatment on aromatic volatile biosynthesis by 'Nanguoli' fruit at different harvest maturity stages. N Z J Crop Hortic Sci 45:191–201. https://doi.org/10.1080/01140671.2016.1272470

Qu G, Liangjie BA, Wang R et al (2022) Effects of melatonin on blueberry fruit quality and cell wall metabolism during low temperature storage. Food Sci Technol 42:e40822. https://doi.org/10.1590/fst.40822

Rahman S, Khan I, Oh DH (2016) Electrolyzed water as a novel sanitizer in the food industry: current trends and future perspectives. Compr Rev Food Sci Food Saf 15:471–490. https://doi.org/10.1111/1541-4337.12200

Rastegar S, Hassanzadeh Khankahdani H, Rahimzadeh M (2020) Effects of melatonin treatment on the biochemical changes and antioxidant enzyme activity of mango fruit during storage. Sci Hortic 259:108835. https://doi.org/10.1016/j.scienta.2019.108835

Razali Z, George DS, Ibrahim NAN (2015). Improvements in quality and antioxidant capacity of guava (*Psidium guajava* L.) exposed to UV-C irradiation. In: Proceedings of 3rd international conference on Chemical, Agricultural and Medical Sciences (CAMS-2015). 10–11 December, Singapore, pp 50–53

Sahu SK, Barman K, Singh AK (2020) Nitric oxide application for postharvest quality retention of guava fruits. Acta Physiol Plant 42:156. https://doi.org/10.1007/s11738-020-03143-8

Salvador A, Sanz T, Fiszman SM (2007) Changes in colour and texture and their relationship with eating quality during storage of two different dessert bananas. Postharvest Biol Technol 43:319–325. https://doi.org/10.1016/j.postharvbio.2006.10.007

Sanchís E, González S, Ghidelli C et al (2016) Browning inhibition and microbial control in fresh-cut persimmon (*Diospyros kaki* Thunb. cv. Rojo Brillante) by apple pectin-based edible coatings. Postharvest Biol Technol 112:186–193. https://doi.org/10.1016/j.postharvbio.2015.09.024

Sayyari M, Babalar M, Kalantari S et al (2009) Effect of salicylic acid treatment on reducing chilling injury in stored pomegranates. Postharvest Biol Technol 53:152–154. https://doi.org/10.1016/j.postharvbio.2009.03.005

Seo HS, Song JT, Cheong J-J et al (2001) Jasmonic acid carboxyl methyltransferase: a key enzyme for jasmonate-regulated plant responses. Proc Natl Acad Sci 98:4788–4793. https://doi.org/10.1073/pnas.081557298

Seymour IJ, Burfoot D, Smith RL et al (2002) Ultrasound decontamination of minimally processed fruits and vegetables. Int J Food Sci Technol 37:547–557. https://doi.org/10.1046/j.1365-2621.2002.00613.x

Shelp BJ, Bozzo GG, Trobacher CP et al (2012) Hypothesis/review: contribution of putrescine to 4-aminobutyrate (GABA) production in response to abiotic stress. Plant Sci 193–194:130–135. https://doi.org/10.1016/j.plantsci.2012.06.001

Shi F, Li X, Meng H et al (2020) Reduction in chilling injury symptoms by hot electrolyzed functional water treatment may function by regulating ROS metabolism in Satsuma orange fruit. LWT 125:109218. https://doi.org/10.1016/j.lwt.2020.109218

Shu P, Min D, Ai W et al (2020) L-Arginine treatment attenuates postharvest decay and maintains quality of strawberry fruit by promoting nitric oxide synthase pathway. Postharvest Biol Technol 168:111253. https://doi.org/10.1016/j.postharvbio.2020.111253

Singh SP, Singh Z, Swinny EE (2009) Postharvest nitric oxide fumigation delays fruit ripening and alleviates chilling injury during cold storage of Japanese plums (*Prunus salicina* Lindell). Postharvest Biol Technol 53:101–108. https://doi.org/10.1016/j.postharvbio.2009.04.007

Singh UB, Malviya D, Singh S et al (2019) *Trichoderma harzianum*- and methyl jasmonate-induced resistance to bipolaris sorokiniana through enhanced phenylpropanoid activities in bread wheat (*Triticum aestivum* L.). Front Microbiol 10:1697. https://doi.org/10.3389/fmicb.2019.01697

Smith MA, Davies PJ (1985) Separation and quantitation of polyamines in plant tissue by high performance liquid chromatography of their dansyl derivatives. Plant Physiol 78:89–91. https://doi.org/10.1104/pp.78.1.89

Sobolev AP, Neelam A, Fatima T et al (2014) Genetic introgression of ethylene-suppressed transgenic tomatoes with higher-polyamines trait overcomes many unintended effects due to reduced ethylene on the primary metabolome. Front Plant Sci 5:632. https://doi.org/10.3389/fpls.2014.00632

Sogvar OB, Saba MK, Emamifar A (2016) Aloe vera and ascorbic acid coatings maintain postharvest quality and reduce microbial load of strawberry fruit. Postharvest Biol Technol 114:29–35. https://doi.org/10.1016/j.postharvbio.2015.11.019

Sohail M, Wills RBH, Bowyer MC, Pristijono P (2021) Beneficial impact of exogenous arginine, cysteine and methionine on postharvest senescence of broccoli. Food Chem 338:128055. https://doi.org/10.1016/j.foodchem.2020.128055

Sonawane SK, Malar T, Patil S (2020) Non-thermal plasma: an advanced technology for food industry. Food Sci Technol Int 26:727–740. https://doi.org/10.1177/1082013220929474

Sukarminah E, Djali M, Andoyo R et al (2017) Ozonization technology and its effects on the characteristics and shelf-life of some fresh foods: a review. KnE Life Sci 2:459. https://doi.org/10.18502/kls.v2i6.1067

Sun Q, Zhang N, Wang J et al (2014) Melatonin promotes ripening and improves quality of tomato fruit during postharvest life. J Exp Bot 66:657–668. https://doi.org/10.1093/jxb/eru332

Tapia R-GMA, Carmona A et al (2008) Use of alginate- and gellan-based coatings for improving barrier, texture and nutritional properties of fresh-cut papaya. Food Hydrocoll 22:1493–1503. https://doi.org/10.1016/j.foodhyd.2007.10.004

Tavallali V (2018) Vacuum infiltration of 24-epibrassinolide delays chlorophyll degradation and maintains quality of lime during cold storage. Acta Sci Pol Hortic Cultus 17:35–48. https://doi.org/10.24326/asphc.2018.1.4

Tijero V, Munoz P, Munne-Bosch S (2019) Melatonin as an inhibitor of sweet cherries ripening in orchard trees. Plant Physiol Biochem 140:88–95. https://doi.org/10.1016/j.plaphy.2019.05.007

Thakur R, Pristijono P, Golding JB et al (2018) Development and application of rice starch based edible coating to improve the postharvest storage potential and quality of plum fruit (*Prunus salicina*). Sci Hortic 237:59–66. https://doi.org/10.1016/j.scienta.2018.04.005

Valero D, Martinez D, Riquelme F, Serrano M (1998) Polyamine response to external mechanical bruising in two mandarin cultivars. Hortic Sci 33:1220–1223. https://doi.org/10.21273/hortsci.33.7.1220

Vinod BR, Asrey R, Meena NK et al (2024) Ozonized water treatment enhances quality and storage life of 'Red Lady' papaya fruit during cold storage. Sci Hortic 328:112859. https://doi.org/10.1016/j.scienta.2024.112859

Vinod BR, Asrey R, Sethi S et al (2023) Recent advances in physical treatments of papaya fruit for postharvest quality retention: a review. eFood 4:e79. https://doi.org/10.1002/efd2.79

Wang Q, Lai T, Qin G, Tian S (2008) Response of jujube fruits to exogenous oxalic acid treatment based on proteomic analysis. Plant Cell Physiol 50:230–242. https://doi.org/10.1093/pcp/pcn191

Wang B, Li Z, Han Z et al (2021) Effects of nitric oxide treatment on lignin biosynthesis and texture properties at wound sites of muskmelons. Food Chem 362:130193. https://doi.org/10.1016/j.foodchem.2021.130193

Wang J, Wang Y, Li Y et al (2023) L-Arginine treatment maintains postharvest quality in blueberry fruit by enhancing antioxidant capacity uring storage. J Food Sci 88(9):3665–3680. https://doi.org/10.1111/1750-3841.16710

Wei Y, Liu Z, Su Y et al (2011) Effect of salicylic acid treatment on postharvest quality, antioxidant activities, and free polyamines of asparagus. J Food Sci 76:S126–S132. https://doi.org/10.1111/j.1750-3841.2010.01987.x

Wildermuth MC, Dewdney J, Wu G, Ausubel FM (2001) Isochorismate synthase is required to synthesize salicylic acid for plant defence. Nature 414:562–565

Wills RBH, Li Y (2016) Use of arginine to inhibit browning on fresh cut apple and lettuce. Postharvest Biol Technol 113:66–68. https://doi.org/10.1016/j.postharvbio.2015.11.006

Wu F, Zhang D, Zhang H et al (2011) Physiological and biochemical response of harvested plum fruit to oxalic acid during ripening or shelf-life. Food Res Int 44:1299–1305. https://doi.org/10.1016/j.foodres.2010.12.027

Wu X, Ren J, Huang X et al (2021a) Melatonin: biosynthesis, content, and function in horticultural plants and potential application. Sci Hortic 288:110392. https://doi.org/10.1016/j.scienta.2021.110392

Wu Z, Dong C, Wei J et al (2021b) A transcriptional study of the effects of nitric oxide on rachis browning in table grapes cv. Thompson Seedless. Postharvest Biol Technol 175:111471. https://doi.org/10.1016/j.postharvbio.2021.111471

Xu X, Tian S (2008) Salicylic acid alleviated pathogen-induced oxidative stress in harvested sweet cherry fruit. Postharvest Biol Technol 49:379–385. https://doi.org/10.1016/j.postharvbio.2008.02.003

Yang J, Sun C, Zhang Y et al (2017) Induced resistance in tomato fruit by γ-aminobutyric acid for the control of alternaria rot caused by *Alternaria alternata*. Food Chem 221:1014–1020. https://doi.org/10.1016/j.foodchem.2016.11.061

Yao H, Tian S (2005) Effects of pre- and post-harvest application of salicylic acid or methyl jasmonate on inducing disease resistance of sweet cherry fruit in storage. Postharvest Biol Technol 35:253–262. https://doi.org/10.1016/j.postharvbio.2004.09.001

Yu J, Engeseth NJ, Feng H (2015) High intensity ultrasound as an abiotic elicitor—effects on antioxidant capacity and overall quality of romaine lettuce. Food Bioprocess Technol 9:262–273. https://doi.org/10.1007/s11947-015-1616-7

Yurdugül S (2016) Effects of edible coating-semperfresh™, ascorbic acid and whey protein treatment on certain microbiological, physical and chemical qualities of peeled bananas "*musa sapientum*." Bulg J Agric Sci 22

Yuting X, Lifen Z, Jianjun Z, Jie S et al (2013) Power ultrasound for the preservation of postharvest fruits and vegetables. Int J Agric Biol Eng 6:116–125

Zalewski K, Lahuta LB, Martysiak-urowska D et al (2019) Effect of exogenous application of methyl jasmonate on the lipid and carbohydrate content and composition of winter triticale

(*Tritico secale* Wittm.) grain and the severity of fungal infections in triticale plants and grain. J Agric Food Chem 67:593–5939. https://doi.org/10.1021/acs.jafc.9b00799
Ze Y, Gao H, Li T et al (2021) Insights into the roles of melatonin in maintaining quality and extending shelf life of postharvest fruits. Trends Food Sci Technol 109:569–578. https://doi.org/10.1016/j.tifs.2021.01.051
Zhai R, Liu J, Liu F et al (2018) Melatonin limited ethylene production, softening and reduced physiology disorder in pear (*Pyrus communis* L.) fruit during senescence. Postharvest Biol Technol 139:38–46. https://doi.org/10.1016/j.postharvbio.2018.01.017
Zhang W, Jiang W (2019) UV treatment improved the quality of postharvest fruits and vegetables by inducing resistance. Trends Food Sci Technol 92:71–80. https://doi.org/10.1016/j.tifs.2019.08.012
Zhang X, Min D, Li F et al (2017) Synergistic effects of l-arginine and methyl salicylate on alleviating postharvest disease caused by *Botrysis cinerea* in tomato fruit. J Agric Food Chem 65:4890–4896. https://doi.org/10.1021/acs.jafc.7b00395
Zhang W, Zhao H, Zhang J et al (2019) Different molecular weights chitosan coatings delay the senescence of postharvest nectarine fruit in relation to changes of redox state and respiratory pathway metabolism. Food Chem 289:160–168. https://doi.org/10.1016/j.foodchem.2019.03.047
Zhang W, Jiang H, Cao J, Jiang W (2021) UV-C treatment controls brown rot in postharvest nectarine by regulating ROS metabolism and anthocyanin synthesis. Postharvest Biol Technol 180:111613. https://doi.org/10.1016/j.postharvbio.2021.111613
Zhang Y, Li S, Deng M et al (2022) Blue light combined with salicylic acid treatment maintained the postharvest quality of strawberry fruit during refrigerated storage. Food Chem X 15:100384. https://doi.org/10.1016/j.fochx.2022.100384
Zheng X, Tian S (2006) Effect of oxalic acid on control of postharvest browning of litchi fruit. Food Chem 96:519–523. https://doi.org/10.1016/j.foodchem.2005.02.049
Zheng X, Jing G, Liu Y et al (2012) Expression of expansin gene, MiExpA1, and activity of galactosidase and polygalacturonase in mango fruit as affected by oxalic acid during storage at room temperature. Food Chem 132:849–854. https://doi.org/10.1016/j.foodchem.2011.11.049
Zhou D, Wang Z, Tu S, Chen S, Peng J, Tu K (2019) Effects of cold plasma, UV-C or aqueous ozone treatment on botrytis cinerea and their potential application in preserving blueberry. J Appl Microbiol 127:175–185. https://doi.org/10.1111/jam.14280
Zhu LQ, Zhou J, Zhu SH (2010) Effect of a combination of nitric oxide treatment and intermittent warming on prevention of chilling injury of 'Feicheng' peach fruit during storage. Food Chem 121:165–170. https://doi.org/10.1016/j.foodchem.2009.12.025

Part II
Advances in Postharvest Technology of Horticulture Crops

Chapter 4
Storage Technologies for Horticulture Crops

Abha Sharma, Agrim Rattan, Abhay Sharma, Ajay Tomar, and Kshitij Parmar

Abstract The techniques for storing food are crucial and can bring in a lot of money for the agricultural and food businesses. Just after China, India is the world's second-largest producer of fruits and vegetables as a result of its unique climatic circumstances. The National Horticulture Database reports that India produced 209.14 and 107.50 million metric tons of fruits and vegetables, respectively, between 2021 and 2022 (Anonymous, Department of Agriculture & Farmers Welfare, MoA & FW, Government of India, India, 2022). With a growing area of 11.37 million hectares for vegetables and 7.06 million hectares for fruits, there is a sizable export potential. However, because most crops are perishable and seasonal, they are extremely vulnerable to changes in the temperature and humidity of the environment, which limits both their availability and quality. As a result, there is an urgent need to transform these products into different forms in order to assure an extended time of availability, as well as to extend the shelf life of these fruits and vegetables through effective storage solutions. Clamps, ventilated buildings, vaults, and evaporative cooling chambers are a few examples of classic, economical storage structures. Contrarily, contemporary storage methods, such as cold or refrigerated storage, modified atmospheric packing, controlled atmospheric storage, and hypobaric storage, are useful in big enterprises and provide incredibly efficient storage options. Both kinds of institutions are equally important for growing nations like India.

Keywords Horticulture · Storage · Postharvest · Technologies

A. Sharma (✉) · A. Tomar · K. Parmar
Amity Institute of Organic Agriculture, Amity University, Noida, Uttar Pradesh, India

A. Rattan
Atmospheric Sciences, IIT Delhi, New Delhi, Delhi, India

A. Sharma
Horticulture Research Station, Seobagh, Kullu, Himachal Pradesh, India

M. Thakur, T. Belwal (eds.), *Advances in Postharvest and Analytical Technology of Horticulture Crops*, https://doi.org/10.1007/978-981-97-7247-6_4

4.1 Introduction

Fruits and vegetables stand as essential dietary supplements, providing vital vitamins, minerals, and fiber necessary for maintaining human health. Nationally, the major vegetables, including potato, onion, tomato, cauliflower, cabbage, bean, cucumber, peas, garlic, and okra, along with major fruits, including mango, grape, apple, apricot, orange, banana, avocado, guava, litchi, papaya, sapota, and watermelon, play a significant role in meeting nutritional needs.

However, the postharvest loss of these valuable commodities remains alarmingly high, occurring before consumption due to deficiencies in harvesting, handling, storage, transportation, and marketing facilities. Developing countries grapple with food shortages, driven by the relentless demands of a growing population. Unfortunately, the increase in food production to address this scarcity results in wastage, given the inadequate storage and preservation facilities available.

India has solidified its global standing in horticultural crop production, encompassing a diverse range of commodities such as vegetables, fruits, coffee, cotton, and spices (Rais and Sheoran 2015). Regrettably, a substantial portion of these valuable products succumbs to losses caused by inadequate storage practices, pest infestation, and improper handling. Postharvest losses for fruits and vegetables are estimated to range from 6% to 18% (Nanda et al. 2012). Mitigating this issue requires the establishment of appropriate storage environments that enhance the quality and marketability of horticultural produce. To achieve an ideal atmosphere for the produce, sophisticated measures involving precise temperature control, adequate air circulation, and relative humidity (RH) management are essential. Furthermore, regulating the air composition within the storage area holds promise. Storage facilities can be broadly classified based on their refrigeration requirements, distinguishing between those necessitating refrigeration and those that do not (El-Ramady et al. 2015). Traditional, cost-effective storage structures such as clamps, ventilated structures, cellars, and evaporative cooling chambers have been conventionally employed. In contrast, modern storage techniques such as cold or refrigerated storage, modified atmospheric packaging, controlled atmospheric storage (CAS), and hypobaric storage find utility in large industries, offering highly effective storage solutions. For developing countries both types of structures hold equal significance (Kale et al. 2016).

Minimizing postharvest losses would not only enhance the farmers' income but also contribute to the overall gross value of the nation's agricultural output. To achieve this, strategic approaches must be employed to store the produce optimally for future use, catering to off-season demands and distant markets lacking access to specific products. Preservation and processing of horticultural produce serve multiple objectives, such as extending the consumption period, adding value, and diversifying products according to consumer preferences.

Modern processing technologies hold immense potential in transforming farm produce into stable forms, facilitating storage and shipment to distant markets throughout the year, thereby extending the availability of processed products while preserving their nutritional value and palatability. The process of transformation can yield more user-friendly and convenient forms of horticultural produce, establishing a strong connection with a broader consumer base.

Demand for processed products such as dehydrated foods, pickles, fruit beverages, fresh and canned mangoes, fruit juices, and salted cashews serves as valuable foreign exchange earners. Consequently, it becomes imperative to enhance and expand the infrastructure for food storage and preservation. By focusing on advanced storage and processing techniques, the shelf life of fruits and vegetables can be significantly extended, ensuring a stable supply of these valuable crops to meet the ever-growing demands of a dynamic population.

4.2 Storage of Fruits and Vegetables

4.2.1 Optimizing Shelf Life and Quality

The vitality of fresh fruits and vegetables lies in their inherent metabolic activity, although postharvest, this process becomes disconnected from the mother plant, devoid of essential substrates for respiration. The crucial supply of mineral nutrients is also severed during harvest, rendering the harvested produce dependent on the recycling of existing nutrients, leading to a shortened lifespan. To ensure an orderly marketing of perishables, storage methods must be employed to strike a balance between harvest and sales fluctuations, or for extended storage to prolong the market availability of these valuable commodities. Therefore, proper storage conditions play a pivotal role in increasing the shelf life of fresh produce by decelerating metabolic rates while safeguarding against complete metabolic stasis.

Regrettably, the lack of adequate infrastructure for processing and postharvest management (PHM) results in a staggering 25–40% wastage of fresh fruits and vegetables, amounting to an annual cumulative loss of approximately Rs. 3000–4000 crores in India (Swaminathan 1981). To address this challenge, storing fruits and vegetables at lower temperatures becomes crucial to regulate the distribution and extend their availability in the market. Storage, in essence, involves preserving produce, either fresh or processed, in a safe condition, minimizing deteriorative changes such as moisture loss, carbohydrate and vitamin depletion, physiological disorders, and susceptibility to pests and diseases, among others, for future use.

The art of storage revolves around controlling respiration, transpiration, ripening, and biochemical changes that contribute to the decomposition and deterioration of produce. Swiftly employing low-temperature storage immediately after harvest effectively curbs the rate of respiration and thermal decomposition, mitigates microbial spoilage, and retains the quality and freshness of the produce for an extended duration (Chadha and Pareek 1993). Central to the storage process is the comprehension of three principles governing fresh fruit and vegetable preservation: they endure better under cold conditions, are vulnerable to damage when subjected to freezing, and tend to wilt or shrivel in dry air. Furthermore, the composition of the storage atmosphere significantly influences the postharvest life of fresh produce.

Various storage systems exist, each with distinct attributes to minimize microbial infection and enhance the storability of produce. The storage life of fresh fruits and vegetables heavily relies on the proper PHM of two crucial factors: respiration and transpiration. Proper control of temperature and RH plays a pivotal role in maximizing storage life and marketable quality.

It is imperative to acknowledge that the socio-economic status of different segments of the developing countries population, diverse food habits across various states, and varied storage and preservation practices in different geographic regions all impact the storage of fruits and vegetables. Generally, low temperature and high humidity are required for fruit and vegetable storage, except for onion and garlic, which necessitate relatively lower RH (70%).

The fundamental objectives of storage can be summarized threefold: to inhibit the biological activity of harvested produce, to curtail drying rates by maintaining appropriate humidity levels in storage spaces, and to slow down the proliferation of microorganisms. Achieving effective storage entails careful consideration of three critical factors: temperature, RH, and atmospheric composition. By optimizing these factors, we can extend the shelf life and enhance the quality of fresh fruits and vegetables, thereby contributing to better food security and reduced wastage.

4.3 Postharvest Losses

4.3.1 A Challenge for Fresh Produce

Fruits and vegetables, being living tissues, remain highly vulnerable to respiration, water loss, and cell softening throughout the postharvest journey. The storage life of these commodities varies significantly, contingent upon factors such as species, variety, and pre-harvest conditions, particularly concerning quality, and maturity. Regrettably, the magnitude of postharvest losses for fruits and vegetables is estimated at a staggering 35–40%, predominantly attributed to inadequate PHM (XI Planning Commission). This loss translates not only to financial investments in the production cycle but also results in the wastage of labor, energy, and valuable inputs. Astonishingly, India discards fruits and vegetables every year, equivalent to the entire annual consumption of the United Kingdom.

The high moisture content inherent in horticultural commodities expedites their reaction response, rendering them highly perishable. To mitigate this challenge, reducing the moisture content to a recommended level through drying or dehydration could improve shelf life, albeit at the cost of compromising the overall texture and impairing the quality of the produce. Lamentably, the dearth of sufficient storage facilities for fruits and vegetables postharvest significantly curtails the quantity of these vital commodities that reach the market. Such limitations have an immediate impact on the distribution and consumption of the required amount for a healthy livelihood, exacerbating food scarcity issues. Urgent attention must be given to improving PHM and storage facilities to curb these losses and ensure a sustainable and resilient food supply chain.

4.3.2 Storage Consideration

4.3.2.1 Maximizing Shelf Life and Quality

The effective storage of fresh produce hinges upon managing two pivotal factors: respiration and transpiration. While it is essential to limit these processes, complete cessation is not ideal. Precise control of temperature and RH emerges as the linchpin for optimizing storage life and ensuring marketable quality. Following harvest, produce quality loss occurs due to a combination of physical, biochemical, physiological, and biological processes, the rates of which are predominantly influenced by the product's temperature at harvesting and the surrounding RH.

For fresh produce, maintaining low temperatures and high RH during storage proves crucial to preserving their quality and extending their shelf life. By meticulously regulating these factors, we can mitigate deterioration and enhance the overall marketability of the harvested commodities. The implementation of strategic PHM practices plays a pivotal role in fostering a sustainable and efficient food supply chain.

4.3.2.1.1 Temperature

4.3.2.1.1.1 The Key to Preserving Freshness

Each type of fruit or vegetable has specific temperature requirements for optimal storage. The deterioration of these products over time is influenced by factors such as temperature, respiration rate, and the stress of harvesting and postharvest handling. To ensure high-quality produce that remains attractive to customers, it is essential to lower the product's temperature quickly after harvest.

Lowering the temperature rapidly offers several benefits:

- Decreases the rate of respiration.
- Reduces water loss.
- Suppresses ethylene production (a gas that speeds up ripening).
- Reduces sensitivity to ethylene.
- Slows down or inhibits the growth of decay-causing microorganisms.

In essence, postharvest cooling not only helps maintain the quality of fruits and vegetables but also provides flexibility in marketing by enabling growers to sell their produce at the most suitable time. By cooling the produce promptly within about 4 h after harvest, we can extend its freshness and appeal, ensuring a positive experience for consumers.

4.3.2.1.2 Relative Humidity

4.3.2.1.2.1 Preserving Freshness and Quality

To preserve the freshness and quality of harvested produce, it is crucial to maintain high humidity levels. This prevents water loss, which can lead to reduced returns due to poor quality, such as wilting or shriveling, and a loss of saleable weight. However, it is essential to strike the right balance and use high humidity in conjunction with low-temperature storage. Combining humidity with warmth can create favorable conditions for the growth of fungi and bacteria, potentially causing spoilage.

Even though the RH of air naturally increases when it is cooled, it is still important to regularly check and ensure that the RH in a cool storage room remains at satisfactory levels. By controlling humidity effectively, we can safeguard the produce from unnecessary deterioration and maintain its appeal, making it more desirable for consumers and better for the market.

4.4 Types of Storage Technologies

(a) Low-cost storage technology
(b) Improved/modern storage methods/high-cost storage technologies

4.4.1 Low-Cost Storage Technology

Certain storage systems offer practical and cost-effective solutions without the need for complex cooling systems. Examples of these low-cost storage technologies include in situ storage, pits or trenches, clamps, cellars, and zero energy cool chambers (ZECCs).

1. **In Situ/On-Site/Field Storage:** The method of on-site storage involves postponing the crop harvest by leaving it in the soil. The commodity is left in the field itself for some time even after maturity and harvesting is delayed till its requirement in the market (Rickard and Coursey 1981). This method is mainly used for root, rhizomes, and tuber crops. A major disadvantage of this method is that land remains occupied where the produce is grown, and new crop plantations might get delayed. This is quite analogous to some fruits which are left on the tree. It should be ensured that the crops are protected from pest and disease

attacks, chilling, and freezing injuries (El-Ramady et al. 2015; Elansari et al. 2019).

2. **Pits or Trenches:** Pits and trenches are simple storage methods involving dug cavities lined with materials such as cut grass, wood shreds, sand, stubble, or soil (Rees et al. 2012). These cavities are usually created at the edge of the field, especially in sloping areas, to facilitate drainage. The process involves lining the pit or trench with straw or other organic materials and then filling it with fresh produce. A layer of organic matter is placed on top, followed by a layer of soil. Ventilation holes are included at the top, which can be covered with snow to allow air passage while preventing rainwater from entering.

   In colder conditions, a 25-cm soil layer is essential to protect crops from extreme temperatures. Conversely, in hot climates, a thicker layer of soil or organic matter is added to prevent excessive heat build-up during the day. This low-cost and natural storage solution helps preserve the freshness and quality of the produce, making it an efficient and sustainable option for farmers in different environmental conditions. Potatoes are stored in pit farms.

3. **Clamps:** Clamps are a time-tested method used in certain parts of the world to store potatoes. Typically built at the edge of the field, these clamps have a width of about 1–2.5 m. The process involves piling up the potatoes in conical heaps and marking out their dimensions. To create a suitable bed for the potatoes, straw is often spread on bare ground. The height of the heap's center is set at around one-third of the clamp's width, determined by the angle of repose of the heap.

   To protect the stored potatoes, rainwater is allowed to run off through the straw, which is bent at the top to prevent spoilage. The compressed straw layer should be about 15–25 cm thick. After about 2 weeks, the clamp is covered with soil to a depth of 15–20 cm, which can vary depending on environmental conditions (El-Ramady et al. 2015; Elansari et al. 2019). In more tropical climates, extra straw is sometimes used instead of soil to enhance ventilation. In extremely low atmospheric temperatures, a second layer of insulation using soil and straw can be added. In areas with high temperatures, chimney-type air outlets may be built at the top of the clamp to ensure proper ventilation. For regions experiencing heavy rainfall, building the clamp under a tree or a roof can prevent rainwater from entering (Kale et al. 2016).

   This traditional method of using clamps for potato storage proves to be effective in preserving the crop while considering the specific environmental conditions of the region. It offers a practical and low-cost solution for farmers in certain geographical areas, ensuring that their harvested potatoes remain fresh and of good quality for an extended period.

4. **Cellars:** These are underground or partly underground rooms, often beneath a house which provide insulation and are protected from excessively low temperatures in cold regions and used for storing apples, cabbage, onion, and potatoes in Britain. The crops are spread out thinly on shelves to ensure good air circulation. The doors are left open at night to maintain a low temperature. Desiccation is common in such cases which can be prevented by wetting the floor of the cellars.

5. **Barns:** Barns offer a straightforward storage solution for onion and garlic. Stakes are driven into the ground, and the tubers are suspended from these stakes using twine. Bulbs are spread over a platform raised with locally available materials such as reeds or bamboo sticks inside a roofed structure. This arrangement allows for the free circulation of air, preventing the accumulation of carbon dioxide produced during respiration. As a result, the storability of bulbs is extended, while they are protected from rain and direct sunlight.
6. **Evaporative Coolers:** Evaporative coolers use the principle of evaporation to provide cooling. In both active and passive systems, the air is passed through a pad of water to achieve the cooling effect. In the passive system, cooling pads are placed over the entrance of the storage area and kept moist. In the active system, a fan draws air into the storage area through a pad constantly kept moist by water pumped over it. The active system is more efficient in cooling but requires a continuous electric supply. Evaporative coolers are an effective means to keep stored produce cool and fresh, ensuring its longevity.
7. **ZECC:** The ZECC is an ingenious cooling system that requires no electricity and is particularly beneficial during the hot summer months. It is constructed using simple and easily available materials such as bricks, river sand, and bamboo. The ZECC is built with a single layer of bricks as the base, and a double-walled rectangular structure is erected on top of it, leaving a 7.5-cm space between the inner and outer brick walls. The dimensions of the chamber are approximately 165 × 115 × 67.5 cm. The cavity between the walls is filled with river sand, and the top is covered with khas khas or gunny cloth within a bamboo frame.

   During operation, the ZECC maintains a cooling effect through direct evaporative cooling. The sides of the cool chamber are double-walled with bricks joined using mud as cementing material, creating a space of 7.5 cm between the walls (Chanbisana et al. 2021). This space is filled with sand that is regularly saturated with water. One significant advantage of using the ZECC for storage is that it helps preserve the firmness of fruits, reducing physiological loss in weight (PLW) and other metabolic processes. As a result, mature green mangoes stored in the cool chamber can have a shelf life extended by 3–4 days compared with ambient conditions. Chanbisana et al. (2021) discussed the storage life of different commodities such as bitter gourd (May to June), carrot (Feb to March), cauliflower (Feb to March), cucumber (May to June), green chilies (May to June), kinnow (Jan to March), ladies finger (May to June), mango (July), peas (Feb to March), plum (June), and spinach (Feb to March) with storage life in ZECC ambient conditions and cool chambers.

   By utilizing ZECC, farmers, especially small and marginal ones, can avoid relying on middlemen or intermediaries in the retail chain. They can store their produce for a few days, giving them better control over market timing and enhancing the overall quality of their fresh products. In support of this sustainable approach, the National Horticulture Board provides farmers with a grant aid of 100% for ZECC implementation (Khan et al. 2017). This environmentally friendly and cost-effective solution presents an excellent opportunity for

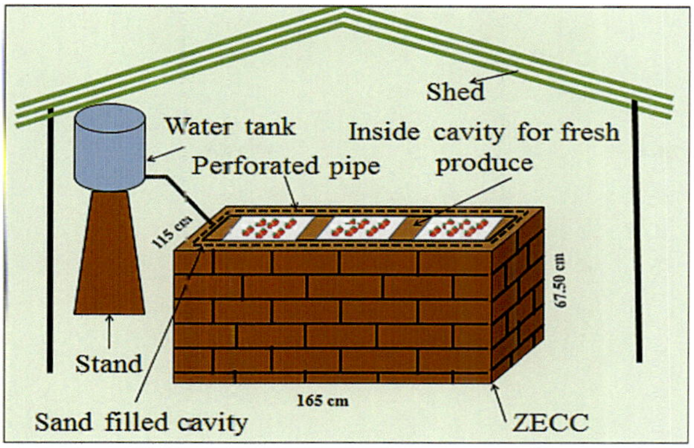

Fig. 4.1 Zero energy cool chambers (Khalid et al. 2020)

small-scale farmers to improve their postharvest handling and increase their income by accessing better markets.

4.4.1.1 Operation of the ZECC (Fig. 4.1)

To ensure effective operation and optimal storage conditions in the ZECC, the following steps should be taken:

1. Keep the sand, top cover, and bricks of the chamber wet with water to maintain a cool and humid environment.
2. Connect a drip line to an overhead water source to ensure a continuous water flow and regulate the required moisture levels.
3. Store fruits and vegetables in perforated plastic crates within the chamber to allow for proper airflow and ventilation.
4. Cover these crates with thin sheets of polythene to protect the produce from direct contact with water and maintain humidity levels.

4.4.1.2 Precautions During ZECC Construction (Fig. 4.2)

1. Choose a location with ample access to airflow to facilitate efficient cooling.
2. Avoid waterlogging by constructing the system at an elevated place.
3. Use good quality, porous bricks for the chamber's structure.
4. Ensure that the sand used contains no organic matter or clay.
5. Both the sand and bricks should be saturated with water before use.
6. Provide a roof over the system to prevent direct exposure to solar heat.
7. Prefer plastic crates over wooden, fiber-board, bamboo baskets, gunny bags, and so on, for storing produce.
8. Prevent water from coming into direct contact with fresh produce.

Fig. 4.2 ZECC, Location: Amity University Noida, Uttar Pradesh

9. Keep the chamber clean and disinfect it regularly using approved fungicides, insecticides, or chemicals to protect against fungi, insects, rodents, pests, reptiles, and so on.

By following these operational guidelines and taking necessary precautions during construction, the ZECC can effectively preserve the freshness and quality of stored fruits and vegetables, providing a sustainable and low-cost solution for farmers and improving their postharvest handling practices.

4.4.2 Improved/Modern Storage Methods/High-Cost Storage Technologies

Various advanced packaging and storage technologies have been developed and commercialized for fruits and vegetables, aiming to enhance their shelf life and quality. Some of these innovative methods are as follows.

4.4.2.1 Refrigerated Storage

Refrigerated storage works on the principle of heat transfer, where heat moves from higher-temperature objects to lower-temperature surfaces. Refrigeration creates a cold environment that absorbs heat through conduction, convection, or radiation. Ammonia is commonly used as a refrigerant in large installations. This technology is employed for storing frozen foods, processed products such as juice, concentrates, and aseptically packaged items.

The main advantages of refrigeration include:

- Reducing the respiration rate, thus extending shelf life
- Minimizing transpiration, which prevents shriveling and water losses
- Slowing down ethylene production, leading to slower ripening
- Enhancing resistance to ethylene's effects
- Reducing microbial activity, preserving freshness
- Maintaining texture, nutrients, and flavor, while also preventing browning
- Delaying ripening and slowing down natural aging processes

However, refrigeration can be expensive, and some tropical and subtropical fruits are sensitive to chilling, which can lead to chilling injury. Careful handling and the development of techniques to extend postharvest life with minimal refrigeration are essential (McGlasson et al. 1979). Cold storages with proper ventilation and humidity control systems ensure the best storage conditions for fresh produce (Sahoo et al. 2019).

Adopting these improved storage technologies for bulk-level applications requires considering factors such as cost-effectiveness, eco-friendliness, product quality, applicability range, storage area to product volume ratio, skill requirements, process control, energy consumption, and safety aspects. By implementing these advanced storage methods, the postharvest life of fruits and vegetables can be significantly extended, reducing food waste and enhancing the overall efficiency of the supply chain (Kitinoja 2013).

4.4.2.2 Hypobaric Storage

Hypobaric storage, also known as low-pressure storage, involves storing fruits under pressures lower than atmospheric levels, typically between 0.2 and 0.5 atmospheric pressure at temperatures ranging from 59 to 75 °F. Humidified air is passed through the storage chambers to remove volatile compounds released by the fruits, as the partial vacuum can cause some drying effects.

The process of hypobaric storage offers several benefits:

- Reduction of ethylene levels in the fruits and the oxygen within the chamber.
- Slowing down the respiration rate and ethylene production due to the reduced oxygen supply.
- Removal of ethylene and other volatile compounds produced by the stored fruits.

- By subjecting fruits and vegetables to hypobaric conditions, the ripening process and natural aging are retarded, leading to an extension of their shelf life.

Studies have shown that hypobaric treatment consistently delays postharvest fungal decay in greenhouse cucumbers without negatively affecting their color, firmness, or nutritional content. Cucumbers stored under hypobaric conditions exhibit higher overall quality compared with traditional storage methods (Zapotoczny and Markowski 2014).

Hypobaric storage presents an innovative and promising approach to extend the postharvest life of fruits and vegetables, improving their marketability and reducing food waste. It offers unique advantages in preserving produce quality and can be a valuable addition to modern postharvest storage technologies.

Hypobaric storage improves the metabolism, antioxidant system, and postharvest quality of Chinese bayberry fruit and provides an effective alternative method to prolong the storage life of this fruit (Hangjun et al. 2013).

4.4.2.3 Controlled Atmosphere Storage

Controlled atmosphere (CA) storage is an advanced and effective method used for preserving fruits. The idea was first suggested by W.R. Phillips of Canada's Department of Agriculture. The primary purpose of CA storage is to control the ripening process, maintain firmness, preserve the green color, and retain organic acids in fruits and vegetables. It is also known to inhibit sprouting and disinfecting produce.

The working principle of CA storage involves creating an insulated chamber where the oxygen (O_2) levels are reduced, and carbon dioxide (CO_2) levels are increased, effectively slowing down the ripening process. The respiratory rate is a key indicator of metabolic activity in fruits. The storage room needs to be well-insulated with adequate vapor barriers and sufficient cooling surfaces to maintain high humidity and ensure proper air circulation for efficient fruit cooling.

To maintain the desired atmosphere inside the room, it is made airtight. Excess CO_2 is removed from the CA storage using hydrated lime, as high levels of CO_2 can be harmful to many commodities. Water scrubbing can also remove CO_2 if levels below 3% are acceptable, along with oxygen concentrations below 5%.

Various substances such as potassium carbonate, ethanolamine, aluminum calcium silicate, and activated carbon are used to absorb CO_2 and water vapor, effectively reducing air humidity. To remove O_2 from the air, it can be burned using propane or natural gas by an atmosphere generator.

The simplest way to establish the desired atmosphere is by adding nitrogen (N_2) gas from tanks at the beginning of the storage process to quickly reduce O_2 levels. Lowering the O_2 level in plant cells reduces the rate of chemical reactions and metabolism, potentially affecting the flavor and aroma of the stored produce. In addition, elevated levels of CO_2 (10–15%) have been found to control gray mold in grapes, but prolonged exposure to these levels can damage the fruit for more than 2 weeks.

CA storage proves to be a highly beneficial and versatile method for extending the postharvest life of fruits, providing a valuable tool for the agricultural industry to ensure produce reaches consumers in excellent condition while minimizing losses and waste (Table 4.1). Different apple varieties used for CA storage in Japan include Starking Delicious, Jonathan, Rallus Janet, and Fuji, maintained in an atmosphere of 3% O_2 and 3% CO_2. However, when CO_2 concentrations fall below 1%, the softening of the flesh accelerates, while levels above 5% increase the browning of the core (Shiko 1972).

4.4.2.4 Modified Atmosphere Storage

In modified atmosphere (MA) storage, the control of gas concentrations surrounding the packaged commodity is less precise compared with CA (CA) storage. Instead, the appropriate atmosphere evolves passively within a sealed package due to the commodity's respiration, which consumes oxygen (O_2) and produces carbon dioxide (CO_2).

To create the desired atmosphere, the packaging film must allow a controlled amount of O_2 to enter the package, balanced by the commodity's O_2 consumption while venting excess CO_2 produced by the commodity. This process needs to occur rapidly without the risk of dangerous levels of CO_2 build-up.

MA packaging provides the necessary altered atmosphere for the packed products. There is a wide variety of flexible plastic films available for packaging, each with different gas permeability, water vapor transmission rates, anti-fogging properties, strength, and stretch-shrink properties. These films are continually evolving and being developed.

Polystyrene is a commonly used polymer with high gas transmission rates and a relatively high CO_2/O_2 permeability ratio. It is extensively used to wrap lettuce and tomatoes and is available as a heat-shrink film. Polystyrene is chemically inert and clear and has high transparency.

Table 4.1 Recommended conditions for controlled atmosphere storage of some horticultural crops (Kaur et al. 2021; Kader 1992)

Fruits and vegetables	Temperature (°C)	Oxygen (%)	Carbon dioxide (%)
Apple	0–5	2–3	1–2
Bananas	12–15	2–5	2–5
Cantaloupe	3–7	3–5	10–15
Kiwifruit	0–5	2	5
Lettuce	0–5	2–5	0
Nuts and dried fruits	0–25	0–1	0–100
Strawberry	0–5	10	15–20

MA storage can be a valuable tool in maintaining the quality of fresh produce throughout the marketing chain, preserving appearance, texture, flavor, and nutritive value (Kader et al. 1989). It offers an efficient way to extend the postharvest life of fruits and vegetables while ensuring they reach consumers in optimal condition, reducing food waste, and enhancing overall supply chain efficiency.

4.5 Conclusion

Postharvest losses of horticultural crops present a significant challenge for farmers and the food industry, primarily due to their short shelf life and inadequate storage practices. Addressing this issue is crucial, especially in developing countries, where substantial losses occur. Traditional storage structures, such as pits, cellars, clamps, and evaporative cooling chambers, have been utilized to some extent, but they have limitations in storing large quantities of produce for extended periods.

To overcome these limitations, improved and modern storage techniques have emerged, including cold stores, hypobaric storage, CAS, and modified atmospheric storage. Although these advanced methods may come with higher costs, they have proven to be effective in extending the shelf life of various horticultural crops by creating optimal storage conditions.

Ongoing research and studies are exploring different storage structures tailored to specific crop species, indicating a vast potential for enhancing horticultural crop storage. By adopting these innovative techniques and continually refining storage practices, we can significantly reduce postharvest losses, boost farmers' returns, and ensure a steady supply of high-quality produce in the market. Ultimately, the development and implementation of improved storage structures hold the key to minimizing postharvest losses, promoting sustainable agriculture, and contributing to food security and economic prosperity in the horticultural sector. Embracing these advancements will pave the way for a more resilient and efficient food supply chain, benefiting farmers, industries, and consumers alike.

References

Anonymous (2022) Department of Agriculture & Farmers Welfare, MoA & FW, Government of India

Chadha KL, Pareek OP (eds) (1993) Advances in horticulture, vol 4. Malhotra Publishing House, New Delhi

Chanbisana C, Prasad H, Saikhom H, Irungbam P, Pandey Y (2021) Storage and processing technology of horticultural crops. In: Advances in horticulture. Akinik, pp 21–46

Elansari AM, Yahia EM, Siddiqui W (2019) Storage systems. In: Postharvest technology of perishable horticultural commodities. Woodhead, pp 401–437

El-Ramady HR, Domokos-Szabolcsy E, Abdalla NA, Taha HS, Fari M (2015) Postharvest management of fruits and vegetable storage. In: Sustainable agriculture reviews. Springer, Cham, pp 65–152
Hangjun C, Hailong Y, Haiyon G, Jie L, Fei T, Xiangjun F, Fang X, Jiang Y (2013) Effect of hypobaric storage on quality, antioxidant enzyme and antioxidant capability of the Chinese bayberry fruits. Chem Cent J 7:4
Kader AA (1992) Postharvest technology of horticultural crops. 2nd Edition, Univ. Calif., Div. of Agr. and Nat. Resources, Publ 3311:15–20
Kader AA, Devon Z, Eduardo LK (1989) Modified atmosphere packaging of fruits and vegetables. Crit Rev Food Sci Nutr 28(1):1–30
Kale SJ, Nath P, Jalgaonkar KR, Mahawar MK (2016) Low-cost storage structures for fruits and vegetables handling in Indian conditions. Indian Hortic J 6(3):376–379
Kaur J, Aslam R, Afthab Saeed P (2021) Storage structures for horticultural crops: a review. Environ Conserv J 22:95–105
Khalid S, Majeed M, Ullah MI, Shahid M, Riasat AR, Abbas T, Aatif HM, Farooq A (2020) Effect of storage conditions and packaging material on postharvest quality attributes of strawberry. J Hortic Posthar Res 3(2):195–208
Khan FA, Bhat SA, Narayan S (2017) Storage methods for fruits and vegetables. Sher-e-Kashmir University of Agricultural Sciences and Technology of Kashmir, Shalimar
Kitinoja L (2013) Use of cold chains for reducing food losses in developing countries. Population 6(1.23):5–60
McGlasson WB, Scott KJ, Mendoza DB Jr (1979) The refrigerated storage of tropical and subtropical products. Int J Refrig 2(6):199–206
Nanda SK, Vishwakarma RK, Bathla HVL, Rai A, Chandra P (2012) Harvest and post-harvest losses of major crops and livestock produce in India. AICRP (ICAR)
Rais M, Sheoran A (2015) Scope of supply chain management in fruits and vegetables in India. J Food Procss Technol 6(3):1–7
Rees D, Westby A, Tomlins K, Van Oirschot Q, Cheema MU, Cornelius E, Amjad M (2012) Tropical root crops. In: Rees D et al (eds) Crop post-harvest: science and technology: perishables. Wiley-Blackwell, Oxford
Rickard JE, Coursey G (1981) Cassava storage part I. Storage of fresh cassava roots. Trop Sci 23(1):1–32
Sahoo K, Bandhyopadhyay B, Mukhopadhyay S, Sahoo U, Kumar TS, Yadav V, Singh Y (2019) Cold storage with backup thermal energy storage system. In: Progress in solar energy technologies and applications. Wiley, pp 181–232
Shiko H (1972) Controlled atmosphere storage of fruits in Japan. Japan Agric Res 6(2):175–180
Swaminathan MS (1981) Chairman's report of the group on perishable agricultural commodities, Govt. of India, Ministry of Agriculture. Dept. of Cooperation
Zapotoczny P, Markowski M (2014) Influence of hypobaric storage on the quality of greenhouse cucumbers. Bulgarian J Agric Sci 20(6):1406–1412

Chapter 5
Drying of Horticultural Produce: Mechanization, Challenges, and Opportunities

Alka Joshi, VikfigonoKsh, Shruti Sethi, Bindvi Arora, Darshaveeta Baruah, Ajay Narola, and Sumit Bhausaheb Uhre

Abstract The drying process plays a crucial role in the preservation of horticultural produce by ensuring extended shelf life. Understanding the mechanisms of heat and mass transfer during drying is essential for optimizing the process and overcoming challenges. This chapter explores the mechanisms involved in heat and mass transfer, drying equipment, and modeling. Additionally, it discusses the challenges faced during the drying of horticultural produce and presents opportunities for improvement in terms of innovative dried products. Keeping all the drying methods in view, economic feasibility, duration, and skilled manpower required, foam mat drying (FMD) and osmodehydration are recommended as the best drying techniques that are easy to understand and use. Moreover, unlike other drying methods, where the main focus is on evaporation of moisture which results in a bland flavor in the dried product, the characteristics and composition of the product can be modified in osmotic drying and FMD.

Keywords Fruits and vegetables · Limitations · Drying models · Dryer · Innovative dried horticultural products

A. Joshi (✉) · VikfigonoKsh · S. Sethi · B. Arora · A. Narola
Division of Food Science and Postharvest Technology, ICAR-Indian Agricultural Research Institute, New Delhi, Delhi, India

D. Baruah
Amity Institute of Food Technology, Amity University, Noida, Uttar Pradesh, India

S. B. Uhre
ICAR-Central Institute of Post-Harvest Engineering & Technology (CIPHET), Ludhiana, Punjab, India

M. Thakur, T. Belwal (eds.), *Advances in Postharvest and Analytical Technology of Horticulture Crops*, https://doi.org/10.1007/978-981-97-7247-6_5

5.1 Introduction

Drying is one of the oldest and most effective interventions employed for food preservation. Drying and dehydration are often used interchangeably. Generally, drying is achieved through natural means, whereas the term dehydration is more related to mechanical drying. Drying and dehydration are also classified as per the resultant moisture retained in the product; if moisture is reduced up to 6–7% (10% max), then it is drying; in other cases, it would be categorized as dehydration. The food industries use different methods for drying a variety of horticultural products. The most suitable technique is chosen based on the industry's requirement, type of raw material, processing time, quality of the end product, and the efficiency of the technique. Dried and dehydrated fruits and vegetables are more concentrated with respect to nutrients. Prolonged heat processing affects their delicate flavor, nutritional quality, and acceptability. Therefore, the new techniques that include osmotic dehydration, foam-mat-drying, modified drum drying, explosion puffing, microwave heating, infrared radiative drying, and slush drying remove water from fruits and vegetables at low or ambient temperature in less time and thus have minimum effect on flavor, nutritional, and thus enhanced acceptability.

5.2 Principle

Most horticultural produce contains enough water to permit their spoilage by the microorganisms and enzymes. Water presents itself in bound and free form. The free water can be removed by drying or dehydration. This free water supports spoilage, biochemical reactions, and microbial growth and is related to water activity. Theoretically, the water activity of any foodstuff is the vapor pressure of that commodity in comparison to pure water at the same temperature and pressure condition. Lower water activity means lower availability of water to support biochemical reactions, such as water activity 0.2–0.3 can support fat oxidation and enzymatic reactions but cannot support microbial damage. Generally, products with water activity less than 0.6 are considered safe from microbial damage. However, a nutrient-dense medium even in very low water activity (less than 0.6) can support microbial growth such as in chocolates (Pandey and Singh 2011).

In the following paragraphs, the different drying and dehydration techniques commonly used in the food industries are discussed.

5.2.1 Tray Drying

It is one of the most common, simplest, and cheap drying equipment used in industries. The materials to be dried are spread over trays and placed inside a cabinet in batch mode. Drying occurs due to heat convection from the hot air flowing across the trays or by conduction from heated trays or by radiation from the heated trays. This method is generally not recommended for high-value and heat-labile crops. In some advanced designs, it is operated under vacuum or controlled humidity conditions. This ensures better end-product quality but adds to the operation cost.

Foam-mat drying (FMD) is a process of drying semi-solid or liquid foods in a foam mat dryer. Similar drying can also be achieved through a tray drier. The product is mixed with a foaming agent and/or stabilizer to produce stable foam followed by a thin layer (1–5 mm) drying in a tray drier. The technology was patented in 1917 by Campbell for the drying of foamed evaporated milk. It is majorly used for viscous or high sugar matrices such as drying of bael, melon, beetroot, cherry, pineapple, and so on, which otherwise are difficult to dehydrate using other conventional drying procedures. This may be due to the physicochemical nature of the commodity and/or cost or feasibility constraints of the drying technology. The enhanced surface area provided by foam and enhanced moisture diffusivity of the evaporation front accelerate the rate of water removal during the course of drying (Thippeswamy et al. 2021).

5.2.2 *Tunnel Drying*

Tunnel dryers are regarded as an upgrade of tray dryers. A series of trays loaded with the material to be dried are conveyed inside a tunnel. Drying occurs as the moist material comes in contact with hot air flowing concurrently (tray and air flowing in the same direction), countercurrently (tray and air flowing in opposite direction), or mixed flow. In places where solar radiation is abundantly available, tunnel dryers are usually powered with solar energy. In one design, transparent flat collector plates were connected in series with a tunnel dryer for heat supply, and heat distribution was achieved using fans (Bala and Mondol 2001). Many horticultural products such as mango, tomato, chili, grapes, fenugreek leaves, and so on are dried in tunnel dryers (Indiarto et al. 2021).

5.2.3 *Roller or Drum Drying*

Drum dryers consist of a hollow drum that is internally heated with steam, hot water, hot air, and so on and rotates on its horizontal axes. The material to be dried is fed to the heated rotating drum surface in thin layers (0.5–2.0 mm), drying occurs due to conduction, and the dried material is scraped off using a scraper or Doctor's knife. The drying temperature and the rotation speed are important parameters and are decided based on the moisture to be removed and the nature of the material. The uniform layer thickness and contact drying ensure uniform and efficient drying (60–90%) at low operation costs (Henríquez et al. 2014). Additionally, the short residence time at elevated temperatures makes it suitable for heat-labile materials. Depending on the number of drums, drum dryers are grouped as single-drum or twin-drum dryers. Commercially, they are employed for drying sticky or viscous materials requiring solubility in cold water post-drying, for example, starches, breakfast cereals, baby food, and instant mashed potatoes. Powdered milk is also produced with drum drying; however, off-flavor and color development at high temperatures restrict its application as ingredients in animal feed or chocolate industries (Courtois 2013).

5.2.4 *Spray Drying*

This technique is employed when powdered end products are desirable from heat-labile liquid feed. In this, liquid feedstock (solvent, emulsion, or suspension) is pumped into an atomizer resulting in the generation of fine droplets (Fig. 5.1). These droplets enter the drying chamber and come in contact with hot air (204–538 °C), leading to heat exchange between the two and resulting in the formation of powder (Fox et al. 2014). The air and powder proceed to the cyclone separator where the air exits from the opening at the top portion of the cyclone separator and the powder settles at the bottom. The powder formed is relatively homogeneous ranging from 10 to 100 μm (Saleem et al. 2017). This method has lower specific energy consumption compared with freeze-drying. Additionally, it involves low operation costs and reduced drying time. Nonetheless, the high temperatures can be a limiting factor for certain feedstocks containing protein, for instance, egg white protein (Liu et al. 2020). Spray drying has a wide scope of applications ranging from the production of instant fruit powders, milk powders, whey protein concentrate, and so on. It is also used in the encapsulation of bioactive compounds from horticultural crops in in food and pharmaceutical matrices (Schuck 2002).

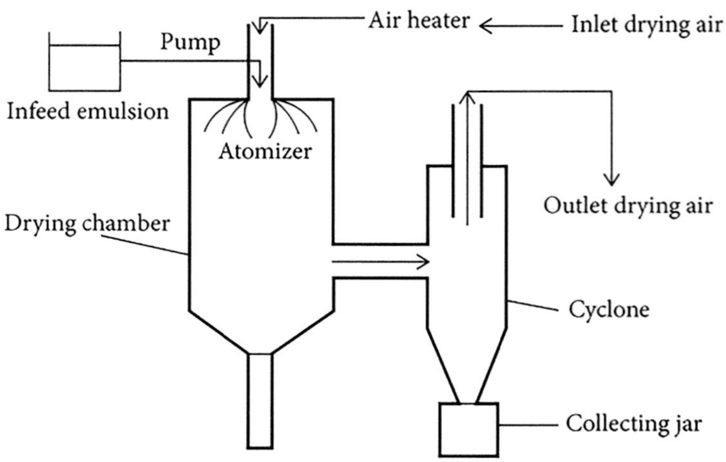

Fig. 5.1 Spray dryer. (Source: Nguyen et al. 2021)

5.2.5 *Freeze Drying or Lyophilization or Cryodesiccation*

Freeze drying can be broken down into three stages: freezing, primary drying, and secondary drying. After the material is frozen, primary drying (≤ -10 °C) commences under vacuum (≤ 2 mmHg) in the form of sublimation, that is, direct conversion from solid (ice) to gas (vapor). As sublimation progresses, a porous dried material is obtained. The latent heat of sublimation is conducted through the dry and frozen layers. In the secondary drying stage, the unfrozen water or bound water (10–35%) is removed by desorption, further reducing the moisture content. It is essential to maintain the right adsorption-desorption balance between the porous material and bound moisture for a high-quality end product (Indiarto et al. 2021). The low drying temperature retards most biochemical activities resulting in high retention of color, flavor, and nutrition. Except for some volatile loss (water-soluble) and minor morphological changes, freeze-dried materials have the best quality compared with those dried using other methods. However, the requirement of a vacuum evacuation system makes it a very expensive method and it is often limited to extremely heat-sensitive high-value materials (Nowak and Jakubczyk 2020). A freeze dryer is also used for encapsulating bioactive compounds. Another type of freeze drying is slush drying in which delicate products such as strawberry pulp and sapota slurry can be dried. In this process, frozen water is partially removed by sublimation and partially through evaporation.

5.2.6 *Fluidized-Bed Drying*

The material to be dried is suspended in a fluid-like state using vertical air flow with heat transfer occurring by convection. In some designs, an optional horizontal airflow is also present for conveying the material. A typical fluidized bed dryer comprises a blower, distribution plate, plenum chamber, drying chamber, and free-board region (Sivakumar et al. 2016). The fluidization medium is dispersed uniformly from the plenum chamber to the drying chamber through the distribution plate such that the material to be dried remains suspended against gravity. Enough free board space above the fluidized bed must be ensured not to permit entrainment of particles thrown up by local high flows of fluidization medium. The materials continue to remain fluidized until a weight loss due to moisture evaporation is observed. This method is suitable for light, granular, and fluidizable materials. Diced vegetables, slurries, pastes, and cereals (rice, soya bean, maize) are dried by fluidized bed drying (Indiarto et al. 2021). Although, owing to high mass transfer rates, it ensures uniform and fast drying, the restriction on the material size limits the applicability of fluidized bed drying.

5.2.7 Microwave Drying

In this method, electromagnetic energy from the microwave is dissipated internally (volumetrically) as heat by conduction (Fig. 5.2). Water molecules are electric dipoles possessing positive and negative charges. As microwaves penetrate into the material, the water molecules rotate in accordance with the changing electrical fields, colliding with other molecules in the process and causing heat generation from within the material. This increases the vapor pressure difference between the inner (high vapor pressure) and outer (low vapor pressure) portions of the material, promoting moisture removal from the material. Other components such as fats, sugars, and ice have lower dipole moments and do not respond well to microwaves. Thus, the drying efficiency depends on the material's composition and wave frequency (Orsat et al. 2017).

Microwave drying is generally used in combination with other drying techniques such as hot air drying, vacuum drying, freeze drying (microwave freeze drying), and so on and opted to reduce drying duration and shrinkage (Indiarto et al. 2021). As heating occurs volumetrically in microwave drying, heat transfer limitations due to case hardening (dry and hot outer surface with humid and hot inner portion), which is a common issue observed in conventional drying techniques, can be addressed. Microwaves at 2450 MHz are employed for both industries and domestic purposes. However, the short penetration depth of microwaves at 2450 MHz may restrict their application. Hence, for industrial drying, 915 MHz (USA), 896 MHz (UK), and 902 MHz (Australia) are used (Radoiu 2020). Innovative microwave-dried potato

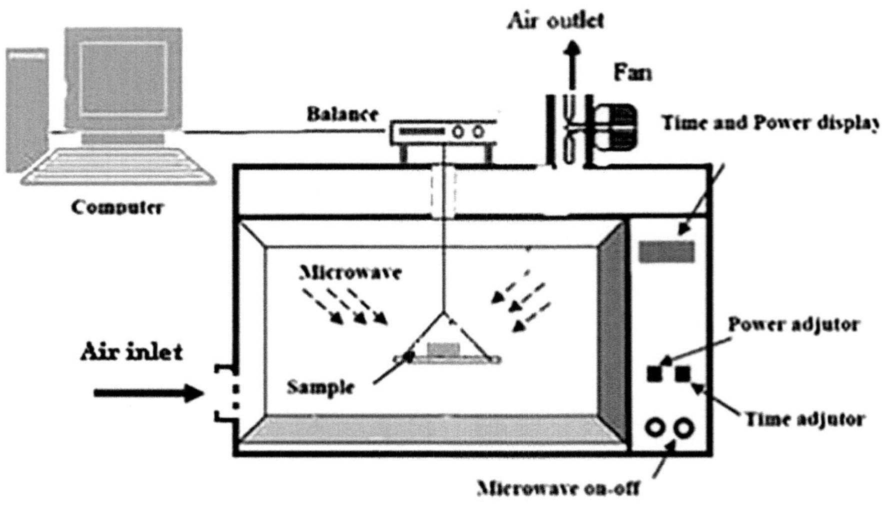

Fig. 5.2 Microwave dryer. (Source: Zarein et al. 2015)

chips have been developed by Joshi et al. (2015) in which the structural, textural, and color attributes of microwaved potato chips were similar to commercial fried potato chips. It was found that at 600 W after 2.5–3.0 min of processing, potato chips gained the fracturability and crispiness index as that of commercial fried chips. Microwave processing was found suitable for processing of potato chips with low-fat content (~3.09% vs. 35.5% in commercial preparation) and with acceptable sensory scores (≥7.6 on a 9.0-point hedonic scale vs. 8.0 in control preparation).

5.2.8 *Infrared Drying*

Infrared radiation is a type of electromagnetic radiation with longer wavelengths than visible light, but shorter than microwaves. Infrared radiation, a form of electromagnetic wave with wavelength ranging from 0.75 to 1000 μm, is also used for drying different agricultural produce. In this type of drying, heat transfer from source to sample occurs by radiation. The incoming irradiation penetrates up to a certain depth and gets absorbed by the molecules within the sample. The molecules start vibrating causing heat generation from within the sample, a temperature rise, and resultant moisture removal from the sample. The sample characteristics (chemical composition, thickness, moisture content, etc.) as well as the radiation wavelength decide the penetration depth (Pawar and Pratape 2017). It is widely used in various industries for drying, heating, and curing processes due to its ability to efficiently transfer heat without the need for direct contact between the heat source and the material being dried. It is employed for drying many food items such as potato slices, carrots, bananas, mushrooms, and so on. However, for heat-labile food items such as raspberries, pomegranates, and cordyceps, infrared drying is generally operated in hybrid mode along with other techniques (hot air drying, vacuum drying, microwave drying, freeze drying, etc.) (Huang et al. 2021). When the emitted infrared radiation comes into contact with the surface of the material to be dried, it interacts with the material's molecules. Some of the infrared energy is absorbed by the material's molecules. As a result, the kinetic energy of these molecules increases, leading to molecular vibration. The vibration of molecules due to the absorbed infrared energy results in the conversion of energy into heat. This localized heating causes the moisture present within the material to heat up. As the moisture within the material heats up, it reaches its boiling point and starts to evaporate. The evaporated moisture escapes from the material in the form of water vapor. The continuous emission of infrared radiation and the subsequent absorption of energy by the material's surface lead to a gradual and controlled drying process. The moisture content decreases over time as more and more moisture evaporates. Infrared drying is energy-efficient because it directly heats the material, rather than heating the surrounding air (as in conventional drying methods). This reduces heat loss and minimizes the energy required for drying.

5.2.9 Heat Pump Dryer

Any dryer that uses convection as the primary mode of heat input to the dryer can be fitted with a suitably designed heat pump. The principal advantage of heat pump dryers emerges from the ability of heat pumps to recover energy from the exhaust as well as their ability to control the drying air temperature and humidity. Successful trial on heat pump drying of strawberries has been conducted in IARI and details of this drying are also available by them (Patel and Kar 2012). The heat pump works like a refrigerator, with an evaporator that absorbs heat and a condenser that emits heat, effectively pumping heat from a low-temperature evaporator to a higher-temperature condenser. Drying food products in a heat dryer offers enormous potential for improvement in product quality due to lower temperatures used in the drying process. Typical drying temperatures in a heat pump dryer are in the range of 30–60 °C. As both the sensible heat and latent heat of the vapor in the air leaving the dryer are recovered and used to heat air being directed to the dryer, heat pump dryers are very energy efficient (Gupta 2012).

5.2.10 Osmotic Dehydration

It is one of the very old but still in use technology for dehydration in fruits and vegetables. Two major simultaneous counter-current flows occur during osmotic drying: water loss from the food and a simultaneous transfer of solute from the solution into the foods. Water loss and solid gain are mainly controlled by the raw material characteristics. The great variability observed among the different fruits is mostly related to tissue compactness (Gupta 2012). In osmotic drying, sugar or salt is used as an osmo-drying agent. Here, sugar mainly acts as a texturizing and preserving agent. The texturizing effect of salt is opposite to that of sugar. However, in both cases, high sugar and salt is the major limitation of osmo-dried products. Nowadays, researchers target to texturize the product with less sugar inputs through the utilization of pectin or other gelling agents. The above-mentioned production of dehydrated fruits and vegetables including those made by solar drying is one answer to the need to produce cheap products due to their low costs of production machinery/equipment and packaging material compared with other methods of processing.

5.2.11 Sun Drying

Sun drying of fruits and vegetables is practical in tropical and subtropical regions, where there is plenty of sunshine and practically little or no rain during the drying season. Equipment essentially consists of drying trays and a few other items such as knives, lye bath, and so on for food preparation prior to drying. It is carried out in the open sun at the mercy of nature and thus has certain disadvantages such as:

- Exposure of produce to dust and atmospheric contamination
- Interference from animals and humans
- Insect infestation
- Dependence of weather

Shade drying and forced air drying are other natural methods of drying that can also give promising results. In curry leaves and pea pod concentrate preparation, shade drying and air drying are found capable of giving excellent product quality as that of freeze drying (Pooja et al. 2023).

Recently, much research has gone into the design of solar dryers for fresh produce in order to overcome these problems. A solar tunnel-type dehydrator (Fig. 5.3) has been developed and tested at I.A.R.I, New Delhi, for drying vegetables such as cauliflower, onions, potatoes, and other leafy vegetables. The dryer, details of which are given below, is capable of safely drying these vegetables to a final moisture content of approximately 5.0%.

5.2.11.1 Common Pre-Treatments Before Drying

(a) **Washing:** Fruits and vegetables are rinsed under running cold water to remove the surface dirt and pesticide spray residues; it is not recommended to wash fruits and vegetables with soap, detergent, or commercial produce wash.

Fig. 5.3 Prototype of dehydrator developed at IARI, New Delhi

(b) **Peeling and Slicing:** Some fruits and vegetables are peeled, trimmed away seeds, core, or damaged portions prior to drying. In apples and root vegetables, peeling is done by methods such as lye, hot brine peels, high-pressure steam for root crops, mechanical knife peelers for apples, and dilute lye for cling-stone peaches. Stone fruits are pitted, sliced, and dried. Cherries, berries, and so on are surface dried as a whole and then pitted.

(c) **Pricking:** Pricking is done to facilitate mass transfer and increase the drying rate. It can be done by stainless steel prickers or forks. Often it is done before pickling fruits and vegetables or osmotic dehydration, freeze-drying blueberries (Ochmian et al. 2020), and solar drying of Amla (Verma and Gupta 2004).

(d) **Soaking/Checking:** Fruits and Vegetables are soaked in water, lime water, brine, or alum for a few hours to a few days before blanching and cooking. Checking is an operation in which fruits, especially raisins and prunes, are dipped for a brief period in a hot solution of alkali. This causes the removal of the waxy layer and it also improves the drying rate.

(e) **Alkaline Dip:** Alkaline dip is used primarily for fruits that are dried as a whole, especially prunes and grapes. A sodium carbonate or lye solution (0.5% or less) is used at a temperature ranging from 93 to 100 °C. It is useful for fruits with a wax layer, as it dissolves the wax layer, enhances the permeability of skin to moisture, and increases drying rate, thus reducing drying time and decreasing quality deterioration of products. Alkali liquor is used for plums, berries, grapes, gold cherry, sour cherry, ginger, tomato, and so on (Deng et al. 2019).

(f) **Acid Dip:** Pectin loosening occurs in an acidic environment and thus can accelerate the drying process. Citric acid is the most commonly used as an anti-browning agent and a texture-modifier of fruits and vegetables. Ascorbic acid or a combination of citric acid and ascorbic acid is also used as an anti-browning agent and a texture modifier. It is used to decrease drying time or improve product color in sweet potato slices, peaches, bananas, and so on. Fruit juices high in vitamin C can also be used. Orange, lemon, pineapple, grape, and cranberry give their own color and flavor to the fruit. An acid solution is used for carrots, tomatoes, quince, apples, peaches, bananas, and so on (Deng et al. 2019).

(g) **Sulfuring**: It prevents oxidation and darkening, acts as a preservative, and prevents vitamin loss. Fruits and Vegetables are exposed to fumes of sublimed sulfur in an enclosed box for 30–60 min. It should be done at a place with adequate air circulation. It is used for dried apricots (Inserra et al. 2017), table grapes, and fresh blueberries (Mahajan et al. 2014).

(h) **Sulfiting**: Sulfite dips can achieve the same long-term anti-darkening effect as sulfuring, but more quickly and easily. Sodium bisulfite, sodium sulfite, or sodium meta-bisulfite solutions are used in low concentration; the use of sulfites in low-concentration solutions can prevent both enzymatic and non-enzymatic browning because sulfite ions react with quinones to stop the activity of polyphenol oxidase activity and reduce the amount of oxygen present. It is used for mushrooms, apples, bananas, carrots, apricots, and tomatoes (Deng et al. 2019).

(i) **Foaming:** Foaming is done as a pre-treatment of FMD. Foam is produced from a pureed or liquid material, whipping with a foaming agent such as methylcellulose (0.25–2%), egg white (3–20%), maltodextrin (0.5–05%), and gum Arabic (2–9%) individually or in combination. It is used for the production of fruit juice powders. Fruits and vegetables that are sticky, heat sensitive, and difficult to dry can be dried using this technique. Foam can be formed by sparging, whipping (beating), and shaking. Under sparging, unlimited air is targeted to be incorporated but under whipping and shaking limited air can be incorporated.

(j) **Syrup Blanching**: Blanching fruit in syrup helps it retain color fairly well during drying and storage. Fruits that can be syrup-blanched include apples, apricots, figs, nectarines, peaches, pears, plums, and prunes.

(k) **Blanching:** The goal of thermal blanching is to inactivate the enzymes involved in the spoilage of fresh agro products, soften the tissues to facilitate drying, remove inter-cellular air from the tissues to prevent oxidation, and impart desired porosity and translucent appearance to the product. It is well described by Deng et al. (2019) and Bassey et al. (2021). Different blanching techniques have been developed such as:

- Conventional hot water blanching used in pumpkin, cherry tomato, and banana
- Steam blanching applied in cauliflower, apple, banana, and potato
- Super-heated impingement blanching practiced for yam, apple, and red pepper
- Ohmic-heat blanching tried in strawberry, artichoke, and grape
- Microwave blanching fruitful in green beans and carrot
- Infrared blanching useful for bitter gourd, mango, and carrot

(l) **Ultrasonic Field:** Ultrasonic pre-treatment on agro-products is a novel technology and is of keen interest as it can be utilized for heat-sensitive compounds. It also enhances the drying rate by altering the microstructure of plant tissue, hence reducing drying time; it increases the extraction ability of compounds. However, it is not yet developed on an industrial scale. It has been used for better retention of pigments in plums and strawberries. Enhanced vitamin C content has been found in quince slices and honeyberry fruits, along with increased phenolic compound content in quince, banana, and sweet potatoes (Llavata et al. 2020).

(m) **Freezing:** It has been used as a pre-treatment for drying fruits and vegetables. Freezing forms ice crystals, which on thawing results in the breakdown of tissues and the formation of porous structure, thus facilitating enhanced drying rate and decreased drying time. It also adds extra operation cost, inability to inactivate enzymes responsible for browning, and nutrition loss during thawing (Deng et al. 2019).

(n) **Edible Film Coating:** Edible films are natural biopolymers used as a surface coating to maintain food safety and quality by protecting against lipid oxidation, enzymatic reaction, mechanical harm, and senescence. The ability of edible films to serve as a protective barrier during drying, especially at high coating concentrations, can ensure minimal oxidation and limit the exposure of bioactive components to heat, thus producing dried fruits and vegetables with improved nutritional and functional properties. For example, coated pineapple exhibited enhanced retention of vitamin C and improved color during drying (Bassey et al. 2021), and calcium-enriched carrageenan coating on fresh strawberries resulted in a decrease in firmness (Ribeiro et al. 2007).

(o) **Carbonic Maceration Pre-Treatment:** In this method, samples to be treated are placed in a closed tank with a carbon dioxide-rich atmosphere. This results in enriched anaerobic conditions. Cytoplasm pH decreases, cell ruptures, cell membrane modifies, and inactivation of key enzymes and extraction of intracellular substances occur. The accumulative effect accelerates the drying rate. It is environment-friendly, safe for food as well as quality; however, carbonic maceration pre-treatment may trigger anaerobic respiration leading to undesirable changes in product texture and aroma such as in dried tomato quarters (Turgut et al. 2018) and dried sweet potatoes (Zhao et al. 2016).

5.3 Mechanisms of Heat and Mass Transfer in Drying of Horticultural Produce

As we know, drying is a complex process that involves the simultaneous transfer of heat and mass within the horticultural produce. Heat is transferred to evaporate the moisture content, while mass transfer occurs as the evaporated moisture moves from regions of higher concentration to lower concentration. This section provides an overview of the mechanisms involved in heat and mass transfer during the drying process (Jayaraman and Gupta 2020).

5.3.1 Heat Transfer Mechanisms

Heat transfer during drying can occur through three main mechanisms: (A) Conduction, (B) Convection, and (C) Radiation

(A) Conduction: Conduction heat transfer involves the transfer of heat through direct contact between particles within the horticultural produce. It occurs when there is a temperature gradient within the material, causing the heat to flow from the hotter regions to the colder regions. Conduction is primarily responsible for heat transfer within the solid matrix of the produce (Zhang et al. 2022).

(B) Convection: Convective heat transfer relies on the movement of fluid, typically air, to transfer heat to the surface of the produce. As the heated air comes into contact with the surface, it absorbs the moisture and carries it away, promoting the drying process. Convective heat transfer is influenced by factors such as air velocity, temperature, and humidity (Ling et al. 2022).

(C) Radiation: Radiative heat transfer occurs through the emission and absorption of electromagnetic waves. In the drying process, thermal radiation is responsible for transferring heat energy from a hotter source to the surface of the produce. The rate of radiation heat transfer is influenced by factors such as surface temperature, emissivity, and the presence of any radiation-absorbing materials (Iranshahi et al. 2022; Delfiya et al. 2022).

5.3.2 *Mass Transfer Mechanisms*

Mass transfer during drying is mainly governed by the principles of diffusion, which involves the movement of moisture from regions of higher concentration to lower concentration (Rurush et al. 2022).

- **Constant Rate Drying**: In the initial stage of drying, the moisture content is relatively high, and the drying rate is primarily controlled by the rate of moisture diffusion within the produce. This stage is referred to as constant rate drying, where the drying rate remains constant, and the moisture content decreases gradually. The moisture movement within the produce occurs due to a moisture concentration gradient (Natarajan et al. 2022; Sharada 2013).
- **Falling Rate Drying**: As the drying process continues, the moisture content decreases, and the drying rate decreases as well. This stage is known as falling rate drying, where the rate of moisture diffusion becomes the limiting factor. Falling rate drying is influenced by factors such as moisture content, temperature, and the material's structural properties. Understanding the falling rate drying stage is crucial for determining the optimal drying time and preventing over-drying or under-drying of the produce (Zhang et al. 2022; Salehi 2020).

5.4 Challenges in Drying Horticultural Produce

Drying horticultural produce presents several challenges that need to be addressed for optimal preservation and product quality. Some of the challenges include (Thippeswamy et al. 2021):

- High moisture concern
- High reducing sugar and therefore, susceptible to browning
- High sugar content
- Low porosity in some produce.
- Rich in heat-labile bioactive compounds
- Drying leads to shrinkage defects

Understanding the mechanisms of heat and mass transfer in drying horticultural produce is crucial for optimizing the drying process and achieving high-quality final products. By addressing the challenges and exploring opportunities for improvement, we can enhance drying efficiency, preserve nutritional quality, and ensure the safety and shelf life of horticultural produce.

- ***Uniform Drying***: Achieving uniform drying across the produce is essential to maintain consistent product quality. Non-uniform drying can lead to uneven moisture content, resulting in variations in texture, color, and microbial growth.
- ***Retention of Nutritional Quality***: The drying process can lead to the loss of certain nutrients, such as vitamins and antioxidants, in horticultural produce. It is crucial to develop drying strategies that minimize nutrient degradation and preserve the nutritional value of the final product.
- ***Control of Microbial Growth***: Drying alone may not be sufficient to eliminate microbial growth in horticultural produce. Controlling microbial activity during drying and subsequent storage is necessary to ensure food safety and prevent spoilage.
- ***Novel drying techniques***: Exploring and implementing innovative drying techniques such as vacuum drying, FMD, freeze drying, and microwave drying can offer improved drying efficiency, reduced drying time, and enhanced product quality as discussed earlier.
- ***Mathematical Modeling and Simulation***: Developing mathematical models and simulation tools can aid in understanding the drying process, optimizing drying parameters, and predicting product quality. These tools enable researchers and practitioners to experiment with different scenarios and evaluate the impact on drying performance.
- ***Energy-Efficient Drying***: Exploring energy-efficient drying technologies and integrating renewable energy sources can contribute to sustainable drying practices. This approach helps reduce energy consumption, operational costs, and environmental impact (Ndlovu et al. 2022; Shelke et al. 2023; Li et al. 2019; Nijhuis et al. 1998).

5.5 Drying Models: Concept and Utility

Drying models play a critical role in understanding and predicting the behavior of the drying process. These models provide valuable insights into the underlying mechanisms, help in optimizing drying conditions, and enhance the drying efficiency of horticultural produce. This chapter explores the concept and utility of drying models, including mathematical, empirical, and computational simulation techniques (Adnouni et al. 2022; Shao et al. 2012). It discusses the benefits of utilizing drying models and their applications in various aspects of the drying process.

Drying models serve as valuable tools for analyzing and simulating the complex phenomenon occurring during the drying process. These models aim to describe the heat and mass transfer mechanisms and predict the changes in moisture content, temperature, and other relevant parameters over time. This section provides an introduction to the concept of drying models and their utility in understanding and improving drying operations.

Mathematical Modeling: Mathematical models involve the formulation of equations based on fundamental principles and physical laws governing heat and mass transfer. These models describe the drying process mathematically and can provide insights into the underlying mechanisms (Nukulwar and Tungikar 2021; Onwude et al. 2016; Sacilik et al. 2006). The utility of mathematical models includes the following:

1. ***Understanding Drying Mechanisms***: Mathematical models allow researchers to analyze the complex interactions between heat and mass transfer during drying. By incorporating relevant equations and boundary conditions, these models help identify the dominant mechanisms and their contributions to the overall drying process.
2. ***Optimization of Drying Conditions***: Mathematical models provide a platform for optimizing drying conditions such as temperature, airflow rate, and humidity. Through simulation and sensitivity analysis, researchers can identify the optimal drying parameters that maximize drying efficiency and product quality.
3. ***Scale-Up and Process Design***: Mathematical models enable the scale-up of drying operations from lab-scale to industrial-scale. By considering the geometrical and operational parameters, these models assist in designing efficient drying systems and predicting the drying behavior in larger-scale applications.

5.6 Empirical Models

Empirical models are based on experimental data and statistical analysis. These models establish relationships between input variables (such as drying time, temperature, and airflow) and output variables (such as moisture content or drying rate) (Kiranoudis et al. 1997). The utility of empirical models includes the following:

1. ***Quick Estimation of Drying Parameters***: Empirical models provide a simple and quick way to estimate drying parameters based on available experimental data. They can be used for preliminary analysis or when detailed mechanistic information is not available (Lingayat et al. 2020; Saikia et al. 2022).
2. ***Process Monitoring and Control***: Empirical models can be employed for real-time monitoring and control of the drying process. By continuously updating the model parameters using measured data, it becomes possible to adjust drying conditions to achieve the desired product quality and drying performance (Destro et al. 2023).

5.7 Computational Simulation Techniques

Computational simulation techniques, such as computational fluid dynamics and finite element analysis, offer advanced tools for analyzing complex drying systems. the utility of computational simulation techniques includes the following:

- ***Detailed Understanding of Drying Processes***: Computational simulation techniques provide a detailed understanding of the drying process by simulating the airflow patterns, heat transfer, and moisture distribution within the horticultural produce. This information can help optimize drying system design and identify potential issues (Capecelatro et al. 2022).
- ***Visualization of Drying Phenomena***: Simulation results can be visualized to observe the spatial and temporal distribution of moisture content, temperature, and other parameters. The visualization aids in identifying drying non-uniformities, hotspots, and potential areas for improvement (Rojano et al. 2014).
- ***Virtual Testing and Optimization***: Computational simulation allows for virtual testing and optimization of drying processes. It enables researchers to explore various scenarios, evaluate the effects of different parameters, and optimize drying conditions without the need for physical prototypes or extensive experimental trials (Kumar et al. 2012).

5.8 Reconstituability

Reconstitution is defined as the rehydration of dried products during which moisture is restored using water. Reconstitution is of interest to both manufacturers and consumers. It has the objective to restore the functional properties of the fresh produce. Reconstitution depends on both operating conditions and material qualities. For dehydrated fruits and vegetables, reconstitution time will depend on the size of the food material. In recent times, the food industry has been intrigued with fruit and vegetable powder having the ability to reconstitute. Reconstitution of food powder generally consists of five steps: *wetting of the powder particles,* i.e., contact between the liquid and particles, penetration of the liquid into the pores due to penetration force during which physiochemical properties of surface play a vital role, *sinking,* i.e., particles sink into the liquid from the surface, *dispersing* during which the bridges between primary particles are dissociated and they disperse in the liquid, and, finally, *dissolution* that involves completely dissolving in the solution, can be rapid or slow if lumps are formed.

For the production of fruit and vegetable powders, dehydration is done by drum drying, FMD, freeze-drying, spray drying, cast-tape drying, and so on. One of the advantages of foam-mat dried powder is the reduction in shipping weight; however, the foaming agent should be such that it does not result in a milky, unnatural appearance. Only spray drying produces particles of equivalent sizes; for the rest, grinding is required followed by sieving through meshes to obtain a powder of equivalent size. Particle size is imperative with regard to the functional, reconstitution, and handling properties. The rehydration ability of powders is related to some of their physical and chemical properties, such as composition, moisture content, particle shape and size and distribution, and morphology, which depends on the conditions the particles have been produced. Agglomeration is a process that involves particle size enlargement carried out to improve the flowability, compaction, and homogeneity of powders. Physical properties of the powder related to its flowability, bulk density, tap density, porosity, compressibility, angle of repose, adhesive, and cohesive nature are important to be known to understand its behavior during storage, handling, and processing. The storage conditions such as relative humidity, temperature, and time also influence the powder rehydration ability. The influence of reconstitution conditions (stirred speed, mixing systems, temperature, and viscosity) is also known to influence reconstitution time. The addition of maltodextrin is commonly used in the production of fruit powders, where it functions as a wall material to envelop spray-drying feed and improve powder yield. It contains sugar with low molecular weight, which functions as a plasticizer and prevents the shrinkage of the powder surface leading to the formation of a smooth surface.

5.9 Reconstitution Ratio

The calculation of water absorbed by dehydrated foods can be made to express the results in terms of "Rehydration Ratio," "Coefficient of Rehydration," and "Percentage of Water in the Rehydrated Material":

- Rehydration Ratio: $\dfrac{\text{Weight of rehydrated sample}}{\text{Weight of dehydrated sample}}$

- Coefficient of Rehydration:

$$= \frac{\text{Weight of the Rehydrated Sample}}{\dfrac{\left(\begin{array}{l}\text{Weight of the dehydrated sample} - \text{Moisture Content}\\ \text{of the dehydrated sample}\end{array}\right)100}{100 - \text{Moisture Content of Initial Sample before dehydration}}}$$

5.10 Innovative Dried Value-Added Products

Figure 5.4 shows some innovative dried products developed by ICAR-IARI, New Delhi. They are mainly infused and osmo-dried products. Their technology is very simple and can be adopted by small and marginal agriprenuers. They are mainly bioactive enriched dried products. Osmo-dehydration and FMD are majorly practiced for fruits and vegetables as they are relatively low-cost techniques and as discussed earlier they have the capability to cater to the challenges of a difficult matrix of horticultural produce such as high sugar, gumminess, juiciness, and heat-labile texture. Therefore, sophisticated instruments and trained manpower are not a mandatory requirement for these cases.

Fig. 5.4 Osmo-dried innovative dried products developed by ICAR-IARI, India

5.11 Conclusion

Innovative drying technologies must complement low-cost drying methods because sometimes comparable benefits can be harnessed with low-cost interventions. As observed by Pooja et al. (2023), pea pod protein concentrate obtained after air drying was at par with the characteristics of the freeze-dried sample. In that case, low-cost technologies can also be suggested as a more sustainable approach. Simultaneously, drying models, including mathematical models, empirical models, and computational simulation techniques, are valuable tools for understanding, optimizing, and improving the drying process of horticultural produce. These models provide insights into the underlying mechanisms, aid in process optimization, and facilitate scale-up and design of drying systems. By utilizing drying models, researchers and practitioners can enhance drying efficiency, improve product quality, and reduce costs and energy consumption.

References

Adnouni M, Jiang L, Zhang XJ, Zhang LZ, Pathare PB, Roskilly AP (2022) Computational modelling for decarbonised drying of agricultural products: sustainable processes, energy efficiency, and quality improvement. J Food Eng 338:111247

Bala BK, Mondol MRA (2001) Experimental investigation on solar drying of fish using solar tunnel dryer. Dry Technol 19(2):427–436. https://doi.org/10.1081/DRT-100102915

Bassey EJ, Cheng JH, Sun DW (2021) Novel nonthermal and thermal pretreatments for enhancing drying performance and improving quality of fruits and vegetables. Trends Food Sci Technol 112:137–148

Capecelatro J, Longest W, Boerman C, Sulaiman M, Sundaresan S (2022) Recent developments in the computational simulation of dry powder inhalers. Adv Drug Deliv Rev 118:114461

Courtois F (2013) Roller and drum drying for food powder production. In: Handbook of food powders: processes and properties. Elsevier, pp 85–104. https://doi.org/10.1533/9780857098672.1.85

Delfiya DA, Prashob K, Murali S, Alfiya PV, Samuel MP, Pandiselvam R (2022) Drying kinetics of food materials in infrared radiation drying: a review. J Food Process Eng 45(6):e13810

Deng LZ, Mujumdar AS, Zhang Q, Yang XH, Wang J, Zheng ZA et al (2019) Chemical and physical pretreatments of fruits and vegetables: effects on drying characteristics and quality attributes–a comprehensive review. Crit Rev Food Sci Nutr 59(9):1408–1432

Destro F, Barolo M, Nagy ZK (2023) Quality-by-control of intensified continuous filtration-drying of active pharmaceutical ingredients. AICHE J 69(2):e17926

Fox B, Bellini G, Pellegrini L (2014) Drying. In: Fermentation and biochemical engineering handbook: principles, process design, and equipment, 3rd edn. Elsevier, pp 283–305. https://doi.org/10.1016/B978-1-4557-2553-3.00014-3

Gupta DKD (2012) Advances in dehydration technologies for vegetables and fruits. In: Continuing eduction programme manual organized by DFRL, Mysore

Henríquez C, Córdova A, Almonacid S, Saavedra J (2014) Kinetic modeling of phenolic compound degradation during drum-drying of apple peel by-products. J Food Eng 143:146–153. https://doi.org/10.1016/J.JFOODENG.2014.06.037

Huang D, Yang P, Tang X, Luo L, Sunden B (2021) Application of infrared radiation in the drying of food products. Trends Food Sci Technol 110:765–777. https://doi.org/10.1016/J.TIFS.2021.02.039

Indiarto R, HodizahAsyifaa A, Citra AngiputriAdiningsih F, Almira Aulia G, RahmaliaAchmad S (2021) Conventional and advanced Food-drying technology: a current review. Int J Sci Technol Res 10:1

Inserra L, Cabaroğlu T, Şen K, Arena E, Ballistreri G, Fallico B (2017) Effect of sulphuring on physicochemical characteristics andaroma of dried Alkaya apricot: a new Turkish variety. Turk J Agric For 41(1):59–68

Iranshahi K, Onwude DI, Martynenko A, Defraeye T (2022) Dehydration mechanisms in electrohydrodynamic drying of plant-based foods. Food Bioprod Process 131:202–216

Jayaraman KS, Gupta DD (2020) Drying of fruits and vegetables. In: Handbook of industrial drying. CRC Press, pp 643–690

Joshi A, Rudra SG, Sagar VR, Raigond P, Dutt S, Singh B, Singh BP (2015) Development of low fat potato chips through microwave processing. J Food Sci Technol 53(8):3296–3303

Kiranoudis CT, Tsami E, Maroulis ZB, Marinos-Kouris D (1997) Drying kinetics of some fruits. Dry Technol 15(5):1399–1418

Kumar C, Karim A, Saha S, Joardder MUH, Brown R, Biswas D (2012) Multiphysics modelling of convective drying of food materials. In: Proceedings of the global engineering, science and technology conference 2012. Global Institute of Science and Technology, pp 1–13

Li K, Zhang M, Mujumdar AS, Chitrakar B (2019) Recent developments in physical field-based drying techniques for fruits and vegetables. Dry Technol 37(15):1954–1973

Ling W, Xing Y, Hong C, Zhang B, Hu J, Zhao C et al (2022) Methods, mechanisms, models and tail gas emissions of convective drying in sludge: a review. Sci Total Environ 845:157376

Lingayat A, Chandramohan VP, Raju VRK, Kumar A (2020) Development of indirect type solar dryer and experiments for estimation of drying parameters of apple and watermelon. Thermal Sci Eng Progr 16:100477

Liu L, Dai X, Kang H, Xu Y, Hao W (2020) Structural and functional properties of hydrolyzed/glycosylated ovalbumin under spray drying and microwave freeze drying. Food Sci Human Wellness 9:80–87. https://doi.org/10.1016/j.fshw.2020.01.003

Llavata B, García-Pérez JV, Simal S, Cárcel JA (2020) Innovative pre-treatments to enhance food drying: a current review. Curr Opin Food Sci 35:20–26

Mahajan PV, Caleb OJ, Singh Z, Watkins CB, Geyer M (2014) Postharvest treatments of fresh produce. Philos Trans Royal Soc Math Phys Eng Sci 372(2017):20130309

Natarajan SK, Elangovan E, Elavarasan RM, Balaraman A, Sundaram S (2022) Review on solar dryers for drying fish, fruits, and vegetables. Environ Sci Pollut Res 29(27):40478–40506

Ndlovu PF, Magwaza LS, Tesfay SZ, Mphahlele RR (2022) Destructive and rapid non-invasive methods used to detect adulteration of dried powdered horticultural products: a review. Food Res Int 157:111198

Nguyen TTT, Le TVA, Dang NN, Nguyen DC, Nguyen PTN, Tran TT, Nguyen QV, Bach LG, Thuy Nguyen Pham D (2021) Microencapsulation of essential oils by spray-drying and influencing factors. J Food Qual 2021. https://doi.org/10.1155/2021/5525879

Nijhuis HH, Torringa HM, Muresan S, Yuksel D, Leguijt C, Kloek W (1998) Approaches to improving the quality of dried fruit and vegetables. Trends Food Sci Technol 9(1):13–20

Nowak D, Jakubczyk E (2020) The freeze-drying of foods—the characteristic of the process course and the effect of its parameters on the physical properties of food materials. Foods 9(10):1488. https://doi.org/10.3390/FOODS9101488

Nukulwar MR, Tungikar VB (2021) A review on performance evaluation of solar dryer and its material for drying agricultural products. Mater Today Proc 46:345–349

Ochmian I, Figiel-Kroczyńska M, Lachowicz S (2020) The quality of freeze-dried and rehydrated blueberries depending on their size and preparation for freeze-drying. Acta Univ Cibiniensis Ser E Food Technol 24(1):61–78

Onwude DI, Hashim N, Janius RB, Nawi NM, Abdan K (2016) Modeling the thin-layer drying of fruits and vegetables: a review. Compr Rev Food Sci Food Saf 15(3):599–618

Orsat V, Raghavan GSV, Krishnaswamy K (2017) Microwave technology for food processing: an overview of current and future applications. In: The microwave processing of foods, 2nd edn. Elsevier, pp 100–116. https://doi.org/10.1016/B978-0-08-100528-6.00005-X

Pandey A, Singh G (2011) Development and storage study of reduced sugar soy containing compound chocolate. J Food Sci Technol 48(1):76–82

Patel KK, Kar A (2012) Heat pump assisted drying of agricultural produce—an overview. J Food Sci Technol 49:142–160

Pawar SB, Pratape VM (2017) Fundamentals of infrared heating and its application in drying of Food materials: a review. J Food Process Eng 40(1):e12308. https://doi.org/10.1111/JFPE.12308

Pooja BK, Sethi S, Bhardwaj R, Chawla G, Kumar R, Joshi A, Bhowmik A (2023) Isoelectric precipitation of protein from pea pod and evaluation of its physicochemical and functional properties. Vegetos 37:1131–1141

Radoiu M (2020) Microwave drying process scale-up. Chem Eng Process Process Intensif 155:108088. https://doi.org/10.1016/J.CEP.2020.108088

Ribeiro C, Vicente AA, Teixeira JA, Miranda C (2007) Optimization of edible coating composition to retard strawberry fruit senescence. Postharvest Biol Technol 44(1):63–70

Rojano A, Espinosa T, Rojano F, Salazar R (2014, August) A computational simulation of forced cactus drying. In: XXIX international horticultural congress on horticulture: sustaining lives, livelihoods and landscapes (IHC2014): 1120, pp 369–374

Rurush E, Alvarado M, Palacios P, Flores Y, Rojas ML, Miano AC (2022) Drying kinetics of blueberry pulp and mass transfer parameters: effect of hot air and refractance window drying at different temperatures. J Food Eng 320:110929

Sacilik K, Keskin R, Elicin AK (2006) Mathematical modelling of solar tunnel drying of thin layer organic tomato. J Food Eng 73(3):231–238

Saikia D, Nayak PK, Krishnan KR, Kondareddy R, Lakshmi DVN (2022) Development of indirect type solar dryer and experiments for estimation of drying parameters of dhekia (Diplazium esculentum). Mater Today Proc 56:774–780

Saleem I, Petkar K, Somavarapu S (2017) Rationale for pulmonary vaccine delivery: formulation and device considerations. In: Micro- and nanotechnology in vaccine development. Elsevier, pp 357–371. https://doi.org/10.1016/B978-0-323-39981-4.00019-1

Salehi F (2020) Recent applications and potential of infrared dryer systems for drying various agricultural products: a review. Int J Fruit Sci 20(3):586–602

Schuck P (2002) Spray drying of dairy products: state of the art. Lait 82(4):375–382. https://doi.org/10.1051/LAIT:2002017
Shao S, Pipattanasomporn M, Rahman S (2012) Development of physical-based demand response-enabled residential load models. IEEE Trans Power Syst 28(2):607–614
Sharada S (2013) Studies on effect of various operating parameters & foaming agents-drying of fruits and vegetables. Int J Modern Eng Res 3(3):1512–1519
Shelke G, Kad V Pandiselvam R, Yenge G, Kakade S, Desai S et al (2023) Physical and functional stability of spray-dried jamun (*Syzygium cumini* L.) juice powder produced with different carrier agents. J Texture Stud 54(4):560–570
Sivakumar R, Saravanan R, Elaya Perumal A, Iniyan S (2016) Fluidized bed drying of some agro products—a review. Renew Sust Energ Rev 61:280–301. https://doi.org/10.1016/J.RSER.2016.04.014
Thippeswamy B Joshi A, Sethi S (2021) Foam mat drying of fruits: concept, opportunities and limitations. Biotica Res Today 3(1):35–37
Turgut SS, Küçüköner E, Karacabey E (2018) Influence of carbonic maceration pre-treatment on functional quality of dried tomato quarters. Food Bioprocess Technol 11:1818–1827
Verma RC, Gupta A (2004) Effect of pre-treatments on quality of solar-dried amla. J Food Eng 65(3):397–402
Zarein M, Samadi SH, Ghobadian B (2015) Investigation of microwave dryer effect on energy efficiency during drying of apple slices. J Saudi Soc Agric Sci 14(1):41–47. https://doi.org/10.1016/J.JSSAS.2013.06.002
Zhang WP, Yang XH, Mujumdar AS, Ju HY, Xiao HW (2022) The influence mechanism and control strategy of relative humidity on hot air drying of fruits and vegetables: a review. Dry Technol 40(11):2217–2234
Zhao D, Wang Y, Zhu Y, Ni Y (2016) Effect of carbonic maceration pre-treatment on the drying behavior and physicochemical compositions of sweet potato dried with intermittent or continuous microwave. Dry Technol 34(13):1604–1612

Chapter 6
Packaging Technologies for Horticulture Crops

S. G. Lekshmi, Shruti Sethi, P. K. Anagha, Bindvi Arora, and Swarajya Laxmi Nayak

Abstract Packaging is one of the crucial elements in the long and complex value chain of horticultural products to deliver them from the production site to the customer. Suitable containers for handling, transporting, and marketing fresh fruit include bags, crates, hampers, baskets, cartons, bulk bins, and palletized containers. Packers, shippers, purchasers, and consumers must thoroughly understand the various packaging alternatives since they represent a considerable expense to the produce sector. Fresh produce or processed products can be delivered in safe and economical condition from the production center to the last customer through the use of packaging. This book chapter goes into great length about the significance, need, and purposes of traditional and modern packaging techniques.

Keywords Packaging · Fruits and vegetables · MAP · CAP · Active packaging · Intelligent packaging

6.1 Introduction

Food packaging is the act of enclosing food to safeguard it against contamination from physical, chemical, or biological sources (Cachon et al. 2019). The ultimate goal of any packaging technology is to prolong the shelf life of the product by preventing physical, chemical, and biological deterioration, guaranteeing product

S. G. Lekshmi · S. Sethi (✉) · B. Arora
Division of Food Science and Postharvest Technology, ICAR-Indian Agricultural Research Institute, New Delhi, Delhi, India

P. K. Anagha
Division of Fruits and Horticultural Technology, ICAR-Indian Agricultural Research Institute, New Delhi, Delhi, India

S. L. Nayak
ICAR-Directorate of Medicinal and Aromatic Plants Research, Anand, Gujarat, India

M. Thakur, T. Belwal (eds.), *Advances in Postharvest and Analytical Technology of Horticulture Crops*, https://doi.org/10.1007/978-981-97-7247-6_6

quality, and safely transporting the item from the producer to the consumer's table. A few decades ago, packaging was merely a necessity, but today, it is both an art and a science. Bags, crates, hampers, baskets, cartons, bulk bins, and palletized containers are all viable means to handle, transport, and sell fresh produce. Once harvested, fresh produce continues to respire, which involves using oxygen and releasing carbon dioxide. The respiration rate varies for each commodity and determines their decay rate, with faster respiration leading to quicker ripening and aging. Lack of oxygen can result in chemical changes, cell breakdown, and spoilage, causing unpleasant flavors and odors. The appropriate packaging for horticultural commodities varies depending on their physical, anatomical, and physiological characteristics, such as transpiration, respiration, and ethylene production rate, as well as their susceptibility to microbial decay. Additionally, post-harvest life is significantly influenced by temperature, relative humidity, and ventilation.

6.1.1 Properties of Packaging

The fundamental prerequisites and features of effective packaging include:

1. Mass manufacturing
2. Reasonable and effective packaging material
3. Appropriate structure and shape
4. Practicality
5. Ease of disposal

6.1.2 Principles of Packaging

Horticultural commodities need packaging that avoids contamination, reactions with the packaging material, decomposition, and other possible problems. The functions of packaging are containment, protection, preservation, convenience, communication, and marketing. As listed below, it can be completed in several layers:

(i) Primary Packaging: This type of packaging is directly in contact with the food, enclosing and securing it.
(ii) Secondary Packaging: Positioned outside the primary packaging.
(iii) Tertiary Packaging: This sturdy outer layer serves for warehouse storage, bulk handling, and transportation/shipping purposes.

6.1.3 Requirements for Effective Food Packaging

- The packaging material must be free from any harmful chemicals.

- It should allow for quick cooling of the products.
- It needs to withstand high humidity and moisture.
- It must meet size, shape, strength, and weight standards for handling and marketing.
- The package should have sufficient strength to safeguard the products.
- It should be able to interlock and stack.
- For easy disposal, the package should be reusable and recyclable.
- It should be cost-efficient. It should open easily.
- It must be suitable for direct contact with food.

6.2 Types of Packaging

6.2.1 Traditional

- Leaf-based packaging
- Wooden (pallets, pallet bins, wire-bound crates, wooden crates and lugs, wooden baskets, and hampers)
- Jute bags
- Paper
- Corrugated fiberboard (CFB) boxes

6.2.1.1 Leaf-Based Packaging

Leaf is an eco-friendly packaging material in primitive society. The commonly used leaves are palm leaves, lotus leaves, palas leaves, banana leaves, pandan leaves, and leaves of Casuarinas and bamboo. Leaf-based packaging has benefits such as being sustainable, biodegradable, and environmentally beneficial and can be easily disposed off. Leaf packaging has religious significance also in India. Some leaves exhibit an antibacterial impact on packaged food. However, lower resistance limits the application of leaf-based packing (Fig. 6.1a).

6.2.1.2 Wooden

Natural wood and commercially produced wood-based sheet materials are used to make wooden boxes. Wood is a versatile and highly effective packing material. Wooden boxes have replaced traditional baskets because they offer superior protection for fresh products against transportation risks. They take up more space and increase the tare weight but have a strong puncture resistance, good tensile strength, and good compression strength. The major drawback of using wooden boxes is that during lengthy transportation, the fruit might get damaged by the nails embedded in

Fig. 6.1 Various packaging materials used for fruits and vegetables: (**a**) leaf packaging of flowers, (**b**) paper pulp container, (**c** and **d**) CFB boxes, (**e**) polyester trays with overwrap, (**f**) plastic punnets, (**g, h**) plastic mesh bag, (**i**) foam sleeve, (**j**) plastic crates, (**k–m**) shrink wrap, (**n**) shrink wrapping machine, (**o**) sleeve, (**p**) flower sleeve, (**q**) vacuum packed pumpkin, and (**r**) vacuum machine

Fig. 6.1 (continued)

Fig. 6.1 (continued)

the box. However, using wooden boxes is no longer recommended because it directly encourages deforestation. The many wooden packaging options for horticultural produce are listed below.

(a) **Pallets**: Most fresh food is transported to consumers on pallets. Most automated pallet handling machines and pallet organizers are made to fit standard-size pallets, which not only saves money because they can be used multiple times but also increases efficiency.
(b) **Pallet Bins:** Nearly all pallet bins are built locally; thus, it is crucial that they maintain consistency in terms of their structure, materials, and, most importantly, size, from lot to lot. A hardwood pallet bin kept outside typically lasts 5 years on average. Pallet bins may be functional for 10 years or more if adequately weatherproofed.
(c) **Wire-Bound Crates:** Wire-bound crates are sewed crates that have a wire under the stitches to provide the container with more strength. The wire also functions as a lid lock and a hinge. Wire-bound crates are strong and rigid, have a significant amount of stacking strength, and are unaffected by moisture.
(d) **Wooden Crates and Lugs:** Once, these were most commonly used for the transportation of apples, tomatoes, stone fruits, and potatoes. Presently, they have been largely substituted by alternative containers because of their higher cost, increased emphasis on tare weight, and advancements in material handling.
(e) **Wooden Baskets and Hampers:** These baskets were once employed for a wide range of crops, including strawberries and sweet potatoes. They can be grouped together for effective transport, and they are robust. However, its application has been severely limited, mostly to small local grower markets, where they may be reused numerous times due to their high cost, disposal issues, and efficiency challenges in palletization.

6.2.1.3 Jute Bags

Jute fiber and other comparable natural materials are traditionally used to make sacks. Most of the jute bags are offered in a plain weave. Materials with a density of no more than 250 g/m^2 are used to convey 1 ton of veggies. Bags made of synthetic materials and paper have frequently replaced those made of natural fibers because they are cheaper, more attractive, have better mechanical qualities, and pose less of a risk of insect infestation and dissemination.

They are efficient due to their low price, excellent strength, reusability, and need for little storage space. However, they offer little defense against impact, vibration, compression, and puncture wounds and are not very stackable. Usually, these sacks are used in conjunction with bamboo baskets and wooden boxes to improve cushioning, lower the risk of losses, and lesser bruising during transit.

6.2.1.4 Paper

Paper is a traditional form of packaging made from cellulose-based materials. Air, water vapor, and gases (oxygen) can pass through it. It has weak tear resistance. Paper is used to create a wide range of bags and boxes for various purposes. Additionally, paper can be utilized to create lightweight cartons that serve as a vibrant outer wrap for goods packed in plastic or metal containers. Cardboard may be used to make a variety of cans for storing items such as nuts, spices, snacks, and even cups for liquids. Paper waste can be recycled, composted, or burned (with energy recovery). Different ways in which the paper can be used as packaging material is as follows:

- **Pulp Containers**: Fresh vegetable packaging for small consumers is usually done in starch-binding and recycled paper pulp containers. Pulp containers are cheaply priced in typical dimensions and are available in a variety of sizes and shapes. Pulp containers have the ability to draw moisture from the surface, which is advantageous for berries and tiny fruits that are susceptible to water damage. The pulp containers are recyclable and made of recyclable and biodegradable substances (Fig. 6.1b).
- **Paper and Mesh Bags**: The only produce items that are still packaged in paper bags for consumer use are potato and onion packs. The utilization of the stronger mesh bag is substantially greater. In addition to potatoes and onions, mesh bags are used to store cabbage, turnips, and citrus. The mesh benefits from free ventilation along with being inexpensive. A healthy airflow particularly benefits onions. Supermarkets employ little mesh bags because they make eye-catching displays that boost sales.

6.2.1.5 CFB Boxes

CFB boxes are widely used as transport or shipping containers for fresh produce. The majority of CFB is formed from paperboard that has been kraft-processed into three or more layers (Fig. 6.1c, d). The significant advantages are:

- Economical compared with its strength and weight
- Effective cushioning abilities
- A smooth, non-abrasive surface
- Suitable for clear printing on the outer board surface
- Simple to assemble and collapsible for storage convenience
- Reusable and recyclable
- Capable of high-volume manufacturing
- Capable of ventilation through hole punching

6.2.2 Plastic Packaging

Many comparable synthetic materials frequently used for food packaging fall under the general category of plastic. It is durable, airtight, lightweight, recyclable, and robust. Plastic bags prolong the product's shelf life and preserve its freshness. Plastic bags can be used to store items that are virtually moisture-free for a long time. We do not necessarily need to open it to see what is inside because it is mostly transparent. Many other kinds of packaging materials, including bags, wraps, bottles, tubs, buckets, containers, re-sealable pouches, and so on, can be made using plastic. Because they are airtight, bags can help keep food from getting soggy in humid environments. Its inability to biodegrade is a drawback. Different forms in which plastics are used as packaging for horticultural produce include the following:

- **Pouches**: Many different styles of pouches are made of high-quality, long-lasting, and environmentally friendly materials. Examples are spout pouches, zipper pouches, printed stand-up pouches, reusable pouches, and so on. Additionally, they have information found on food labels, such as the production date, expiry date, nutrient content, logos, and slogans.
- **Flexible Plastic Films**: The packaging of horticulture products uses a variety of flexible plastic films, including cellulose acetate, polypropylene (PP), polyvinyl chloride, and LDPE (low-density polyethylene). Most of the time, these films are employed as pouches with breathing holes placed at regular intervals. They can be designed to manage the ambient gases inside the pouch and come in a wide range of grades and thicknesses. The most often-used substance is LDPE.
- **Trays With Overwrap:** The trays that are utilized are often constructed of plastic materials, including extruded polystyrene, polyvinylchloride (PVC), and polypropylene (PP) or molded pulp trays. To prevent bruising and abrasion during shipping, the product is put into separate compartments. The trays additionally cushion the vegetables. The overwrap film is a transparent, see-through, odorless plastic material that, when stretch-wrapped around a product, adheres to it. It is possible to apply this film without using heat. It is typically constructed of PVC, LDPE, or linear LDPE (LLDPE). The films allow for the exchange of gases necessary for the product to breathe and are semi-permeable (Fig. 6.1e).
- **Plastic Punnets**: These are sturdy, adaptable, transparent, bright containers with holes for ventilation to keep the fruit fresh. They also enhance product visibility. These containers are good-looking, food-grade, odorless, low in weight, stackable, and recyclable. They are made of either PET (polyethylene tertrapthalate), PVC, or PP (Fig. 6.1f).
- **Plastic Net Bags (Extruded and Woven):** Stretching and accommodating all product sizes and shapes is a property of plastic net bags. With a stretch width of 200–400 mm, these bags are offered in roll form or precut lengths. These net bags extend the freshness and shelf-life of the food by allowing air to circulate around and inside the produce. Additionally, they get rid of pack condensation, which stops waste and spoilage. They enhance the natural color of fresh fruit and

create a bright point-of-sale display by making the contents visible. These are often made of polyamide or HDPE (high-density polyethylene) (Fig. 6.1g, h).
- **Foam Sleeve:** This is a plastic tubular film with varying colors, diameters, and lengths manufactured of polyethylene foam. It fits snugly over the individual fruits and is simple to slip on. Fresh food is shielded from abrasion and scrapes while in transportation, thanks to its cushioning effect. It is non-toxic, odorless, and hygienic (Fig. 6.1i).
- **Light-Weight Plastic Crates**: These compact crates can be transported without an exterior pack because they are lightweight. The product is kept fresh by the ventilation provided by the perforations. Their strong compression strength and stackability ensure that the fresh product within is adequately protected. These crates are sanitary, spotless, reusable, and recyclable. They can be created from PP or HDPE (Fig. 6.1j).
- **Plastic Crates:** These are typically injection-molded from HDPE or PP. While polypropylene is more scratch-resistant and has lower ultraviolet radiation degradation, polyethylene has a higher impact strength. The addition of antioxidants and UV filters (for sun protection) can enhance the efficacy of both materials.

The advantages of plastic crates are as follows:

- They possess strength and rigidity.
- Suitable for extended transportation, resulting in a relatively low cost per trip.
- Available in various sizes and shapes.
- Simple to clean and disinfect.
- Strong and weather-resistant, making them applicable in humid regions.

These boxes may be collapsed, stacked, or nested together. The most expensive ones are plastic collapsible crates, followed by stack-nest and stackinates. The use of foldable crates lowers the cost and amount of storage space needed to move empty containers. The typical capacity ranges from 20 to 40 kg.

- **Shrink Wrap**: This is one of the newest packaging innovations for fresh food wherein individual produce or sorted fruits are trayed and wrapped. Individual shrink wrapping's primary benefit is its capacity to limit moisture loss. The coating creates a barrier that boosts protection against water vapor by lowering the transpiration rate and keeping the fruit solid (Sharma et al. 2013). By utilizing selectively permeable plastic sheets, it is possible to reduce the transpiration rate by a factor of 5–20. A flexible film is used to seal each piece of fruit loosely. These packs are then put through a heat-shrink tunnel where they are briefly (a few seconds) exposed to hot blowing air before the film is compressed securely around the fruit. The fresh produce is then quickly ventilated to chill it. LDPE or LLDPE films are the most often utilized materials (Fig. 6.1k–n).
- **Sleeve Packing:** Sleeve packaging of flowers refers to the practice of using protective sleeves or covers to package and transport flowers. The floral industry frequently uses this technique to maintain the freshness of cut flowers and improve their appearance. Sleeve packaging plays a crucial role in the floral business by guaranteeing the secure transport and aesthetic appeal of cut flowers. These sleeves, which are intended to fit around individual flower stems or bou-

quets, are often made of translucent or semi-transparent materials, such as plastic or cellophane. Sleeve packing is primarily used to safeguard flowers during shipping, guarding against damage to delicate petals, foliage, and stems. The sleeves serve as a barrier, protecting the flowers from any outside elements including temperature fluctuations, humidity, and physical damage (Fig. 6.1o, p).

- **Corrugated Boxes/Cartons:** Many fruits, including mangoes, apples, grapes, and so on, are packed either in corrugated boxes made of paper board or polymers like polypropylene in compact packs of 2–4 kg. These cartons/boxes are lightweight with strong compression. They can be printed to look good on the shelf.

6.2.3 Modern Packaging Technologies

In today's world, traditional packaging methods fall short of meeting our needs. While packaging has traditionally served as a container for food, facilitating transportation and preventing damage, it can no longer ensure the necessary level of efficiency and safety for consumers. The need for more advanced packaging solutions has become increasingly apparent (Ozdemir and Floros 2004). Packaging aims to shield food from external elements such as oxygen, moisture, light, and potential contaminants, both chemical and biological. Unfortunately, perishable items may sometimes face shipment delays or temperature fluctuations during transportation, resulting in food spoilage or over-ripening and, ultimately, financial loss. Active packaging is not like traditional packaging that sits there. It is a system designed to improve packaging by engaging with the food or environment (Lim 2011). Additionally, an intelligent packaging system has emerged that can assist in monitoring the state and quality of products using specialized indicators like freshness indicators and radio frequency identification tags (Siddiqui et al. 2018). Modern packaging has been developed to cater to the specific needs of the commodity by managing the internal gas composition inside the pack as discussed below.

6.2.3.1 Reduced Oxygen Packaging

It is a wide classification in which fruits and vegetables are packaged with less oxygen than in normal atmospheric conditions. In all these systems, the respiration of the commodity is manipulated.

6.2.3.1.1 Vacuum Packaging

Vacuum packaging is a common method for altering the atmosphere within a package (Fig. 6.1q, r). It involves sealing the food in an airtight package to prevent gas permeation and removing most of the oxygen surrounding the food resulting in modified atmosphere (MA) conditions (Meena et al. 2017). When food is vacuum-packaged, it slows down the process of oxidative reactions and prevents the growth of aerobic microorganisms. This, in turn, helps to prevent the degradation of food during storage (Martens and Farber 1995).

6.2.3.1.1.1 Advantages of Vacuum Packaging

(i) An uncomplicated method for packaging items needing protection from oxygen.
(ii) Effective management of the moisture levels in the produce.
(iii) Prevents the proliferation of aerobic spoilage bacteria.
(iv) Less expensive compared to rigid containers.
(v) Extended shelf life for goods.

6.2.3.1.1.2 Disadvantages of Vacuum Packaging

(i) Removing all the oxygen from food is almost impossible as there will always be small quantities trapped within food cells.
(ii) This tool is not suitable for items with sharp edges or those that need to be crispy.
(iii) Needs packaging material with strong barrier properties to sustain a vacuum.
(iv) Anaerobic conditions may trigger the promote the growth of *Clostridium botulinum* and *Listeria monocytogenes*.
(v) Microorganisms that are not impacted by oxygen are not effectively controlled through vacuum packing.
(vi) Capital intensive.

6.2.3.1.2 Modified Atmosphere Packaging

Modified atmosphere packaging (MAP) is based on altering the levels of O_2 and CO_2 to form the desired atmosphere, which can lower the respiration rate, ethylene production, and the growth of organisms that cause decay (Madhav and Sethi 2016). In MA storage, the atmospheric composition is initially set to a pre-determined gas composition, but changes may happen due to the product's physiological activity and the physical environment's characteristics. It is important to note that the atmospheric composition is not actively controlled in passive MAP (such as the composition of the gas atmosphere, temperature, and relative humidity). The respiring commodity inside the package itself changes the composition of gases within the pack. The MAP system is a dynamic one where respiration and permeation occur simultaneously. Factors affecting both of these must be considered when designing a package.

The respiration of the product within its packaging is impacted by several factors:

- Amount of the produce
- Ripeness stage
- Temperature

- Presence of ethylene gas
- Intensity of light

Meanwhile, the permeability of the package is affected by:

- Type and characteristics of the material
- Thickness and surface area of the material
- Temperature and humidity levels
- Differences in partial pressure of O_2 and CO_2

A combination of these factors works together to establish balanced levels of O_2 and CO_2 within a sealed package. The choice of film, or a combination of films (laminate), is determined by the expected transpiration and respiration rates of the produce. Typically, the selected film should allow for more CO_2 permeability than O_2 for most fresh produce.

6.2.3.1.3 Gas Packaging (Active MAP)

Gas packaging is a form of modified atmospheric packaging, wherein air is removed from the package and is replaced with a singular gas or combination of gases. Gas packaging can be achieved in two fundamental ways. Air can be replaced with a gas or mixture either mechanically or through the use of atmosphere modifiers, such as oxygen absorbents. There are two methods for mechanically replacing air in a package:

Gas Flushing: This method involves using a continuous gas stream to remove air from the packaging before sealing. However, this technique is not very effective in eliminating air from the pack, which can result in residual oxygen levels ranging from 2% to 5%. Consequently, gas flushing is not recommended for food products that are sensitive to oxygen. With the increasing popularity of minimally processed fruits and vegetables, active MAP seems to be a sustainable alternative to maintain their quality and safety (Kumar and Sethi 2021).

Compensated Vacuum: The first stage of the compensated vacuum method involves removing air from the pack by creating a vacuum. However, achieving a complete vacuum is not always possible as reduced pressures can cause water to boil. The second stage of this method involves gas flushing, where the pack is filled with a modified gas mixture. While gas flushing is effective, evacuating air from the pack results in even lower residual oxygen levels, making this method ideal for packing products that are sensitive to oxygen. Unfortunately, the two-stage process of the compensated vacuum method results in a slower packaging rate than gas flushing.

6.2.3.1.4 Edible Packaging of Fruits and Vegetables

Fresh fruits and vegetables have natural skins that protect them from water loss, browning, and microbial deterioration. The natural bloom on the commodity may get damaged during handling and transportation. Applying edible coatings/films on their surface can offer additional protection. With the growing demand for fresh-like items, the usage of edible films and coatings is rapidly increasing. Moreover, due to the increasing sale of freshly cut goods, the edible coatings industry is becoming more economically important. Edible packaging comes in several forms, such as films, sheets, coatings, and pouches. Edible films (thickness < 254 μm) or sheets (thickness >254 μm) are independent structures that are produced separately from the food itself and then placed on or between food components or sealed into edible pouches. This is in contrast to edible coatings, which are thin layers of edible substances formed directly onto the outside layer of the food (Janjarasskul and Krochta 2010). Edible packaging provides a barrier that selectively slows down the migration of moisture, gas, oil, fat, and solutes. Additionally, it enhances the mechanical handling properties of foods, preserves their integrity, and retains volatile flavor compounds. Edible packaging can also serve as carriers of food additives such as antioxidants, antimicrobials, colorants, and so on (Nayak et al. 2019; Sethi et al. 2021).

The major advantages of edible packaging are as follows:

(i) It enhances shelf life without altering the sensory properties of the produce.
(ii) It enhances the barrier qualities such as water vapor permeability and gaseous transmission and guards the product against microbiological, chemical, and physical harm.
(iii) **They can be consumed directly without the need to unwrap or dispose of the covering.**
(iv) **These items are eco-friendly as they are crafted from renewable resources.**

6.2.3.1.4.1 Components of Edible Packaging

Based on the type of biopolymer, the edible packaging is categorized as hydrocolloid-based, protein-based, lipid-based, and composites.

- Hydrocolloid-Based Edible Packaging: It is made up of protein (gelatin, collagen, gluten, whey isolates, casein, pea protein) or polysaccharide (cellulose, hemicellulose, chitosan, starches, seaweeds, and gums). It possesses good mechanical and gaseous permeability properties but has poor permeability to water vapor.
- Protein-Based Edible Packaging: They are good oxygen barriers, but they are susceptible to moisture. Common examples include gelatin, zein, and aqueous ethanol solutions of wheat gluten. The addition of plasticizers such as glycerin or sorbitol can improve the mechanical properties of the film.
- Lipid-Based Edible Packaging: It is made up of waxes (carnauba, paraffin, and bee wax), fatty acids, resins, and essential oil (rosemary, peppermint, clove, cin-

namon) and possesses excellent barrier properties to water vapor but lacks mechanical properties.
- Composites: To integrate the qualities of these three diverse kinds of materials and create edible films and coatings with increased permeability and mechanical capabilities for particular applications, hydrocolloids, proteins, and/or lipids are combined with one another. Heterogeneous films can include soy protein isolates with polylactic acid, gelatin and soluble starch, whey isolate and lipids, maize zein, methylcellulose, and lipids.

The efficacy of any edible coating in preserving postharvest quality and extending the storage life of horticultural produce depends upon the nature of the coating (Kumar et al. 2018b), concentration (Edirisinghe et al. 2014), time of application (Kumar et al. 2017), commodity physiological characteristics (Kumar et al. 2018a), and the active ingredients added into the formulation (Kumar and Sethi 2021).

6.2.3.1.4.2 Disadvantages of Edible Coatings

- The respiration may be restricted by thick coatings, which might lead to excessive ethanol accumulation and off-flavor development in the product. An increase in coating or film thickness not only has negative effects on anaerobic fermentation by decreasing internal oxygen and raising carbon dioxide concentration but also taints the product's natural flavor and taste. It might stop the condensation of water vapor, which could lead to microbiological deterioration in fruit and vegetable packaging. Inadequate water vapor barrier properties in the coatings may cause the product to lose moisture or weight. Good gas barrier qualities in films might lead to anaerobic respiration and obstruct natural ripening. To prevent anaerobic conditions, the coating or film should permit a specified amount of oxygen permeability.
- For coatings like whey protein, which acts as a food for microbial development in damp environments, the spoiling might be quick. This issue can be resolved by adding antimicrobials such as potassium sorbate to the coatings.
- Basic information on the make-up, properties, methods of application to food surfaces, and evidence of the effectiveness of film coatings is lacking. It takes a lot of research to apply edible coatings to food, especially to just-cut fruits and vegetables.

By choosing the right type and thickness of coating, avoiding treating immature, flavorless fruit, and avoiding storing coated fruits at high temperatures, those disadvantages can be avoided. According to Regulation 2.4.5 of the Food Safety and Standards (Packaging and Labeling) regulations (FSSAI 2011), the Food Safety and Standards Authority of India (FSSAI) in India allows fresh fruits to be coated with bee wax (white and yellow), carnauba wax, or shellac wax at a level that does not exceed good manufacturing practices when properly declared on the label.

6.2.3.2 Active Packaging

Active packaging refers to a packaging system that intentionally incorporates additional components within the packaging material or the package headspace to improve its overall performance (Madhav and Sethi 2016). Active packaging systems are designed to alter the package conditions, thereby enhancing food safety, prolonging the shelf life, preserving sensory properties, and ensuring product quality. Such systems typically involve removing or absorbing undesirable substances, such as oxygen, carbon dioxide, ethylene, flavor/odor, and excessive moisture, to modify the environmental or physiological conditions inside the food package. Some active packaging systems use sachets, labels, or films to add or release compounds such as carbon dioxide, antioxidants, and preservatives into the package headspace (Bodbodak and Rafiee 2016).

Active packaging systems can be classified as follows:

- Active scavenger systems (O_2 scavengers, CO_2 absorbers, ethylene absorbers, drip absorbers)
- Active release systems (CO_2 emitters)
- Controlled release systems (sulfur dioxide, antimicrobial, humidity, and condensation controllers)

There are two mechanisms of active packaging:

1. Placing the active element inside the package along with the product to be packed (sachets and labels)
2. Making the active element part of the materials that form the package itself (film)

6.2.3.2.1 Systems Used in Active Packaging

(a) Oxygen Scavengers

Among active packaging, oxygen scavengers are the most often utilized commercially. They are used in the form of a little sachet that contains several iron-based powders together with an effective catalyst that may lower oxygen levels in the pack to around 0.01%. Organic reducing agents such as ascorbic acid, ascorbate salts, or catechol are used by non-metallic scavengers. Additionally, glucose oxidase or ethanol oxidase, which can be included in sachets, sticky labels, or immobilized on packaging film surfaces, are employed in enzyme oxygen scavenging systems.

(b) Ethylene Scavengers

There are two types of produced ethylene that can affect the ripening process of fruits and vegetables, depending on the source: endogenous ethylene, which the plant produces itself by biological pathway, and exogenous ethylene, which originates from adjacent crops, plastics, and smoke. It is important to remove both types of ethylene when trying to slow down the ripening process of fruits and vegetables, as they will ripen faster regardless of which type, they are exposed to. However, it is crucial to keep in mind that ethylene can significantly impact plant growth and development, even at a concentration as low as 0.1 μL/L, due to its sensitivity to plants. Fruit ripening, softening, and aging can be accelerated by ethylene. The most commonly used method to scavenge ethylene involves packaging with sachets containing either potassium permanganate or activated carbon with a metal catalyst. There are also plastic film-based products that remove ethylene, consisting of PE infused with finely dispersed minerals such as clays, zeolites, and carbon.

6.2.3.3 Intelligent Packaging/Smart Packaging

Intelligent packaging systems offer valuable information to users about the food or its surroundings, such as temperature and pH. This advanced packaging serves as an extension of traditional packaging and can detect, sense, and record changes in the product's environment, effectively communicating with consumers. In general, radiofrequency identification (RFID) systems, sensors, and indicators are the three types of intelligent systems.

Sensors: A sensor is a device that is used to identify, locate, or measure matter or energy. It provides a signal for the measurement or identification of a chemical or physical characteristic that the instrument reacts to.

Sensors used in intelligent packaging are as follows:

- *Bioreceptors* may be used for the detection of biological or organic substances, such as microorganisms, hormones, nucleic acid, enzymes, and antigens. The transducers might be electrochemical, acoustic, or optical, for example, Food Sectinel System® (SIRA Technologies Inc.)—food pathogen and ToxinGuard® (Toxin Alert, Canada)—*Escherichia coli* and *Listeria* sp.
- *Gas sensors* are used for detecting the presence of gaseous analyte in the package. It includes oxygen sensors, carbon dioxide sensors, water vapor sensors, ethanol sensors, metal oxide semiconductor field effect transistors, organic conducting polymers, and piezoelectric crystal sensors.
- *Printed electronics* is an emerging technology that utilizes electrically functional inks on flexible substrates. One of the key advantages of this technology is the unique properties of the resulting electronic sensors, which include being lightweight, bendable, rollable, portable, and foldable.
- *Chemical sensor* is a chemical selective coating capable of detecting the presence, activity, composition, and concentration of a particular chemical or gas through surface adsorption.
- *Electronic noses* are systems used to imitate the mammalian olfactory system within an instrument intended to acquire reproducible readings and enable the identification and categorization of aroma mixes contained in the odor.

6.2.3.3.1 Indicators

Indicators are defined as substances that can reveal the presence, absence, or concentration of another substance. They can also indicate the level of reaction between two or more substances by producing a characteristic change, often in color.

- *Freshness indicators* offer valuable information about the quality of a food product by detecting any microbial growth or chemical changes. The indicators within the packaging interact with the metabolites produced by the microbes, providing a visual indication of the product's microbial quality, for example, bromophenol blue (BPB) was immobilized onto bacterial cellulose membrane-Guava, Sensorq® (DSM NV and Food Quality Sensor International).
- *Time-temperature indicator* is designed to show if a certain temperature limit has been surpassed over a period of time and to estimate how long a product has been exposed to temperatures above that limit (known as time-temperature history). These labels display clear visual cues of temperature history during transportation and storage, for example, Monitormark™ (Minnesota), Checkpoint® (Vitsab), and Thermax (Thermographic Measurements Ltd.).
- *Integrity Indicators:* Preventing leaks is crucial when producing and distributing packaged food. Integrity indicators are used to guarantee the food's integrity at every step of the distribution chain, for example, Ageless Eye® (Mitsubishi Gas Chemical Company), Novas® (Insignia Technologies Ltd.).
- *RFID:* RFID technology is an advanced system that enables the automatic identification of items through the use of wireless sensors. It operates without human intervention, relying on tags and readers to communicate. Typically, RFID tags contain unique identification numbers, which the reader can access from a database to gather relevant information and take appropriate action, for example, Intelligent Box (Mondi Plc), CS8304 (Convergence Systems Ltd.), Temptrip (Temptrip LLC).

6.2.3.4 Nanotechnology-Based Packaging

The application of nanotechnology in synthesizing nanoparticles has generated a great deal of attention in the food packaging sector. It has the ability to produce food packaging with improved qualities that might increase the longevity of food items. Improved, active, and intelligent packaging that helps preserve food quality and traceability along the supply chain is made possible by the use of nanoparticles. Nanoparticles possess various properties—such as antimicrobial activity, oxygen scavenging ability, and UV impermeability—that make them invaluable for the production of nanocomposites. Various types of nanoparticles, including organic, inorganic, and combined varieties, such as nanocellulose, silver nanoparticles, nanostarch, protein nanoparticles, chitosan nanoparticles, carbon nanotubes, and nanoclay, are used in the development of effective food packaging.

References

Bodbodak S, Rafiee Z (2016) Recent trends in active packaging in fruits and vegetables. In: Eco-friendly technology for postharvest produce quality. Academic Press, pp 77–125

Cachon R, Girardon P, Voilley A (eds) (2019) Gases in agro-food processes. Academic Press

Edirisinghe M, Ali A, Maqbool M, Alderson PG (2014) Chitosan controls postharvest anthracnose in bell pepper by activating defense-related enzymes. J Food Sci Technol 51:4078–4083

FSSAI (2011) Food Safety and Standards Authority of India, Ministry of Health and Family Welfare, The Gazette of India: Extraordinary. Part III: Sec. 4

Janjarasskul T, Krochta J (2010) Edible packaging materials. Annual Rev Food Sci Technol 1:415–48

Kumar P, Sethi S (2021) Influence of modified atmospheres on shelf life and quality of fresh-cut apples. J Packaging Technol Res 5:209–216

Kumar P, Sethi S, Sharma RR (2017) Combined effect of edible coatings and low temperature on plum fruit quality. Int J Curr Microbiol Appl Sci 6:4210–4218

Kumar P, Sethi S, Sharma RR (2018a) Inhibition of browning in fresh-cut apple wedges through edible coatings and antibrowning agents. Ind J Hortic 75:517–522

Kumar P, Sethi S, Sharma RR, Varghese E (2018b) Influence of edible coatings on physiological and biochemical attributes of Japanese plum (*Prunus salicina* Lindell) cv. Santa Rosa Fruits 73:31–38

Laurile E, Ahvenainen R (2002) Minimal processing in practice; fresh fruits and vegetables. In: Ohlsson T, Bengtsson N (eds) Minimal processing technologies in the food industry. CRC Press, Boca Raton, pp 219–244

Lim LT (2011) Active and intelligent packaging materials. In: Moo-Young M (ed) Comprehensive biotechnology, 2nd edn. Elsevier, San Diego, pp 629–644

Madhav JV, Sethi S (2016) Effect of modified atmosphere packaging on irradiated and minimally processed garlic cloves. Curr Biot 9(4):322–327

Martens T, Farber MJ (1995) Current status of sous vide in Europe. In: Principles of modified-atmosphere and sous vide product packaging. Routledge, pp 37–68

Meena MK, Chetti MB, Nawalagatti CM, Naik MC (2017) Vacuum packaging technology: a novel approach for extending the storability and quality of agricultural produce. Adv Plants Agric Res 7(1):221–225

Nayak SL, Sethi S, Sharma RR, Prajapati U (2019) Active edible coatings for fresh fruits and vegetables. In: Gutierrez TG (ed) Polymers for agri-food applications. Springer Nature, Cham, pp 417–432

Newton P, Gillespie A (2010) A practical guide to vacuum packing. Lava Vacuum Packing, Warwickshire

Ozdemir M, Floros JD (2004) Active food packaging technologies. Crit Rev Food Sci Nutr 44(3):185–193

Sethi S, Joshi A, Nayak SL, Sharma RR (2021) Edible coatings for quality and disease management of fruits and vegetables. In: Singh D, Sharma RR, Devappa V, Kamil D (eds) Postharvest handling and diseases of horticultural produce. CRC Press, Boca Raton

Sharma RR, Pal RK, Singh D, Samuel DVK, Sethi S, Kumar A (2013) Evaluation of heat shrinkable films for shelf life, and quality of individually wrapped Royal Delicious apples under ambient conditions. J Food Sci Technol 50:590–594

Siddiqui MW, Rahman MS, Wani AA (2018) Innovative packaging of fruits and vegetables: strategies for safety and quality maintenance. CRC Press, Boca Raton

Chapter 7
Extraction Technologies for Horticultural Crops

P. M. Sindhu, Shruti Sethi, Suma H. Budnimath, Alka Joshi, and H. R. Raghavendra

Abstract Fruits and vegetables play a vital role in horticultural trade due to their high value. They are consumed in both raw and processed form and provide beneficial components that promote health. These commodities serve as crucial raw materials for agro-based processing industries. During the processing of fruits and vegetables, a significant amount of waste is generated, primarily consisting of peels, seeds, and rags. These wastes are rich in bioactives which after extraction are valuable additives for the food and pharmaceutical industries. Additionally, effective utilization of waste can help combat environmental pollution issues. This chapter focuses on extracting bioactive compounds from horticultural crops using various conventional and non-conventional methods, such as maceration, percolation, soxhlet extraction, supercritical fluid extraction, pressurized hot water extraction, pressurized liquid extraction, ultrasound-assisted extraction, microwave-assisted extraction, pulsed electric field-assisted extraction, and enzyme-assisted extraction. This comprehensive approach will enable readers to harness appropriate technologies for extracting the beneficial components present in horticultural crops.

Keywords Extraction · Horticulture crops · Bioactive compounds · Conventional extraction methods · Non-conventional extraction methods

P. M. Sindhu · S. Sethi (✉) · A. Joshi
Division of Food Science and Postharvest Technology, ICAR-Indian Agricultural Research Institute, New Delhi, Delhi, India

S. H. Budnimath
Department of Horticulture, College of Agriculture, Kerala Agricultural University, Padannakkad, Kerala, India

H. R. Raghavendra
ICAR-Indian Agricultural Research Institute, Gogamukh, Assam, India

M. Thakur, T. Belwal (eds.), *Advances in Postharvest and Analytical Technology of Horticulture Crops*, https://doi.org/10.1007/978-981-97-7247-6_7

7.1 Introduction

Horticulture crops encompass a wide range of plants cultivated for their fruits, vegetables, flowers, and medicinal or aromatic properties. These crops hold immense economic, nutritional, and medicinal value, making it essential to optimize the extraction process to harness their full potential. Extraction technologies play a crucial role in the horticulture industry, enabling the efficient and effective recovery of valuable compounds from different crops. Extraction is the first step to isolate the desired natural compounds from the raw materials. The extraction of bioactive compounds from horticulture crops involves the separation and recovery of desired components, such as essential oils, pigments, flavors, antioxidants, and other phytochemicals. These compounds find applications in various industries, including food and beverages, cosmetics, pharmaceuticals, and nutraceuticals. Extraction technologies serve as the key link between the raw agricultural produce and the value-added products derived from them.

Over the years, numerous extraction technologies have been developed and advanced, providing researchers with diverse methods to extract and isolate target compounds. These technologies vary in their principles, mechanisms, efficiency, and suitability for different horticultural crops. The choice of extraction technology depends on factors such as the nature of the desired compounds, the properties of the raw material, economic considerations, and the intended application of the extracted products. Traditional extraction methods, such as maceration, steam distillation, and solvent extraction, have been used for centuries to obtain valuable compounds from horticulture crops. However, advancements in technology have led to the emergence of modern extraction techniques that offer enhanced efficiency, selectivity, and sustainability. These include techniques such as supercritical fluid extraction (SFE), microwave-assisted extraction (MAE), ultrasound-assisted extraction (UAE), and enzyme-assisted extraction (EAE), among others. Each extraction technology brings its own set of advantages and limitations, making it essential for researchers and industry professionals to stay updated with the latest developments. Factors such as extraction yield, extraction time, product quality, energy consumption, environmental impact, and scalability must be considered when selecting an appropriate extraction method for a particular horticulture crop. This chapter aims to explore the various extraction technologies available for horticulture crops, highlighting their principles, applications, and recent advancements. By understanding these technologies, researchers can optimize their extraction processes, improve product quality, and unlock the full potential of horticulture crops in diverse industries.

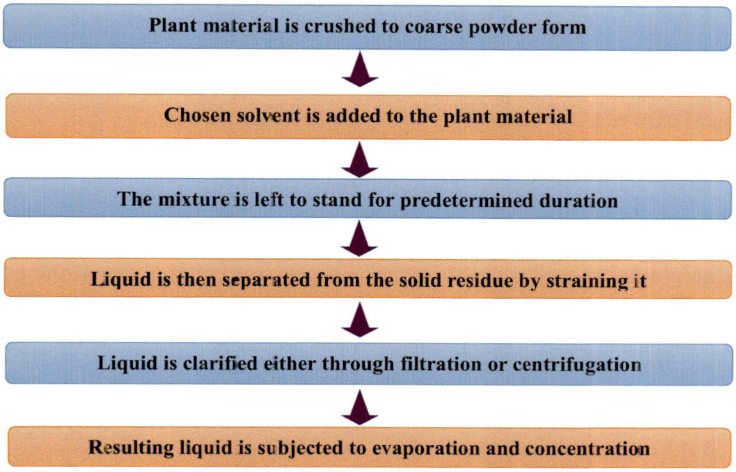

Fig. 7.1 Methodology of extraction of bioactive compounds using maceration

7.2 Maceration

Maceration is a conventional extraction method in which plant material is immersed in a liquid solvent for a specified duration to facilitate the transfer of soluble compounds from the plant matrix into the solvent (Fig. 7.1). During maceration, the solvent gradually permeates the plant material, dissolving and extracting the desired bioactive compounds. This process is driven by diffusion, as the concentration gradient between the plant material and the solvent allows for the transfer of compounds through the cell walls and into the solvent (Mujumdar 2019).

7.2.1 Principle

The principle of maceration is rooted in the concept of solubility and diffusion, allowing for the transfer of compounds from the plant matrix into the solvent. It is a relatively passive process that does not involve external forces such as heat or mechanical agitation. The extraction efficiency in maceration depends on factors such as the solubility of the target compounds, the solvent choice, the contact time between the solvent and the plant material, and the permeability of the plant matrix (Bhambra and Singh 2013).

7.2.2 Advantages of Maceration

- Maceration is an easy extraction method that requires minimal equipment and technical expertise.
- It is a cost-effective extraction technique that can be used to extract a wide range of bioactive compounds from various plant materials.
- It is a gentle extraction method that minimizes the degradation and alteration of sensitive compounds. The extraction process occurs at ambient or low temperatures, helping to preserve the structural integrity and bioactivity of the extracted compounds.
- The process allows for the use of different solvents, including water, ethanol, methanol, and other organic solvents.

7.2.3 Disadvantages of Maceration

- Compared with other extraction techniques, maceration may have lower extraction efficiency. The passive diffusion process during maceration can be slow, resulting in incomplete extraction and lower yields of target compounds.
- Maceration typically requires a longer extraction time compared with other extraction methods. The extended soaking period is necessary for the solvent to penetrate the plant material and extract the bioactive compounds fully.
- The choice of solvent in maceration can significantly impact the extraction efficiency and selectivity.
- Maceration does not provide precise control over extraction parameters such as temperature, pressure, and agitation.

7.3 Percolation

Extraction by percolation involves the flow of a solvent through a solid matrix to extract desired compounds. It is commonly used in the extraction of bioactive compounds from plant materials and has been employed for centuries in various industries. A commonly used tool for this process is a percolator, which is a narrow, cone-shaped container with open ends. The solid ingredients are moistened with a specific amount of the prescribed liquid and left undisturbed for 4 h in a tightly sealed container. Afterward, the mixture is tightly packed and the upper part of the percolator is closed. A little more liquid is poured on top to create a shallow layer above the mixture, and the closed percolator is allowed to steep for 24 h. The outlet of the percolator is then opened, and the liquid inside is allowed to drip out slowly. The remaining solid residue is pressed to extract any liquid, which is combined with the collected liquid. More liquid is added to reach the desired volume, and the mixed liquid is clarified either through filtration or by allowing it to settle and then pouring off the clear liquid (Manousi et al. 2019).

7.3.1 Principle

Percolation is based on the principle of solvent flow through a solid matrix, typically a packed bed of plant material. The solvent is allowed to pass through the matrix, dissolving and extracting the desired compounds as it percolates through the solid material. The process relies on the solubility and permeability of the target compounds, as well as the diffusion and mass transfer mechanisms (Guimarães et al. 2013).

7.3.2 Advantages of Percolation

- Percolation is a versatile extraction method and allows for the selective extraction of specific compounds by controlling the choice of solvent and the extraction conditions while minimizing the extraction of unwanted substances.
- The process can achieve efficient extraction due to the continuous flow of the solvent through the solid matrix.
- The process can be easily scaled up for industrial applications.
- It offers good reproducibility as the extraction parameters, viz., solvent composition and flow rate, can be controlled precisely.

7.3.3 Disadvantages of Percolation

- It can be a time-consuming process, particularly when extracting compounds with low solubility or when using a large solid matrix.
- There is a need for specialized equipment, such as percolators or column systems, to maintain the flow of solvent through the solid matrix.
- Inadequate packing or uneven distribution of the plant material can result in channeling, where the solvent flows through preferential pathways, leading to incomplete extraction.
- The extraction requires a larger volume of solvent compared with other extraction methods potentially increasing the cost and environmental impact of the process.

7.4 Soxhlet Extraction

Soxhlet extraction (SE) has been the standard extraction technique used for over a century. It was originally used for the determination of fat in milk (De Castro and Garcıa-Ayuso 1998). In the conventional SE method, the sample is placed in a thimble that is progressively filled with condensed solvent from a distillation flask. The solvent + extracted analytes are transferred into the distillation flask containing the bulk liquid below by a siphon when the liquid exceeds the overflow level. The solute remains in the distillation flask, while the solvent returns to the flask. This procedure is repeated till the complete extraction is accomplished.

7.4.1 Advantages

- The sample is repeatedly brought in contact with a fresh volatilized fraction of the extractant, assisting in the displacement of the mass transfer equilibrium.
- There is no need for filtering following the leaching process.
- Basic extraction assembly is affordable and simultaneous parallel extraction systems attached in series can boost sample throughput.

7.4.2 Disadvantages

- The lengthy extraction process and the significant volume of extractant waste are not only expensive to dispose of but can also generate further environmental issues.
- Samples are often extracted at the extractant's boiling point for an extended length of time and the risk of thermal breakdown of thermolabile chemicals is possible.
- Because of the significant volume of solvent utilized, an evaporation/concentration phase is required following extraction.

To overcome these drawbacks and also to preserve the benefits associated with SE, the technique has now been applied in combination with ultrasound, microwave, or electric heating.

7.5 Pressurized Hot Water Extraction

Pressurized hot water is the term used to designate the condensed phase of water which lies between the temperature range of 100–374 °C, that is, between the boiling point of water and the critical point of water, respectively. It is also referred to as extraction using "hot compressed water" or "subcritical water." In the case of pressurized hot water extraction (PHWE), the density of water remains almost constant over this range of temperatures so that the pressure effect on the properties of water is minimal (Meyer 1993). The use of pressurized water as an extraction fluid at high temperatures was initially described by Hawthorne et al. (1994) for the extraction of various polar and non-polar analytes from soil samples. Moderate pressures (15 bar at 200 °C and 85 bar at 300 °C) are required during the extraction to maintain a condensed phase of water. Superheated steam is generated when the pressure falls below the boiling point. PHWE is a promising green solvent extraction technique as it uses pressurized water at extreme temperatures and regulated pressure, thus reducing the usage of organic solvents. Several studies have demonstrated that at certain temperatures and pressures, the polarity of water may be modified similarly to that of alcohol. As a result, it can dissolve a wide spectrum of analytes with medium and low polarity. The viscosity of water is reduced while its diffusivity is improved, enabling better penetration through the particles of the matrix. Introducing fresh water continuously during dynamic extraction in PHWE enhances mass transfer, resulting in an increased extraction rate. The high temperature and pressure can disrupt the surface equilibria. Elevated temperature can overcome the interaction between solute and matrix caused by Van der Waal forces, hydrogen bonding, dipole attraction of solute molecules, and active sites in the matrix. As a result, the supplied thermal energy can disrupt both cohesive (solute-solute) and adhesive (solute-matrix) interactions by reducing the activation energy required for the desorption process. The transfer of analytes from the matrix to pressurized hot water occurs through diffusion and convection processes. However, compounds that are sensitive to heat may degrade at higher temperatures. To achieve this, sufficient pressure must be applied to the water when temperatures exceed its boiling point. The presence of pressure facilitates the extraction from samples where analytes are trapped in the pores of the matrix. This pressure ensures that water reaches areas of the matrix that would not typically be covered if atmospheric pressure water were used (Teo et al. 2010).

PHWE has progressively gained popularity as a method for isolating bioactive and nutritious components from plants and food items. It is a direct approach for recovering analytes that does not require any cleaning. This saves money because the analytes obtained are safe for further processing, testing, and human ingestion.

7.5.1 Advantages of PHWE

- Reduced consumption of organic solvents.
- Water is readily accessible, non-toxic, and environmentally friendly when recycled or disposed off.
- It improves the mass transfer effects and solubility.
- It enhances the disruption of surface equilibria.

- The capacity of the fluid to solubilize analytes increases when the characteristics of water change at high temperatures.

7.5.2 Disadvantages of PHWE

- The process involves high temperatures, which can cause the degradation of thermally labile compounds.
- PHWE may lack selectivity, as it can extract a wide range of compounds simultaneously. This can result in the co-extraction of unwanted substances, leading to complexity in analysis and potentially affecting the accuracy of target compound quantification.
- Operating the extraction process at elevated temperatures and pressures requires significant energy input.
- Working with pressurized systems carries inherent safety risks.
- Extracts obtained through PHWE can be complex, containing a wide variety of compounds. Analyzing such complex matrices may require sophisticated analytical techniques and instruments, increasing the complexity and cost of the analysis process.

7.6 Pressurized Liquid Extraction

Pressurized liquid extraction (PLE) is a method that involves utilizing liquid solvents at increased temperature and pressure to enhance extraction efficiency compared with techniques performed at room temperature and atmospheric pressure (Andreu and Pico 2019). The advantage of employing solvents at temperatures higher than their boiling points at atmospheric pressure lies in the improved solubility and mass transfer properties. The introduction of PLE took place in 1995 at the Pittcon Conference by Dionex Corporation, where it was initially presented as Accelerated Solvent Extraction (ASE®) Technology. This technique is also referred to as PLE, pressurized solvent extraction, accelerated solvent extraction, and enhanced solvent extraction. When water is used as the extraction solvent, the technique is specifically known as PHWE, sub-critical water extraction, or superheated water extraction (Waldeback 2005). Figure 7.2 gives the schematic diagram of the PLE system.

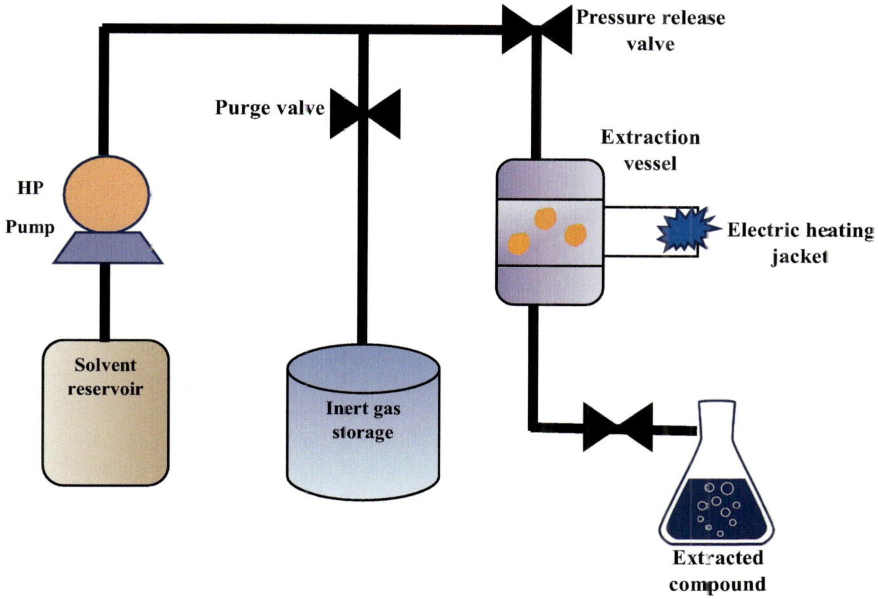

Fig. 7.2 Schematic diagram of pressurized liquid extraction system

7.6.1 Principle

The general principle guiding solvent selection is "like dissolves like" suggesting that polar solvents should be used for polar analytes, while nonpolar analytes are best dissolved in nonpolar solvents. When choosing a solvent, important factors to consider include the solubility characteristics of the desired analyte, its diffusivity in the solvent, and the sample's properties (Santoyo et al. 2006). Ideally, the goal is to achieve an extract with high purity and selectivity, which means that the analyte of interest should have high solubility in the solvent, while other compounds should have minimal or no solubility. When extracting analytes present at low concentrations, the extraction rate is influenced not by the analyte concentration itself but by the rate of mass transfer. Therefore, selecting a solvent with appropriate chemical properties is crucial to facilitate solvation and release of the analyte.

7.6.2 Advantages

PLE has demonstrated notable benefits compared with other techniques in terms of time efficiency, solvent usage, automation, and overall effectiveness. One advantage of PLE over MAE is the elimination of an additional filtration step. This is possible because any matrix components that remain undissolved in the extraction solvent can be retained within the sample extraction cell. This convenience is particularly advantageous for the automation and integration of extraction and separation techniques in online applications (Carabias-Martínez et al. 2005).

7.7 Supercritical Fluid Extraction

SFE is the method of separating the desired component from the matrix by employing supercritical fluids as the extracting solvent. Extraction is often performed on a solid matrix, although it may also be performed on liquids. SFE is a green technology because the use of supercritical CO_2 is non-toxic, non-flammable, noncorrosive, and simple to handle, allowing supercritical operation at low pressures and close to room temperature. A "supercritical fluid" is a dense fluid having characteristics that are intermediate between those of a gas and liquid at temperatures and pressures above its critical point. It has a density comparable to that of a liquid and a viscosity and diffusivity identical to that of a gas. As a result, a supercritical fluid can operate as a solvent in the same way that a liquid does, but with improved mass transfer kinetics (Hicks et al. 2016). SFE has emerged as a superior alternative technique for extracting bioactive compounds from natural products due to its shorter extraction time, lower consumption of organic solvents, suitability for thermo-sensitive substances, and environmental friendliness. It is a "Generally Regarded As Safe" technique (Sanchez-Camargo et al. 2014). Currently, SFE is widely used not just in the food and medicine industries but also in textiles, polymers, petrochemicals, environment toxicology, and chemistry, among other fields.

7.7.1 Principle

The goals of this technique include high selectivity, quick extraction periods, better pollution prevention, and the use of harmless organic solvents. The solid feed is introduced into the extractor in the column, while the mobile phase is formed by mixing solutes with supercritical fluid. To form supercritical fluid, gaseous CO_2 is compressed into liquid followed by extraction. The mobile phase is exposed to a temperature (ambient to 300 °C) and pressure (50–500 atm) close to or above the critical point for increasing mobile phase solvating power. Then the dissolved solute is isolated by precipitation (Wang and Weller 2006). The easiest way to remove a solute from a supercritical fluid is to modify the pressure of the fluid such that it is no longer supercritical or the solute is no longer soluble in the supercritical fluid. Figure 7.3 gives a schematic representation of the SFE process.

Despite the fact that CO_2 is the preferred extraction solvent (for extracting non-polar chemicals), the polarity of supercritical SC-CO_2 can be improved by using a miscible polar component (such as ethanol) as a modifier (Ahmad et al. 2019). The

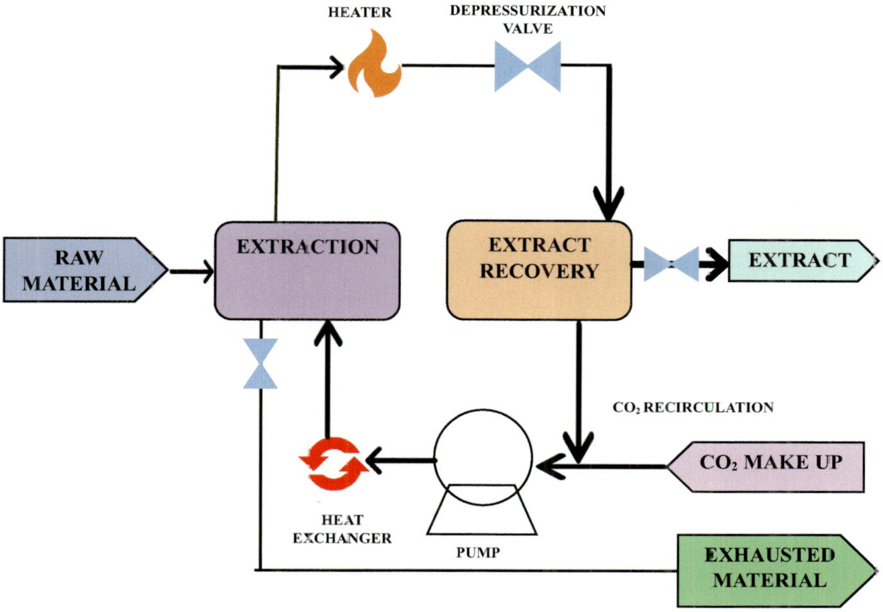

Fig. 7.3 Schematic representation of supercritical fluid extraction

SFE process involves selectivity; therefore, the extracts produced using this method have a low concentration of undesirable chemicals. Furthermore, following depressurization, SC-CO_2 becomes gaseous and can be readily removed from a flow system. In addition, SC-CO_2 is used to destroy microorganisms without affecting nutritional content or organoleptic properties, making it a promising alternative for pasteurizing bioactive compounds in food and medicine that would otherwise be destroyed by conventional thermal processes (Hicks et al. 2016).

7.7.2 Advantages

- Supercritical fluids, such as carbon dioxide, can be adjusted to have specific solvent properties, allowing selective extraction of target compounds while leaving undesirable substances behind.
- SFE operates at relatively low temperatures, which helps preserve the integrity and quality of heat-sensitive compounds.
- Supercritical fluids can easily be separated from the extracted compounds by simply depressurizing the system, leaving behind a solvent-free extract.
- SFE often employs carbon dioxide, which is non-toxic, non-flammable, and readily available. It eliminates the need for traditional organic solvents, reducing environmental impact.

7.7.3 Disadvantages

- SFE requires specialized equipment designed to handle high pressures and temperatures. The initial investment for such equipment is high.
- SFE is not suitable for all types of compounds. It works best for non-polar or slightly polar substances, while polar compounds may not be efficiently extracted.
- In some cases, SFE may not achieve the same level of extraction efficiency as other methods. This is particularly true for compounds with low solubility in supercritical fluids.
- The extraction conditions, including temperature, pressure, and flow rate, need to be carefully optimized to obtain desired extraction results. The process may require extensive experimentation and expertise to achieve optimal extraction conditions.

7.8 Ultrasound-Assisted Extraction

UAE is a technique that utilizes ultrasound energy to enhance the extraction of compounds from solid or liquid samples. UAE offers several advantages over traditional extraction methods, making it a promising and efficient technique (Chemat et al. 2017). Ultrasound plays a crucial role in promoting sustainable green chemistry. It has a notable impact on the speed of various processes in the chemical and food industries. It has gained significant attention in various fields, including food, pharmaceuticals, nutraceuticals, and natural product industries, and is also utilized for extracting bioactive compounds from various sources such as plants, fruits, algae, fungi, bacteria, and even animals, in both laboratory and industrial settings (Anastas and Zimmerman 2003).

Ultrasound energy offers several benefits for extraction, such as improved mixing and micromixing, enhanced energy and mass transfer, reduced thermal and concentration gradients, lower extraction temperatures, selective extraction, compact equipment requirements, quicker process control, faster initiation, increased production, and elimination of certain process steps. The utilization of ultrasound during extraction is thought to be influenced by cavitation phenomena and the augmentation of mass transfer.

7.8.1 Principle

The principle behind UAE is the phenomenon known as acoustic cavitation. When ultrasound waves pass through a liquid medium, they create alternating high-pressure and low-pressure cycles. During the low-pressure cycles, small bubbles or cavities form in the liquid and as the pressure increases, these bubbles collapse violently, creating localized hotspots and shockwaves. This phenomenon generates intense mixing, increased mass transfer, and disruption of cell walls, facilitating the extraction process.

Typically, ultrasonic systems used in laboratories consist of baths, probes, and other extraction devices like cup-horn sonoreactors. These systems can be customized by various companies to meet the specific requirements of clients, ranging from 30 to 1000 Lin volume and with power outputs between 500 and 16,000 W (Pingret et al. 2013). Tanks in industrial ultrasonic systems are filled and emptied using pump systems. Laboratory systems are typically operated in batch mode, while industrial systems often operate in flow mode. Ultrasound systems utilize electromechanical transducers to generate ultrasonic energy by converting mechanical or electrical energy. Piezoelectric transducers are the most commonly used due to their versatility and efficiency. Currently, the most popular piezoceramic material is lead zirconate titanate (PZT), although piezoelectric composites and polymers are being introduced as functional materials for ultrasonic transducers, as they offer advantages such as ease of production and usage at lower pressures and temperatures. The frequency of the transducer is determined by its thickness, with the obtained wavelength generally being twice the thickness, requiring the thickness to be half of the desired wavelength. These transducers can achieve a frequency range from 20 kHz to 5 MHz.

7.8.2 Bath System

The essential components of an ultrasonic bath include the tank, the electronic generator (which supplies power to the transducer), and the transducer itself. Typically, multiple transducers are attached to the tank either on the bottom or the sides using epoxy resins or brazing. Thermostatically controlled heaters are often included in the baths for temperature regulation. In industrial applications, these systems can incorporate additional units for stuffing and destuffing, transportation, washing, rinsing, drying, and solvent recovery.

For extraction purposes, the raw material and extractant are placed inside the bath vessel, resulting in indirect sonication. Sufficient ultrasound power is required to induce cavitation within the extraction vessel. Ultrasonic baths are available in a frequency range of 20 to 500 kHz. When higher frequencies ranging from 0.5 to 5 MHz are used, they are referred to as megasonic baths. The frequency of the bath determines the occurrence of cavitation and the energy released. Most commercial ultrasonic baths operate at a single frequency, typically around 40–45 kHz, suitable for cleaning, degassing, and extraction. However, there are also multi-frequency baths that incorporate transducers with different frequencies arranged to avoid interference. When the system is activated, all frequencies are simultaneously utilized, such as 40 and 80 kHz, providing combined effects (Kim and kwon 2015). When using a conventional ultrasonic bath for extraction purposes, the placement of the extraction vessel inside the bath is crucial. In the simplest configuration, with a single transducer located at the bottom of the bath, the extraction vessel should be positioned just above the transducer. Temperature control enhances reproducibility and the uniformity of the bath can be improved by sweeping the generator frequency, allowing adjustable power and frequency control.

7.8.3 Probe System

Probe systems consist of a generator, transducer, and horn. The generator serves as the source of alternating electrical frequency, typically around 20 kHz. The transducer and horn work together to amplify and concentrate ultrasound energy, resulting in more effective sonication compared with bath systems (up to 100 times greater efficiency). The generator allows for amplitude regulation and can be used in pulsed mode, which helps control medium cooling. The upper horn is made of titanium, while the detachable horn can be constructed from various materials, although a titanium alloy is commonly used. Glass and ceramic probes are also available, but they have limitations in cavitation production and are fragile. By combining the upper and detachable horns, the entire probe assembly does not need to be replaced due to tip erosion caused by cavitation.

For extraction purposes, the probe is immersed directly into the vessel containing the raw material and extractant, enabling direct sonication. The choice of detachable horn depends on the volume to be sonicated, with various diameters and tip configurations available. Stepped probes are the most common due to their achieved amplitude. The majority of the energy is transmitted near the tip, resulting in stronger sonochemical effects in that region (Crum et al. 1998). Spiral probes have also been introduced more recently as the sonication power is distributed along the probe. The shape of the vessel is important, as dead zones should be minimized. With continued use, tip erosion can occur, leading to potential contamination of the medium with metal particles from the tip and a decrease in efficiency.

While ultrasonic baths are commonly utilized for extraction purposes, it is generally acknowledged that probes offer superior performance due to their ability to transmit higher energy to the medium. This results in shorter extraction times under gentle conditions and improved reproducibility. However, using ultrasonic probes for extracting bioactive compounds may present certain challenges. Typically, probes are less productive when processing a single sample at a time. Nonetheless, at the laboratory scale, multi-element probes can be employed to simultaneously process multiple samples, ranging from 4 to 96 (such as microplates), thereby increasing productivity. Additionally, there is a risk of losing volatile compounds during ultrasonic extraction with probes due to the enhanced degassing effect. The use of probes generates heat, leading to elevated temperatures in the medium, which can affect its characteristics and cavitation. Therefore, cooling the sonication vessel, such as with an ice bath or refrigerated vessel, or utilizing pulsed mode is often necessary. It is crucial to carefully control the sonication conditions to prevent degradation of bioactive compounds (Chen et al. 2011). Moreover, there is a higher potential for cross-contamination when using ultrasonic probes. To mitigate these challenges, sonoreactors with indirect sonication, possessing greater power than baths, can be advantageous devices for UAE.

7.8.4 Applications

7.8.4.1 Dietary Fibers

Dietary fibers are commonly found in by-products of fruits and vegetables, and the use of UAE has been found to improve their extraction efficiency from plant matrices. Studies on UAE of dietary fiber from citrus peels and apple pomace have demonstrated higher yields (6.62% and 16.4%), shorter extraction times, and lower temperature requirements compared with conventional solvent extraction methods (Zhang et al. 2017).

7.8.4.2 Polyphenols

Studies on UAE of polyphenols have focused on investigating the impact of various extraction parameters, including temperature, solvent concentration, frequency, and extraction time. For example, Pan et al. (2012) conducted research on the extraction of flavonoid compounds from hawthorn seed and identified ultrasound time, ethanol concentration, and extraction temperature as the most significant variables affecting the extraction yield of flavonoid compounds. On the contrary, ultrasound temperature, solid-liquid ratio, and extraction time had a negligible effect on the extraction of flavonoids (Rodrigues et al. 2008). Raghavendra et al. (2022) optimized conditions for the UAE of phenolics from potato peel waste and could achieve a 2.84-fold increase in extraction yield over the conventional process.

7.8.4.3 Polysaccharides

Chen et al. (2011) conducted a study in which they extracted water-soluble polysaccharides from litchi seeds using UAE and optimized the extraction. They confirmed the structural characteristics and bioactivity, including antioxidant activity, of the polysaccharides obtained through UAE. Raza et al. (2017) utilized sijiaoling stems to extract antioxidant polysaccharides and compared the results obtained from UAE with those from hot water extraction. They found that the polysaccharides obtained through UAE exhibited higher DPPH and ABTS radical scavenging activity, stronger reducing power, and greater total antioxidant capacity compared with the polysaccharides obtained through hot water extraction.

7.8.4.4 Pectin

UAE has proven to be effective in obtaining pectin yields above 25% from various plant by-products such as pomegranate peel, grape pomace, tomato waste, grapefruit peel, orange peel, and eggplant peel. The yield of pectin depends on ultrasonic parameters like frequency, power, duty cycle, power intensity, and sonication time, as well as extraction variables including temperature, liquid-to-solid ratio (LSR), and pH of the solvent, either individually or in combination. Several studies have demonstrated an increase in pectin yield and a reduction in extraction time through UAE. For example, Wang et al. (2015) compared chemical extraction with UAE of pectin from grapefruit peel and found a higher yield of 16.34% and a 37.78% reduction in extraction time using UAE.

7.8.4.5 Flavors, Aromas, and Biocolorant

Traditional methods of extracting flavors and fragrances such as steam distillation and hydrodistillation involve passing steam through a flask containing aromatic plants, causing the aromas to evaporate (Adahchour et al. 2003). They have significant drawbacks, including low yields and the formation of by-products due to the instability of the desired compounds. The high temperatures and extended extraction times associated with these methods can lead to chemical alterations of the aromatic components and the loss of highly volatile molecules. To overcome these limitations and obtain high-quality extracts from aromatic plant materials, numerous studies have explored the advantages of utilizing ultrasonic energy for extraction (Rostagno et al. 2003). Before the UAE process, several parameters including the moisture content of the plants, particle size, ultrasonic device design, temperature, and choice of solvent need to be optimized. For example, in order to extract phycocyanin from *Spirulina platensis* cells, the selection of ultrasonic frequency was important (Furuki et al. 2003). The purity of phycocyanin in its crude extract depended on ultrasonic frequency. For example, phycocyanin was extracted with higher purity at 28 kHz than at 20 kHz, due to the selective extraction of active components at these frequencies. The aroma was extracted from garlic by using US bath (35 kHz) with a temperature of 25 °C and a time of 10 min by using diethyl ether, hexane, and ethyl acetate as solvents. Another example is extracting flavor volatile compounds from spearmint by using US probe (200 W) with extraction conditions like amplitude = 25%, solvent of ethyl alcohol/water, 70/30 (V/V), time of 5, 10, and 15 min, and temperature less than 40 °C. Aroma compounds were extracted from tea by using 40 kHz and 250 W US bath and other conditions like time of 40 min and temperature of 60 °C for 40 min and using water as solvent. By carefully adjusting these factors, researchers aim to enhance the efficiency and effectiveness of the extraction process, leading to improved extraction yields and preservation of the desired aromatic compounds (Iriti et al. 2006). Recently, Pooja et al. (2023) developed an ultrasound-assisted protocol for the effective utilization of garden pea waste as a source of natural chlorophyll pigment wherein pigment yield was enhanced by 37.4% over the conventional process.

7.9 Microwave-Assisted Extraction

Vij et al. (2013) first mentioned the use of microwave energy, employing domestic ovens in the laboratory to treat biological samples for metal trace analysis. Subsequently, they introduced the concept of extracting organic compounds through microwave irradiation. The first patent for using microwaves to extract a natural product was filed by Pare (1995). Initially employed for digestion purposes in various sample types, such as environmental, biological, and geological matrices, microwave extraction has evolved and gained significant acceptance for extracting natural products from plant materials. In recent years, it has garnered growing interest due to its ability to rapidly extract solutes from solid matrices. The extraction efficiency is comparable to classical techniques, but it offers several advantages, including reduced solvent quantity, minimized solvent waste, lower release of solvents into the environment, and reduced human exposure. As the demand for green technology increases, MAE has emerged as one of the most promising methods. It has become a major technique for extracting valuable compounds from vegetable materials, and advancements in the process have led to solvent-free extraction, eliminating the use of solvents in the system. MAE holds great potential as an eco-friendly and efficient approach for obtaining valuable compounds from natural sources.

7.9.1 *Principle*

MAE utilizes non-ionizing electromagnetic waves within the frequency range of 300 MHz to 300 GHz to disrupt or modify the cellular structure of a sample. This process, as described by Leonelli et al. (2012), involves the transfer of energy through two primary mechanisms: ionic conduction and dipole rotation, which often occur simultaneously in various MAE applications. The mechanism of ionic conduction pertains to the movement of ions within a solution when exposed to electromagnetic waves. This movement leads to uniform heating of the solvent due to the resistance encountered by the ions during their migration, as elucidated by Veggi et al. (2012). On the contrary, dipole rotation refers to the reorientation of molecules in response to the electric field's direction. This reorientation causes thermal agitation, which persists even after the electromagnetic waves are removed, as the molecules return to their disordered state.

During MAE, the primary focus of microwave heating is the moisture content present in the sample matrix. As the microwave energy causes water to evaporate, the pressure within the cells increases, resulting in the rupture of cell walls. This phenomenon, as highlighted by Sosa-Hernandez et al. in 2018, enhances the leaching of valuable compounds.

In conventional extraction processes, heat is transferred from the heating medium to the interior of the cells, while the movement of solutes occurs in the opposite direction. In contrast, MAE harnesses both heat and mass transfer gradients that operate in the same direction, facilitating the outward movement of heat and solutes from the cells. This unique characteristic enables MAE to enhance the extraction yields of high-value compounds while reducing the overall extraction time, as emphasized by Veggi et al. in 2012.

7.9.2 Instrumentation

There are two commercially available types of instruments for MAE, each employing distinct approaches. The more prevalent method entails extraction within a closed vessel, carefully controlling pressure and temperature. Alternatively, an alternative approach involves an open extraction vessel operating at atmospheric pressure. It is crucial to emphasize that using a household microwave oven for laboratory purposes is strongly discouraged. The application of microwave energy to highly flammable organic solvents can pose severe hazards. Additionally, the lack of field homogeneity in a domestic device may result in poor reproducibility. Therefore, it is strongly recommended to employ only equipment specifically approved for MAE.

Closed Vessel: Closed vessel systems are commonly recommended for digestions, acid mineralizations, or extractions under severe conditions, as they allow solvents to be heated to approximately 100 °C above their atmospheric boiling point. This approach significantly enhances the extraction speed and efficiency. To address the hazards associated with heating highly flammable solvents, modern security techniques are employed, such as high-capacity exhaust fans, solvent vapor detectors, and pressure-burst safety membranes installed on each vessel. Closed vessel systems offer the flexibility to work with various solvents, and the pressure inside the vessels largely depends on the volume and boiling point of the solvents used. The temperature can be precisely set by adjusting the microwave power. Teflon is a commonly used material for constructing cells in closed vessel systems. The maximal power delivered in closed vessel systems typically ranges from 600 to 1000 W. However, it is essential to carefully select the power level to avoid excessively high temperatures that could lead to potential solute degradation and overpressure issues. Before opening the vessel, it is crucial to cool it to room temperature, particularly when volatile solutes are present, as they may partition into the head space. This cooling step is essential for ensuring safety and minimizing risks during the process.

Open Cell: Open cell systems consist of quartz vessels equipped with a vapor condenser. Operating at atmospheric pressure, these systems rely on the boiling point of the solvent to determine the maximum attainable temperature. In this configuration, the solvent is heated and refluxed through the sample, while the microwaves are concentrated on the sample inside the vessel, ensuring homogeneous and highly efficient heating. To simplify the extraction process, the sample can be placed in a Soxhlet-type cellulose cartridge, eliminating the need for additional filtration steps. Alternatively, the sample can be directly immersed in the solvent. Open-cell extractions offer improved safety in sample handling and enable larger sample sizes to be processed. This system operates with percentage power increments ranging from 10% to 100%, corresponding to a maximum power output of 250 W.

7.9.3 Advantages

- MAE enables the adoption of environmentally friendly non-fossil fuel sources (e.g., solar, wind, and hydroenergy sources) because MAE is reliant upon electrical energy and that electrical energy can be taken from these nonfossil fuel sources which promotes a greener and more sustainable approach to extraction processes.
- Another advantage of MAE is the enhanced safety and improved working conditions for operators. MAE systems have the ability to instantly switch off the energy supply, establishing safe conditions promptly in case of any emergencies or operator concerns.
- Additionally, MAE allows for fully automated operation, eliminating the need for manual intervention during the extraction process. This automation reduces human error and increases process efficiency.
- One of the key features of MAE is its precise control over the energy absorbed by the extraction mixture. This control enables accurate and adjustable application of energy, ensuring optimal extraction conditions for the target compounds. By fine-tuning the energy input, MAE allows for improved extraction efficiency and selectivity.
- Microwave-assisted heating enhances the effectiveness and selectivity of the heating process. The use of microwaves allows for the development of solvent-free extraction processes with ease.
- Microwave heating accelerates energy transfer, resulting in quicker start-up times and rapid response to heating control. This rapid energy transfer enables efficient heating of the extraction medium, reducing the overall extraction duration.
- By utilizing microwaves, thermal gradients are decreased, leading to smaller equipment sizes and fewer operation units. This reduction in thermal gradient contributes to a more uniform and controlled heating process.
- Furthermore, microwave-assisted heating allows for the attainment of high power intensity. The focused application of microwave energy results in the generation of intense heat within the extraction medium, facilitating efficient extraction of bioactive compounds.

- The process enables the development of solvent-free processes, accelerates energy transfer, decreases thermal gradients, reduces equipment size, and allows for the attainment of high power intensity.

7.10 Pulsed Electric Field-Assisted Extraction

Pulsed electric field (PEF) treatment shows potential as a viable alternative to conventional methods of cell disintegration. When biological cells are exposed to intense electric field pulses, the structure of the cell membrane can be altered. The external electric field causes the membrane to become charged, leading to a rearrangement of the phospholipid molecules when the transmembrane voltage reaches a sufficiently high level (~0.5–1 V) (Frey et al. 2013). Consequently, the membrane loses its barrier function and becomes permeable, a phenomenon commonly referred to as "electroporation" or "electro-permeabilization."

PEF technology involves the application of short-duration electrical treatment, ranging from nanoseconds to milliseconds, using electric field strengths varying from 100–300 V/cm to 20–80 kV/cm (Koubaa et al. 2015). When employed at high electric fields (>20 kV/cm), PEF can serve as an alternative to conventional thermal processing for the inactivation of harmful microorganisms and enzymes associated with product quality (Sanchez-Vega et al. 2015). PEF induces the electrical puncturing of the biological membrane, resulting in a temporary or permanent loss of its semi-permeability (Deng et al. 2015). This phenomenon enables the selective extraction of high-value compounds from various matrices. Moreover, PEF shows promise as a technology to enhance drying and freezing processes.

As per existing literature, the primary electrical factors that impact the efficiency of electro-permeabilization are the strength of the external electric field, pulse duration, pulse shape, number of pulses administered, and the specific treatment energy. Depending on the electrical conditions utilized, the resulting membrane permeabilization can be reversible or irreversible. When aiming for an efficient extraction of valuable intracellular components, it is imperative to achieve an irreversible permeabilization of the cell membrane. The disruption of the cell membrane facilitates the release of intracellular contents and enables the penetration of solvents into the cell. PEF serves as a relatively gentle method for cell disintegration, typically conducted at ambient temperatures without introducing additional impurities to the process. This helps prevent undesired alterations in the target material (Ben Ammar et al. 2011). PEF can be carried out in both batch and continuous modes, making it a scalable process. For instance, an industrial-scale PEF device was developed by Sack et al. (2010) for the continuous treatment of sugar beets, capable of processing 10 tons of beets per hour. Furthermore, PEF has been employed not only to enhance the extraction of soluble compounds from plant tissue but also to expedite the extraction of vegetable oils, such as olive oil and maize oil (Guderjan et al. 2005).

7.10.1 Advantages

- It has the ability to retain or minimally alter the sensory, nutritional, and health-promoting properties of liquid food products.
- PEF allows for the selective extraction of target compounds from different matrices, enabling the recovery of high-added value components.
- PEF is a non-thermal technology, meaning that it can process products at lower temperatures compared with traditional thermal processing methods.
- PEF can effectively inactivate harmful microorganisms and enzymes associated with food spoilage, extending the shelf life of products and improving food safety.
- PEF can reduce the need for chemical additives, such as preservatives and antimicrobial agents, due to its antimicrobial effects. This can lead to cleaner label products and reduced reliance on synthetic additives.

7.10.2 Disadvantages

- PEF requires specialized equipment capable of generating high-intensity electric field, which can be expensive to acquire and maintain.
- PEF is most effective on thin or liquid products, as the electric fields may have limited penetration depth in thicker or solid materials. This can restrict its application in certain food processing scenarios.
- Optimizing PEF parameters (electric field strength, pulse duration, etc.) for specific products and desired outcomes can be complex and require thorough experimentation and understanding.
- PEF as a food processing technology may be subject to regulatory scrutiny and approval in some regions, requiring compliance with specific standards and regulations.

7.11 Enzyme-Assisted Extraction

In the quest to harness the therapeutic potential of natural compounds, researchers and industries are constantly seeking innovative and sustainable extraction technologies. Among these, EAE has emerged as a promising approach that leverages the power of enzymes to enhance the extraction of bioactive compounds from plant sources. Enzymes are biocatalysts that can break down complex molecules into simpler forms, facilitating the release of desired compounds from plant tissues (Streimikyte et al. 2022).

In EAE, plant material is first pretreated to enhance the accessibility of target compounds. This may involve processes such as grinding, drying, or steam treatment. The pretreated material is then mixed with a specific enzyme or a combination of enzymes that have the ability to hydrolyze the cell walls and cellular components of the plant material (Nadar et al. 2018). During the extraction process, the enzymes break down the structural barriers of the plant cells, allowing the bioactive compounds to be released and solubilized. The extraction can be carried out under various conditions, such as specific temperature, pH, and incubation time, depending on the characteristics of the enzymes and the target compounds. Once the extraction is complete, the mixture is typically subjected to further processing steps, such as filtration or centrifugation, to separate the liquid extract containing the bioactive compounds from the solid residue. The extract can then be further purified or concentrated using various techniques.

7.11.1 Advantages

- Enzymes can be highly specific in their action, allowing for the targeted extraction of specific compounds from the source material. This selectivity helps to enhance the purity and quality of the extracted compounds.
- EAE typically operates under mild temperature and pH conditions, which helps preserve the integrity and activity of heat-sensitive compounds.
- Enzymes are biodegradable and derived from natural sources, making EAE a more environmentally friendly alternative to traditional extraction methods that may rely on chemical solvents or high temperatures.
- Enzymes can break down complex cellular structures, such as cell walls, facilitating the release of target compounds (Table 7.1). This can lead to higher extraction yields and reduced processing time compared with conventional methods.

7.11.2 Disadvantages

- Enzymes may have limited substrate specificity, meaning they may not effectively act on all types of compounds or plant materials. This can restrict the range of compounds that can be efficiently extracted using enzyme-assisted methods.
- Enzymes used for extraction can be costly, adding to the overall production expenses. Additionally, the optimization of enzyme concentration and reaction conditions may require additional resources and expertise.
- Enzymes may have specific storage and stability requirements, such as temperature control and protection from microbial contamination. Proper handling and storage practices are necessary to maintain enzyme activity and prevent degradation.
- EAE may involve additional steps, such as enzyme activation, pretreatment of the plant material, and subsequent enzyme inactivation, which can increase the overall complexity and time required for the extraction process.

Table 7.1 Enzymes used in the extraction of bioactive compounds from horticultural sources

Plant sources	Enzymes	Target compounds	References
Grape pomace	Cellulase	Polyphenols	Serea et al. (2021)
Papaya leaves	Protease	Papain	Tigist et al. (2016)
Ginger rhizome	Zingiberene	Gingerol	dos Santos Reis et al. (2020)
Broccoli	Myrosinase	Sulforaphane	Deng et al. (2015)
Pineapple	Bromelain	Bromelain	Salve and Ray (2020)
Soybean	Protease	Isoflavones	de Queiros et al. (2021)
Green tea	Polyphenol oxidase	Catechins	Heemann et al. (2019)
Pineapple peel	Bromelain	Bromelain	Ketnawa et al. (2011)
Grape skins	Cellulase	Anthocyanins, Resveratrol	Costa et al. (2020)
Ginger	Cellulase	Gingerol, Shogaol	Phomkaivorn and Meesc (2020)
Turmeric	Amylase	Curcuminoids	Zhang et al. (2019)
Apple pomace	Pectinex	Antioxidant phenols	Oszmiański et al. (2011)
Cauliflower	Viscozyme	Phenolics	Huynh et al. (2014)
Citrus peel	Viscozyme	Phenolics	Nishad et al. (2022)
Grape marc seeds	Pectinase	Phenolics	Stambuk et al. (2016)

7.12 Conclusion

The extraction technologies discussed in this chapter offer significant opportunities for harnessing the valuable bioactive compounds present in fruits, vegetables, and medicinal plants. Through various conventional and non-conventional extraction methods, such as maceration, SE, UAE, and more, we can effectively retrieve these compounds and maximize their utilization. This not only contributes to the valorization of waste pomace, peels, seeds, and so on but also provides a prevailing source of beneficial additives for the food and pharmaceutical industries. By embracing advanced extraction techniques, we can fully exploit the potential of horticultural crops and unlock their valuable components, ultimately benefiting both the industry and consumers.

References

Adahchour M, van Stee LL, Beens J, Vreuls RJ, Batenburg MA, Udo A (2003) Comprehensive two-dimensional gas chromatography with time-of-flight mass spectrometric detection for the trace analysis of flavour compounds in food. J Chromatogr A 1019:157–172

Ahmad T, Masoodi FA, Rather SA, Wani SM, Gull A (2019) Supercritical fluid extraction: a review. J Biol Chem Chron 5:114–122

Anastas PT, Zimmerman JB (2003) Peer reviewed: design through the 12 principles of Green engineering. Environ Sci Technol 37(5):94A–101A

Andreu V, Pico Y (2019) Pressurized liquid extraction of organic contaminants in environmental and food samples. TrAC Trends Anal Chem 118:709–721

Ben Ammar J, Lanoiselle JL, Lebovka NI, Van Hecke E, Vorobiev E (2011) Impact of a pulsed electric field on damage of plant tissues: effects of cell size and tissue electrical conductivity. J Food Sci 76:E90–E97

Bhambra AS, Singh R (2013) Review on maceration, its techniques and advances in plant cell maceration. Pharm J 5:123–128

Carabias-Martínez R, Rodriguez-Gonzalo E, Revilla-Ruiz P, Hernandez-Mendez J (2005) Pressurized liquid extraction in the analysis of food and biological samples. J Chromatogr A 1089:1–17

Chemat F, Rombaut N, Sicaire AG, Meullemiestre A, Fabiano-Tixier AS, Abert-Vian M (2017) Ultrasound assisted extraction of food and natural products. Mechanisms, techniques, combinations, protocols and applications. A review. Ultrason Sonochem 34:540–560

Chen D, Sharma SK, Mudhoo A (2011) Applications of ultrasonication. In: Handbook on applications of ultrasound: sonochemistry for sustainability, vol 5. Elsevier, pp 101–123

Costa JR, Tonon RV, Cabral L, Gottschalk L, Pastrana L, Pintado ME (2020) Valorization of agricultural lignocellulosic plant byproducts through enzymatic and enzyme-assisted extraction of high-value-added compounds: a review. ACS Sustain Chem Eng 8:13112–13125

Crum LA, Mason TJ, Reisse JL, Suslick KS (1998) Sonochemistry and sonoluminescence, vol 524. Springer Science & Business Media

De Castro ML, Garcıa-Ayuso LE (1998) Soxhlet extraction of solid materials: an outdated technique with a promising innovative future. Anal Chim Acta 369:1–10

de Queiros LD, Dias FFG, de Avila ARA, Macedo JA, Macedo GA, de Moura JMLN (2021) Effects of enzyme-assisted extraction on the profile and bioaccessibility of isoflavones from soybean flour. Food Res Int 147:110–474

Deng Q, Zinoviadou KG, Galanakis CM, Orlien V, Grimi N, Vorobiev E, Barba FJ (2015) The effects of conventional and non-conventional processing on glucosinolates and its derived forms, isothiocyanates: extraction, degradation, and applications. Food Eng Rev 7:357–381

dos Santos Reis N, de Santana NB, de Carvalho Tavares IM, Lessa OA, dos Santos LR, Pereira NE, Franco M (2020) Enzyme extraction by lab-scale hydrodistillation of ginger essential oil (*Zingiber officinale* Roscoe): chromatographic and micromorphological analyses. Ind Crops Prod 146:112–210

Frey W, Gusbeth C, Schwartz T (2013) Inactivation of *Pseudomonas putida* by pulsed electric field treatment: a study on the correlation of treatment parameters and inactivation efficiency in the short-pulse range. J Memb Biol 246:769–781

Furuki T, Maeda S, Imajo S, Hiroi T, Amaya T, Hirokawa T, Nozawa H (2003) Rapid and selective extraction of phycocyanin from *Spirulina platensis* with ultrasonic cell disruption. J Appl Phycol 15:319–324

Guderjan M, Topfl S, Angersbach A, Knorr D (2005) Impact of pulsed electric field treatment on the recovery and quality of plant oils. J Food Eng 67:281–287

Guimarães R, Barros L, Dueñas M, Carvalho AM, Queiroz MJR, Santos-Buelga C, Ferreira IC (2013) Characterisation of phenolic compounds in wild fruits from Northeastern Portugal. Food Chem 141:3721–3730

Hawthorne SB, Yang Y, Miller DJ (1994) Extraction of organic pollutants from environmental solids with sub-and supercritical water. Anal Chem 66:2912–2920

Heemann ACW, Heemann R, Kalegari P, Spier MR, Santin E (2019) Enzyme-ssisted extraction of polyphenols from green yerba mate. Braz J Food Technol 22:201–222

Hicks M, Regalado E, Tan F, Gong X, Welch C (2016) Supercritical fluid extraction: principles and practice. J Biomed Anal 117:316–324

Huynh NT, Smagghe G, Gonzales GB, Van Camp J, Raes K (2014) Enzyme-assisted extraction enhancing the phenolic release from cauliflower (*Brassica oleracea* L. var. botrytis) outer leaves. J Agric Food Chem 62:7468–7476

Iriti M, Colnaghi G, Chemat F, Smadja J, Faoro F, Visinoni F (2006) Histo-cytochemistry and scanning electron microscopy of lavender glandular trichomes following conventional and microwave assisted hydrodistillation of essential oils: a comparative study. Flavour Fragr J 21:704–2006

Ketnawa S, Chaiwut P, Rawdkuen S (2011) Extraction of bromelain from pineapple peels. Int J Food Sci Technol 17(4):395–402

Kim YY, Kwon YE (2015) Review of magnetostrictive patch transducers and applications in ultrasonic nondestructive testing of waveguides. Ultrasonics 62:3–19

Koubaa M, Roselló-Soto E, Sic Zlabur J, RezekJambrak A, Brncic M, Grimi N, Barba FJ (2015) Current and new insights in the sustainable and green recovery of nutritionally valuable compounds from *Stevia rebaudiana* Bertoni. J Agric Food Chem 63:6835–6845

Leonelli C, Veronesi P, Cravotto G (2012) Microwave-assisted extraction: an introduction to dielectric heating. In: Microwave-assisted extraction for bioactive compounds: theory and practice. Springer, Boston, pp 1–14

Manousi N, Sarakatsianos I, Samanidou V (2019) Extraction techniques of phenolic compounds and other bioactive compounds from medicinal and aromatic plants. In: Engineering tools in the beverage industry. Woodhead Publishing, pp 283–314

Meyer CA (1993) Steam tables: thermodynamic and transport properties of steam. American Society of Mechanical Engineers, New York

Mujumdar AS (2019) Drying kinetics for fruits and vegetables. In: Handbook of industrial drying, vol 1. Elsevier, pp 501–598

Nadar SS, Rao P, Rathod VK (2018) Enzyme assisted extraction of biomolecules as an approach to novel extraction technology: a review. Food Res Int 108:309–330

Nishad J, Joshi A, Sethi S, Rudra SG, Varghese E, Shankhwar N, Bhatia A, Saurabh V, Kaur C (2022) Functionalization of ash gourd: infusion of citrus peel polyphenols through vacuum impregnation. Food Biosci 50:102095. https://doi.org/10.1016/j.fbio.2022.102095

Oszmiański J, Wojdyło A, Kolniak J (2011) Effect of pectinase treatment on extraction of antioxidant phenols from pomace, for the production of puree-enriched cloudy apple juices. Food Chem 127:623–631

Pan G, Yu G, Zhu C, Qiao J (2012) Optimization of ultrasound-assisted extraction (UAE) of flavonoids compounds (FC) from hawthorn seed (HS). Ultrason Sonochem 19:486–490

Pare JJ (1995) Microwave assisted natural products extraction. Patent no. 519588 (5002784)

Phomkaivorn M, Meeso N (2020) Enzyme-assisted extraction of gingerol and shogaol from ginger rhizomes. J Food Sci Technol 57:130–140

Pingret D, Fabiano-Tixier AS, Chemat F (2013) Ultrasound-assisted extraction. In: Natural product extraction: principles and applications, vol 21. Royal Society of Chemistry, p 89

Pooja BK, Sethi S, Joshi A, Varghese E, Kaur C, Kumar R, Shridhar (2023) Ultrasound assisted extraction of chlorophyll from pea pod waste: optimization, kinetics and stability study. Food Anal Methods 16:1358–1369. https://doi.org/10.1007/s12161-023-02502-8

Raghavendra HR, Sethi S, Bhowmik A, Varghese E, Joshi J (2022) Phenolics from potato peel and its extraction intensification using response surface methodology and genetic algorithm approach. Ind J Hortic 79:83–90

Raza A, Li F, Xu X, Tang J (2017) Optimization of ultrasonic-assisted extraction of antioxidant polysaccharides from the stem of *Trapa quadrispinosa* using response surface methodology. Int J Biol Macromol 94:335–344

Rodrigues S, Pinto GA, Fernandes FA (2008) Optimization of ultrasound extraction of phenolic compounds from coconut (*Cocos nucifera*) shell powder by response surface methodology. Ultrason Sonochem 15:95–100

Rostagno MA, Palma M, Barroso CG (2003) Ultrasound-assisted extraction of soy isoflavones. J Chromatogr A 1012:119–128

Sack M, Sigler J, Frenzel S, Eing C, Arnold J, Michelberger T, Müller G (2010) Research on industrial-scale electroporation devices fostering the extraction of substances from biological tissue. Food Eng Rev 2:147–156

Salve RR, Ray S (2020) Comprehensive study of different extraction methods of extracting bioactive compounds from pineapple waste—a review. J Pharma Innov 9:327–340
Sanchez-Camargo AP, Mendiola JA, Ibanez E, Herrero M, Reedijk J (2014) Reference module in chemistry, molecular sciences and chemical engineering. Elsevier, Waltham
Sanchez-Vega R, Elez-Martínez P, Martín-Belloso O (2015) Influence of high-intensity pulsed electric field processing parameters on antioxidant compounds of broccoli juice. Inno. Food Sci Emerg Technol 29:70–77
Santoyo S, Herrero M, Senorans FJ, Cifuentes A, Ibanez E, Jaime L (2006) Functional characterization of pressurized liquid extracts of *Spirulina platensis*. Eur Food Res Technol 224:75–81
Serea D, Rapeanu G, Constantin OE, Bahrim GE, Stănciuc N, Croitoru C (2021) Ultrasound and enzymatic assisted extractions of bioactive compounds found in red grape skins *babeascaneagra (Vitis vinifera)* variety. Ann Univ Dunarea Jos Galati Fascicle VI Food Technol 45:9–25
Sosa-Hernandez JE, Escobedo-Avellaneda Z, Iqbal HM, Welti-Chanes J (2018) State-of-the-art extraction methodologies for bioactive compounds from algal biome to meet bio-economy challenges and opportunities. Molecules 23:2953
Stambuk P, Tomašković D, Tomaz I, Maslov L, Stupic D, Karoglan Kontic J (2016) Application of pectinases for recovery of grape seeds phenolics. 3 Biotech 6:1–12
Streimikyte P, Viskelis P, Viskelis J (2022) Enzymes-assisted extraction of plants for sustainable and functional applications. Int J Mol Sci 23:2359
Teo CC, Tan SN, Yong JWH, Hew CS, Ong ES (2010) Pressurized hot water extraction (PHWE). J Chromatogr A 1217:2484–2494
Tigist M, Getnet B, Beza K, Endalamaw M, Lulit M, Tamirat D, Tinsae M (2016) Extraction and purification of papain enzyme from papaya leaf and the phytochemical components of the leaf. Int J Mol Sci 45:23–59
Veggi PC, Martinez J, Meireles MAA (2012) Fundamentals of microwave extraction. In: Microwave-assisted extraction for bioactive compounds: theory and practice. Springer, Boston, pp 15–52
Vij NN, Patil DR, Saudagar RB (2013) Microwave assisted extraction of phytoconstituents used in herbal cosmetics. Res J Top Cosmetic Sci 4:54–58
Waldeback M (2005) Pressurized fluid extraction: a sustainable technique with added values. Doctoral dissertation, Acta Universitatis Upsaliensis. Int J Mol Sci 45:233–592
Wang L, Weller CL (2006) Recent advances in extraction of nutraceuticals from plants. Trends Food Sci Technol 17:300–312
Wang W, Ma X, Xu Y, Cao Y, Jiang Z, Ding T, Liu D (2015) Ultrasound-assisted heating extraction of pectin from grapefruit peel: optimization and comparison with the conventional method. Food Chem 178:106–114
Zhang W, Zeng G, Pan Y, Chen W, Huang W, Chen H, Li Y (2017) Properties of soluble dietary fiber-polysaccharide from papaya peel obtained through alkaline or ultrasound-assisted alkaline extraction. Carbohydr Polym 172:102–112
Zhang QA, Zhang L, Cui L, Luo L (2019) Enzyme-assisted extraction of curcuminoids from turmeric (*Curcuma longa* L.) rhizomes. J Agric Food Chem 67:5558–5566

Chapter 8
Natural Bio-stimulants as an Integrated Management Strategy for Horticulture Crops

Naleeni Ramawat, Naresh Kumar, and Deepshikha Thakur

Abstract The sustainability of farming techniques has been negatively impacted by the misuse of chemical fertilizers, which has resulted in soil deterioration, water pollution, and a reduction in biodiversity. This has highlighted the critical need for new environmentally friendly substitutes that can sustain agricultural ecosystems' long-term health. Biostimulants, a broad category of compounds sourced from natural sources, have shown promise in addressing this issue. These chemicals, which include products of hormones, amino acids, and humic compounds, have demonstrated the ability to improve nutrient uptake, stress tolerance, and plant growth. Growing interest has been shown in the application of biostimulants in horticulture in recent years, as they are being investigated as a way to improve product quality and increase plant resistance to environmental stresses. Horticulturists may be able to lessen their need for chemical inputs by utilizing the special qualities of biostimulants, which would encourage more environmentally friendly agricultural methods. However, more study and development are required to completely understand the mechanisms of action of biostimulants and optimize application techniques in order to fully realize their potential in horticulture. Furthermore, legal frameworks that guarantee the safe and efficient use of biostimulants will need to be developed in order for their successful integration into horticulture operations. Researchers, legislators, and industry participants will work together to create regulations for the manufacturing, labeling, and promotion of biostimulant products. The advantages of biostimulants should also be explained to farmers and producers, and they should be given the assistance and training they need to integrate these products into their

N. Ramawat (✉)
Agriculture University Jodhpur, Jodhpur, Rajasthan, India

College of Agriculture, Agriculture University, JODHPUR, India

N. Kumar
Agriculture University Jodhpur, Jodhpur, Rajasthan, India

College of Agriculture, Agriculture University, JODHPUR, India

D. Thakur
Amity University Uttar Pradesh, Noida, Uttar Pradesh, India

M. Thakur, T. Belwal (eds.), *Advances in Postharvest and Analytical Technology of Horticulture Crops*, https://doi.org/10.1007/978-981-97-7247-6_8

operations. In conclusion, biostimulants have a lot of potential to replace chemical fertilizers in horticulture in a sustainable manner.

Keywords Biostimulants · Chemical fertilizers · Horticulture produce · Immune system

8.1 Introduction

Biostimulants represent a new category of agricultural tools that can enhance crop productivity and bolster resistance to environmental stresses. This is due to the significant correlation between their production and application (Rouphael and Colla 2020). Most biostimulants are derived from raw materials rich in bioactive molecules, with botanical sources being a promising avenue for industrial biostimulant synthesis. The ongoing challenge of increasing food production within the constraints of limited land resources, coupled with a growing population, has led to excessive use of fertilizers and pesticides, resulting in environmental pollution and health risks.

Recognizing the impact on both land and human health, various technologies have emerged as alternatives to chemical inputs and fertilizers. Among these, natural biostimulants derived from plant materials have gained traction. They promote plant growth, increase crop yield, improve nutrient utilization, and enhance stress tolerance (Rouphael and Colla 2020). Thus, biostimulants are substances that not only stimulate plant growth but also improve nutrient utilization when applied in small amounts. They facilitate plant growth, enhance nutrient availability, and reduce reliance on chemical fertilizers.

The use of biostimulants has become more prevalent in modern agricultural practices, particularly for high-value crops such as fruits and vegetables. However, there is still limited understanding of their effectiveness and mechanisms of action. Agricultural biostimulants (ABs) support plant growth from germination to maturity by restoring natural processes within plants. They can be sourced from synthetic or organic materials. While biostimulants have a long history in fruit cultivation, they have recently been integrated into broader crop production systems to enhance growth, yield, and quality across various crops, including cereals, corn, oilseeds, and soybeans (Naik et al. 2020; Rouphael et al. 2018). Any substance or microbe that improves nutritional efficiency, stress tolerance, or overall plant quality while increasing nutrient content qualifies as a plant biostimulant. This category includes commercially available products containing combinations of such substances and/or microbes (Calvo et al. 2014; Du Jardin 2015). Du Jardin (2012) defines plant biostimulants as a diverse range of materials, including microbial and complex organic compounds, humic substances (HS), and industrially derived materials, such as those from sewage sludge and urban waste. These materials contain components beneficial for plants, such as inorganic compounds and Co, Al, Si, and Se. Among organic plant biomass or biostimulants, seaweed extracts (SWEs) are particularly notable and widely used in horticulture and agriculture. Additionally, microalgae have been proposed as a renewable source of plant biostimulants (Chiaiese et al. 2018).

8.2 Classification of Natural Biostimulants

Biostimulants come in diverse formulations, each with its unique components. Generally, they can be categorized into three main groups based on their composition and origin. These groups include products containing humic compounds, hormones, and amino acids. Hormonal products, like extracts from marine algae, are rich in active plant growth regulators such as auxins and cytokinins, along with their derivatives and other microbes that can significantly influence plant vigor (Du Jardin 2015). It is worth noting that the effectiveness of these plant biostimulants may vary depending on the plant species.

Despite recent efforts to clarify the regulatory status of biostimulants, there is no universally accepted legal or regulatory definition for plant biostimulants worldwide, including within the European Union (EU) and the United States. This current situation hinders a comprehensive listing and classification of the compounds and microorganisms covered by this concept. However, certain key categories that encompass both chemicals and microorganisms are broadly recognized by scientists, regulatory bodies, and stakeholders (Calvo et al. 2014; Du Jardin 2012; Halpern et al. 2015). Microorganisms like beneficial fungi and bacteria, particularly plant growth-promoting rhizobacteria (PGPRs), are notable examples. These microorganisms can exist as endosymbiotic, free-living, or within the rhizosphere. Various researchers have outlined these categories, as summarized in Table 8.1 for a concise overview.

8.3 Use of Biostimulants as an Integrated Management Strategy

8.3.1 Biostimulant for Plant Defenses Against Abiotic Stresses in Horticultural Crops

Abiotic stress in agriculture often leads to significant reductions in crop yield, primarily due to adverse weather conditions that exacerbate soil erosion. The application of biostimulants, including protein hydrolysates, fulvic and humic acids, and microbial inoculants, has shown promising results in enhancing plant growth and resilience to various environmental challenges (Xu and Leskovar 2015). These interventions are crucial for minimizing yield losses. Notably, studies have demonstrated that the introduction of Rhizobium and Azospirillum species inoculation can enhance plant tolerance to salt stress. In crops such as corn, soybean, lettuce, and onions, mycorrhizal symbioses have been found to improve drought resistance (Augé 2001; Calvo et al. 2014). The application of humic acid to soil has been shown to reduce its electrical conductivity, thereby enhancing salt tolerance in salinized environments. Furthermore, humic acid application has been found to mitigate the negative impacts of saline conditions on Chrysanthemum growth and

Table 8.1 Biostimulants classification

Ikrina and Kolbin (2004)	Kauffman et al. (2007)	Du Jardin (2012)	Halpern et al. (2015)	Du Jardin (2015)	La Torre et al. (2016)
Microorganisms (e.g., bacteria, fungi)	Humic substances	Humic substances	Humic substances	Humic and fulvic acids	Humic substances
Plant material (land, freshwater, and marine)	Hormone-containing products (seaweed extracts)	Complex organic materials	Protein hydrolysate and formulations	Protein hydrolysate; and amino add formulations	Seaweed extracts
Humate- and humus-containing substances	Amino acid-containing products	Beneficial chemical elements	Seaweed extracts	Seaweed extracts and botanicals	Hydrolyzed proteins and amino acids
Vegetable oils		Inorganic salts (such as phosphite)	Plant-growth-promoting microorganisms (including mycorrhizal fungi)	Chitosan and other biopolymers	Inorganic salts
Natural minerals		Seaweed extracts			Microorganisms
Water (activated, degassed, thermal)		Chitin and chitosan			
Resins		Antitranspirants			
Other raw materials (oil and petroleum fractions, shale substances)		Free amino acids and other N-containing substances			

flowering stages (Mazhar et al. 2012). In saline water, humic materials have been observed to promote pistachio growth while reducing proline synthesis and ABA levels (Moghaddam and Soleimani 2012).

8.3.2 Microbial-Based Biostimulants for Plant Growth and Development

The utilization of microbial agents to promote sustainable agriculture is on the rise. The recognized benefits of symbiotic associations, which enhance the nutritional uptake of macronutrients (especially phosphorus) and micronutrients, serve as a driving force for this trend. Moreover, these interactions are vital for shielding plants from biotic and abiotic stresses and for regulating water balance. Numerous studies have shown that biostimulants, which comprise collections of microorganisms, enhance the growth and yield attributes of fruits and vegetables (refer to Table 8.2).

Table 8.2 Studies using microbial biostimulants in crops and their effects on horticultural crops

Crop	Bio-formulations	Impacts on the crop	References
Daucus carota	*Trichoderma harzianum* Rifai T-22, Beta-Chikol, Timorex Gold 24 EC, Fungicide Zaprawa Nasienna T 75 DS/WS	Development and health status of carrot plants against *Alternaria, Fusarium, Haematonectria, Rhizoctonia,* and *Sclerotinia*	Patkowska et al. (2020)
Phaseolus vulgaris	*Pseudomonas putida* UW4 + *Rhizobium* PGPR	Enhanced plant growth under stress	Remans et al. (2007)
Solanum lycopersicum	*Azospirillum brasilense (BNM65)*	Increased plant height and dry weight	Romero et al. (2014)
	Achromobacter piechaudii	Improved fresh and dry weights of tomato seedlings increased P and K use efficiencies	Mayak et al. (2004)
	Trichoderma viride	Increase the plant biomass	Lee et al. (2016)
Cucumis sativus	*Ascophyllum nodosum*	Improved fruit yield	Demir et al. (1999)
Vicia faba/ Pisum sativu	*Rhizobium leguminosarum* (GRA19–GRL19)	Improved plant growth and yield characteristics	Del Pilar et al. (1999)
Pisum sativum	*Pseudomonas* spp. (*P. putida; P. fluorescens*)	Improved plant growth and yield characteristics	Arshad et al. (2008)
Brassica oleracea var. italica	*Ascophyllum nodosum* + amino acids	Increased chlorophyll content	Kałuzewicz et al. (2017)
Ocimum basilicum	*Pseudomonades, Bacillus lentus, Azospirillum brasilens*	Improved CAT and Chlorophyll content	Heidari and Golpayegani (2012)
Capsicum annuum	*Azospirillum brasilense/ Pantoea dispersa*	Improved growth attributes and growth rates	Del Amor and Cuadra-Crespo (2012)
Cucurbita pepo	*Bacillus species, Bacillus pumilis, Trichoderma harzannum, Paenibacillus azotoformans*	Increased fruit weight	Yildirim et al. (2006)
Lactuca sativa	*Azospirillum brasilense*	Increase plant growth attributes, ascorbic acid, and more plant survival after transplant	Barassi et al. (2006); Fasciglione et al. (2015)
Cucumis melo	*Trichoderma aggressivum, T. saturnisporum, T. longibrachiatum*	Alleviate the stress produced by salinity, showed antagonistic activity against *Pythium ultimum*	Sánchez-Montesinos et al. (2019)
Allium cepa	*Azospirillum brasilense, Burkholderia ambifaria, Gluconacetobacter diazotrophicus,* and *Herbaspirillum seropedicae*	Enhanced plant growth and soil fertility status	Pellegrini et al. (2021)

Although biostimulants do not contain nutrients and therefore do not qualify as fertilizers, they can still play a crucial role in nutrient acquisition for plants. This is achieved through various mechanisms, such as the bacterial fixation of atmospheric nitrogen or the expansion of the rhizosphere's surface area by mycorrhizal growth.

8.4 Use of Biostimulants as Plant Immune System Elicitor

Despite having less sophisticated immune systems than animals, plants can exhibit resistance against harmful microorganisms. In addition to stimulating the pathogen response, many biostimulants have been reported to activate plant gene-regulators and elicitors (e.g., chitosan, laminarin, some PGPRs, etc.). However, regulators and stakeholders increasingly agree that biostimulation and biocontrol should remain distinct from a regulatory perspective. Throughout their life cycle, plants display innate immunity through physical and chemical barriers (Vargas-Hernandez et al. 2017). This immunity is divided into two categories: non-specific resistance, which provides different plants with effective defense against various pathogen species and strains, and specific resistance, which provides protection against a particular pathogen. Plants develop acquired immunity in response to disease and pest attacks. According to Kunkel and Brooks (2002) and Abdul Malik et al. (2020), this immunity is either triggered in cells surrounding the infection site (locally acquired immunity) or develops later in distant areas of plants as induced systemic resistance (ISR) and systemic acquired resistance (SAR). Localized pathogen attacks initiate SAR, which transduces the signal throughout the plant via the salicylic acid (SA) pathway. Microorganisms that promote plant growth and are not harmful, such as rhizobacteria or fungi (PGPFs and PGPRs), cause ISR. To transduce the defensive signal across the entire plant, ISR depends on ethylene (ET) and jasmonic acid (JA) (Vallad and Goodman 2004; Van Wees et al. 2008; Fu and Dong 2013).

Cell extracts (microalgae, plants, etc.), microorganisms that stimulate plant growth, and natural substances such as SA, JA, fragmented DNA, chitin, and chitosan are a few examples of biostimulant and elicitor agents that have been used in agriculture (Vázquez-Hernández et al. 2019). When applied foliar or irrigation during cultivation in economically significant species such as tomato and chili, these chemicals have demonstrated the effects described (Mejía-Teniente et al. 2019; Solano-Alvarez et al. 2021). Elicitors are naturally occurring substances that are isolated from algae or crustaceans, such as hepta-β-glucoside or chitosan. By binding to receptor molecules in the plant plasma membrane, they can mimic the plant–pathogen interaction and trigger defense mechanisms in plants (Chandra et al. 2015; Iriti and Varoni 2015; Abdul Malik et al. 2020).

Because of their "auxin-like, gibberellin-like" hormonal modulation properties, biostimulants are becoming a safer alternative to some chemical fertilizers. They come in an assortment of formulas with different components. Hormone-containing products (HCP), amino acid-containing products (AACP), and HS are the three main categories into which biostimulants are typically divided. Auxins, cytokinins, and their derivatives are among the active plant growth chemicals that are abundant in SWEs, which are classified as HCP (Du Jardin 2015). By interacting with plants' physiological and biochemical processes, these growth hormones increase agricultural productivity (Calvo et al. 2014; Hernández-Herrera et al. 2014). The presence of organic molecules like phytohormones and betaines has been associated with the benefits of biostimulants (Blunden et al. 2010; Battacharyya et al. 2015).

8.5 Biostimulants for Enhancing Quality Parameters of Horticultural Crops

Plant biostimulants, also known as ABs, are organic substances that enhance nutrient uptake, stimulate plant growth, and reduce the need for mineral fertilizers (Kunicki et al. 2010). They are organic materials that promote plant growth and help plants withstand harsh environments when applied in small quantities (Chen et al. 1994). Biostimulants are known to improve fruit size, appearance, and other quality characteristics by regulating fruit growth and development, affecting crop load, enhancing plant vigor, and modifying plant architecture (Looney 1993; Weber et al. 2018 and Basile et al. 2020).

Various types of plant protection agents are used as a preventive approach, as both biotic and abiotic factors can affect crop quality and yield (Drobek et al. 2019; Pandey et al. 2022). As consumer awareness of healthy food increases, organic farming must utilize natural biostimulants (Kyriacou and Rouphael 2018). These plant stimulants have a wide range of effects, and their activity and performance in various environmental conditions are influenced by both the crop and the type of biostimulant used (Tarantino et al. 2018; Bulgari et al. 2019).

Scientific research indicates that biostimulants can increase plant production. According to Mugnai et al. (2008), applying marine bioactive compounds enhances ammonium and potassium root influxes, benefiting the growth of *Vitis vinifera* plants. In vegetable production, biostimulants can be used to increase yield, productivity, and plant health, as well as resistance to abiotic stress. Fruits and vegetables are known for their antioxidant activity. Research has shown that the use of biostimulants in plant breeding can affect antioxidant qualities by altering enzyme activity, which in turn affects the system. Many antioxidants have properties that help prevent oxidative stress, such as phenolic substances, ascorbic acid, and lycopene (Table 8.3).

8.6 Biostimulants as Effective Post-Harvest Management Tools

Biostimulants play a crucial role in postharvest management by:

- Enhancing the nutritional and functional quality of fruit
- Reducing postharvest losses

Table 8.3 Biostimulants for enhancing quality parameters of horticultural crop

Biostimulants with dosage	Crops	Beneficial effect	References
Humic and fulvic acids (500 g/100 L)	Apricot (*Prunus armeniaca*)	Increased yield	Tarantino et al. (2018)
All amino acids (3 L/ha)	Mango (*Mangifera indica*)	Increased growth and yield	Morales-Payan (2015)
6-benzyladenine (100 mg/L)	Blueberry (*Vaccinium corymbosum*)	Increase fruit yield by 0.5 kg/tree	Milic et al. (2018)
Humic acids (3 g/L)	Cucumber (*Cucumis sativus*)	Increase in vegetable diameter	Fawzy (2012)
Nitrogen, amino acids, auxins (0.45 cm/L)	Cucumber (*Cucumis sativus*)	Lengthening vegetables	Fawzy (2012)

- Increasing the shelf life of fruit by modulating catabolic processes
- Inhibiting microbial growth on the surface of fruit
- Delaying ripening and senescence by binding with key enzymes
- Reducing chilling and freezing injuries during cold storage

An imbalance in nutrient concentration in fruit is often the primary cause of postharvest diseases in apple fruits (*Malus domestica* Borkh.), including bitter pit, water core, senescence breakdown, Jonathan spot, and blossom-end rot (Bangerth 1979; Dris and Niskanen 1999). Calcium deficiencies in fruit are frequently associated with postharvest diseases. The combination of calcium with biostimulant chemicals may enhance the effectiveness of calcium treatments. According to Soppelsa et al. (2020), applying biostimulants derived from SWEs and inorganic compounds such as zinc and silicon, along with calcium prior to harvesting, can improve the aesthetic quality of apple fruits by intensifying their skin color and reducing the likelihood of storage disorders.

8.7 Implementation Challenges in Horticultural Crops

The application of biostimulants has been shown to significantly reduce the quantity of chemical inputs, benefit ecosystems and farmers through energy conservation, and create opportunities for sustainable fertilization (Chandler et al. 2011; Wezel et al. 2014). The agricultural system of today is entirely dependent on agrochemicals, and farmers still do not favor biostimulants. The integrated management system, in which biostimulants are crucial to sustainable agriculture, must replace this agrochemical-dependent agricultural system (Vallad and Goodman 2004; Wezel et al. 2014).

Implementing biostimulants can present technological, scientific, or regulatory obstacles (Du Jardin 2015). Because of how biostimulants interact with plant physiology, the majority of the scientific problems surrounding them are physiological in nature. Biostimulants have a positive impact on a plant's growth and development by influencing its physiological processes. As these impacts affect the plant's core metabolism and homeostatic controls, which are subject to fluctuations in response to environmental cues, they can be intricate and difficult to understand. The three-way interactions between the biostimulant, the plant, and the environment must be addressed for biostimulants to be implemented successfully. In the event of a deficiency, for instance, the plants' exudates may harbor PGPRs, which in turn may assist the plant through a variety of mechanisms including phosphate solubilization and iron mobilization. It is also challenging to assess nutrient utilization efficiency (Adhikari et al. 2023), and quantifiable plant features in the field barely support the application of a biostimulant that would target this property (Du Jardin 2015).

The formulation of the biostimulant, its mixing and combining with other fertilizers or biocontrol agents, and its application are the main technical challenges that occur in the application of biostimulants. In the event that the biostimulant is to be used in conjunction with other fertilizers, further research into the interaction of the two materials is necessary, as well as optimization. Technical challenges also arise when selecting if, when, and how to apply biostimulants and when to monitor crops. One more obstacle to the biostimulant market is the regulatory issues. Premarket evaluation, intellectual property, and product classification are the main areas of concern (Rouphael et al. 2020).

The growing need for biostimulants will require a number of regional and national efforts. There are several challenges and problems in correctly applying biostimulants in horticulture crops. The nation's limited production setup availability is one of the primary challenges that may affect the timely distribution of biostimulant commodities in the planned amounts in the future. The fact that farmers are generally unaware of the usage of biostimulants is another cause for concern. In spite of the intense efforts at both the national and grassroots levels, farmers are generally uninformed about and distrustful of the usage of biostimulants. Farmers should be urged to profit from short- and long-term advantages including resource savings and ecosystem services. National and international efforts to increase awareness of and promote the use of biostimulants for sustainable farming are still woefully lacking.

8.8 Screening Biostimulant and Biocontrol Products

The primary challenge in screening biostimulants and biocontrol products lies in demonstrating their efficacy in real-world settings, despite their effectiveness being demonstrated in laboratory settings (Jindo et al. 2020). According to studies, some of the most potent PGPR strains available may exhibit decreased ability in vitro (Araújo 2013). Before being introduced as commercial products, biostimulants, including elicitors, SWEs, semiochemicals, and PGPRs, must undergo extensive testing for specific crops and growing environments (Antón-Herrero et al. 2023). Certain screening failures may arise from the differences between laboratory and field conditions (Herrmann and Lesueur 2013), leading to an inability of attributes to be expressed in vitro. Sometimes, the ability of large-field roots to absorb nutrients differs from that of pot cultures grown on a smaller scale.

The most common method to screen a biocontrol agent or an efficient PGPR inoculant is by isolating strains from pathogen-suppressive or plant growth-promoting soil (Mendes et al. 2011). Similarly, elicitor screening is typically conducted in a controlled environment before field testing. Screening procedures are tailored to a specific disease and the plant that needs to be protected (Zehra 2021). It is possible to test different pathogenic infectious strains against different genotypes of plants that exhibit varying degrees of disease susceptibility. Along with the method of administration, the quantity of treatments, and the stage of plant development, the elicitors' dosage and location must be optimized (Walters et al. 2013). Subsequently, the signaling pathways implicated in the elicitation process will be examined through a variety of techniques, including molecular biology studies that measure the expression of genes linked to plant defense mechanisms or biochemical studies that quantify the quantity of compounds related to plant defense (plant hormones, phytoalexins, enzymatic activity, ROS) (Worrall et al. 2012; Walters et al. 2014). Finally, the field trials look into the impact of several environmental parameters such as temperature, relative humidity, and luminance (Walters et al. 2014).

8.9 Formulation and Regulatory Framework

The formulation of biostimulants and biocontrol products is a critical multistep process that involves incorporating one or more strains of microorganisms into a suitable carrier. This carrier provides a safe environment to protect the microorganisms from harsh storage conditions, ensuring their survival and establishment in soils. Quality control at every stage of the process is crucial for the development and manufacture of formulations (Herrmann and Lesueur 2013). The intended applications and composition of biostimulants significantly influence their formulation. The formulation and application process is arguably one of the most important factors in determining the effectiveness of biostimulant and biocontrol products. There are several characteristics that the ideal biostimulant should possess. The composition should be user-friendly and retain its ability to effectively promote plant development or act as a biocontrol agent (Bashan et al. 2014; Walters et al. 2014). The situation with biostimulants is far more complex than that of other agrochemicals, whose suite of practices is typically optimized.

Formulating biofertilizers is particularly challenging due to two important factors: the interactions between the native microbiota of the rhizosphere and the microbiological components of biostimulant mixtures, and the biostimulant inoculant. An even more challenging task is creating a biostimulant formulation that includes microbial consortia. This is because the numerous bacteria in the consortia must be screened for interactions between the target plant, the local rhizomicrobial community, and the strains of the biostimulant product. Recent developments in molecular biology and biotechnology have made PGPR screening easier (Gowtham et al. 2022). Various formulation techniques, including peat, coir dust, charcoal, and polymer-like alginate, have been developed and summarized for PGPR inoculants (Bashan et al. 2014). Additionally, they provided an overview of several useful methods for applying inoculants and achieving production goals.

Semiochemicals, a class of biostimulants, have certain limitations because they are volatile by nature (Barbosa-Cornelio et al. 2019). Consequently, the formulation needs to ensure a gradual, regulated release of semiochemicals. According to Heuskin et al. (2011), slow-release devices should ideally meet various requirements, such as efficiency over a specified time period, repeatability in dispenser output, and sufficiently high aerial concentrations for detection. It is advisable to use semiochemical dispensers early in the growing season, when insect infestation is at its lowest, as most of them have a decreasing release rate over time (Witzgall 2001). Optimal concentration and density of semiochemicals are critical for efficient manipulation of biological control agents. Rodriguez-Saona et al. (2012) state that combining volatile substances can also make them more appealing.

The main obstacle to biostimulant marketing is the unclear regulatory environment surrounding them. Biostimulants are a novel class of biological inputs, in contrast to previous biological inputs such as biopesticides and biofertilizers (Sible et al. 2021). They are intended to boost plant growth and yield through the use of abiotic stress tolerance, nutrient availability, and other techniques. The most closely comparable category is plant regulators. The term "biostimulant" has no established legal meaning at the national or international levels because biostimulants can be classified in various ways depending on how they work and the outcomes they are intended to achieve (Sible et al. 2021). There is no clear regulatory structure in place for biostimulants, and many nations specify them in different ways.

The notion of biostimulants was included for the first time in the EU's new fertilizer legislation, UE 2019/1009 (Antón-Herrero et al. 2023). Until recently, there has been no official regulation of biostimulants, but gradually many nations have begun to regulate their promotion.

Under the Fertilizer Control Order (FCO) of 1985, biostimulants are classified as both organic and biofertilizers in India. Products with carrier-based living microorganisms (solid or liquid) that improve soil and/or crop yield by nitrogen fixation, phosphorus solubilization, or nutrient mobilization are referred to as biofertilizers. The Indian government has taken action to provide precise guidelines for biostimulant registration in response to the country's expanding biostimulant business and to maintain the highest standards for biostimulant quality.

Because there is not a widely accepted definition of biostimulants in China, they are categorized as fertilizers. November 20, 2021, saw the issuance of the agricultural standard "NY/T 3831-2021," which deals with "General Regulations for Organic Water-Soluble Fertilizers." This standard classifies organic water-soluble fertilizers according to the components that make them up. These come from free amino acids, SWEs, humic acid made from minerals, chitosan, polyglutamic acid, polyaspartic acid, molasses, low-quality fish, and the fermentation products of those fish. Furthermore, it is acceptable to include microelement supplements, secondary elements, and other organic compounds.

Fertilizer Regulation (2017) defines Biostimulants as belonging to Fertilizer Group 3 in South Africa. This category includes materials or organisms, manmade or natural, that improve soil properties, plant development, or yield. Seaweed, organic acids, PGPR, biofertilizers, fertilizer coatings, and moisture-absorbing goods are all on the inclusion list. The 2004 law that is now in effect in Mexico makes no specific mention of biostimulants. On the contrary, they might be grouped as non-synthetic fertilizers or non-synthetic PGŔ, or as microbial inoculants for direct soil application or seed treatments (Caradonia et al. 2019). Ley 21-349 in Chile defines biostimulants as compounds, combinations of substances, or microorganisms that can be applied to seeds, plants, or the rhizosphere. They seek to activate natural plant feeding mechanisms to enhance nutrient utilization efficiency, abiotic stress tolerance, qualitative features, and the availability of immobilized nutrients in the soil or rhizosphere.

The EU employs a range of tactics to promote advancements in biostimulant and biocontrol technology in agriculture. Harmonizing the procedures for licensing plant protection items across the EU is the goal of Regulation 1107/2009. Biostimulants have different regulatory statuses depending on whether they are registered under the REACH regulation, classified as fertilizers under national laws, treated as pesticides under European legislation, or allowed under frameworks for organic production. The laws governing biostimulants are scattered over many regulatory frameworks, frequently without clear nomenclature, making it difficult to create thorough product registries and accurate usage data. Many gaps in the control of biostimulants must be filled in order for them to function properly. If the input sector, political leaders, and research organizations have a thorough understanding of the implications and effects of the current legislation toward the adoption of the input, as well as the implications of the revision of norms and regulations included in this work, they will be supported in obtaining appropriate norms and technical procedures that can enable regulatory safety and product validation (Moraes and Azevedo 2016). The lack of broad legal recognition of the biostimulant idea is the reason for the restricted availability and unreliability of market statistics. Having a clear process in place for registering active substances with the potential to be used in two applications is essential.

Increasing the legitimacy of biostimulants in the market requires proactive backing of industry standards. The Biological Products Industry Alliance and the EPA are working together to overcome some regulatory concerns related to biostimulants. To boost consumer confidence, a definition of biostimulants that is officially recognized needs to be developed together with an open, predictable, and consistent introduction strategy into the market. There is an urgent need for global standardization of criteria and practices. Most importantly, clear and accepted biostimulant claims must be established.

Because of their relationship to fertilizers, biostimulants have rather varying market positioning in different countries, which presents another difficulty in standardizing the biostimulant sector. Regulators may also be able to enhance data integration with existing databases and protocols—such as those outlined in the REACH regulation (EC No 1907/2006), which addresses chemical registration, evaluation, authorization, and restriction—by creating a novel approach for biostimulants. Biostimulants present distinct challenges for intellectual property protection. It can be difficult to protect patents and stop illegal duplication or reverse engineering of biostimulant products. It can be difficult to ascertain whether a product is innovative and qualifies as an invention, both of which are requirements for patent eligibility.

As a result, companies routinely patent the industrial processes involved in creating biostimulants. Protecting intellectual property could be improved by enforcing mandatory biostimulant registration in conjunction with data protection protocols. The adoption of data-sharing policies for the registration of biostimulant chemicals or microorganisms may encourage market expansion for biostimulants by fostering cooperative ties and partnerships within the industry. While the REACH law has created a precedent for this kind of innovation in Europe, it is crucial to keep in mind that the regulation only pertains to chemicals and does not address microorganisms, a significant subset of biostimulants.

The recent global agreement to restrict the use of pesticides would boost the market for biostimulants. Because national regulatory systems are not harmonized, providers operating in diverse market situations are currently engaged in unfair competition. Uniform biostimulant regulations are necessary to prevent market fragmentation (Caradonia et al. 2019).

8.10 Conclusion

Plant biostimulants represent a novel and distinct category of sustainable, environmentally friendly agricultural inputs that can be employed as part of an integrated management strategy for the cultivation of horticultural crops, alongside synthetic fertilizers. These biostimulants are widely recognized for their ability to influence the morphological, physiological, and qualitative characteristics of plants, thereby enhancing their growth, yield, and overall quality while simultaneously mitigating the adverse effects of abiotic stress. Despite their potential benefits, the use of biostimulant compounds in the horticulture industry has not received sufficient attention, necessitating further in-depth research.

In an integrated approach to enhancing agricultural productivity and quality, fortifying plant resilience to salinity, water stress, cold, and high-temperature stress, and improving crop nutrition and water utilization efficiency, biostimulants may prove to be a valuable tool. The input industry would greatly benefit from a comprehensive understanding of the implications and impacts of the current legislation governing the use of biostimulants, as well as the adaptation of standards and regulations to promote sustainable crop production solutions.

References

Abdul Malik NA, Kumar IS, Nadarajah K (2020) Elicitor and receptor molecules: orchestrators of plant defense and immunity. Int J Mol Sci 21(3):963

Adhikari S, Anuragi H, Chandra K, Hanmantrao Tarte S, Dhaka SR, Hanuman Singh Jatav HS, Hingonia K (2023) Molecular basis of plant nutrient use efficiency—concepts and challenges for its improvement. In: Aftab T, Hakeem KR (eds) Sustainable plant nutrition. Academic Press, pp 107–151

Antón-Herrero R, García-Delgado C, Antón-Herrero G, Mayans B, Delgado-Moreno L, Eymar E (2023) Design of a hydroponic test to evaluate the biostimulant potential of new organic and organomineral products. Sci Hortic 310:111753. https://doi.org/10.1016/j.scienta.2022.111753

Araújo AEDS (2013) Response of traditional upland rice varieties to inoculation with selected diazotrophic bacteria isolated from rice cropped at the northeast region of Brazil. Appl Soil Ecol 64:49–55

Arshad M, Shaharoona B, Mahmood T (2008) Inoculation with pseudomonas spp. containing ACC-deaminase partially eliminates the effects of drought stress on growth, yield, and ripening of pea (*Pisum sativum* L.). Pedosphere 18:611–620

Augé RM (2001) Water relations, drought and vesicular arbuscular mycorrhizal symbiosis. Mycorrhiza 11:3–42. https://doi.org/10.1007/s005720100097

Bangerth F (1979) Calcium-related physiological disorders of plants. Annu Rev Phytopathol 17:97–122

Barassi CA, Ayrault G, Creus CM, Sueldo RJ, Sobrero MT (2006) Seed inoculation with azospirillum mitigates NaCl effects on lettuce. Sci Hortic 109:8–14

Barbosa-Cornelio R, Cantor F, Coy-Barrera E, Rodríguez D (2019) Tools in the investigation of volatile semiochemicals on insects: from sampling to statistical analysis. Insects 10(8):241. https://doi.org/10.3390/insects10080241

Bashan Y, de Bashan LE, Prabhu SR, Hernandez JP (2014) Advances in plant growth-promoting bacterial inoculant technology: formulations and practical perspectives (1998–2013). Plant Soil 378:1–33

Basile B, Rouphael Y, Colla G, Soppelsa S, Andreotti C (2020) Appraisal of emerging crop management opportunities in fruit trees, grapevines and berry crops facilitated by the application of biostimulants. Sci Hortic 267:109330

Battacharyya D, Babgohari MZ, Rathor P, Prithiviraj B (2015) Seaweed extracts as biostimulants in horticulture. Sci Hortic 196:39–48

Blunden G, Morse PF, Mathe I, Hohmann J, Critchley AT, Morrell S (2010) Betaine yields from marine algal species utilized in the preparation of seaweed extracts used in agriculture. Nat Prod Commun 5(4):581–585

Bulgari R, Franzoni G, Ferrante (2019) Biostimulants application in horticultural crops under abiotic stress conditions. Agronomy 9:306. https://doi.org/10.3390/agronomy9060306

Calvo P, Nelson L, Kloepper JW (2014) Agricultural uses of plant biostimulants. Plant Soil 383:3–41

Caradonia F, Battaglia V, Righi L (2019) Plant biostimulant regulatory framework: prospects in Europe and current situation at international level. J Plant Growth Regul 38:438–448. https://doi.org/10.1007/s00344-018-9853-4

Chandler D, Bailey AS, Tatchell GM, Davidson G, Greaves J, Grant WP (2011) The development, regulation and use of biopesticides for integrated pest management. Philos Trans R Soc B 366(1573):1987–1998

Chandra S, Chakraborty N, Dasgupta A, Sarkar J, Panda K, Acharya K (2015) Chitosan nanoparticles: a positive modulator of innate immune responses in plants. Sci Rep 5(1):51–95

Chen Y, Magen H, Riov J (1994) Humic substances originating from rapidly decomposing organic matter. Proc Int Meet 1992:427–443

Chiaiese P, Giandomenico C, Giuseppe C, Kyriacou CM, Youssef R (2018) Renewable sources of plant biostimulation: microalgae as a sustainable means to improve crop performance. Front Plant Sci 9(1782):1–6

Del Amor FM, Cuadra-Crespo P (2012) Plant growth-promoting bacteria as a tool to improve salinity tolerance in sweet pepper. Funct Plant Biol 39:82–90

Del Pilar Cordovilla M, Berrido SI, Ligero F, Lluch C (1999) Rhizobium strain effects on the growth and nitrogen assimilation in *Pisum sativum* and *Vicia faba* plant growth under salt stress. J Plant Physiol 154:127–131

Demir K, Günes A, Inal A, Alpaslan M (1999) Effects of humic acids on the yield and mineral nutrition of cucumber (*Cucumis Sativus*, L.) grown with different salinity levels. Acta Hortic 492:95–104

Dris R, Niskanen R (1999) Calcium chloride sprays decrease physiological disorders following long-term cold storage of apple. Plant Food Hum Nutr 54:159–171

Drobek M, Magdalena FA, Cybulska J (2019) Plant biostimulants: importance of the quality and yield of horticultural crops and the improvement of plant tolerance to abiotic stress—a review. Agronomy 9:335. https://doi.org/10.3390/agronomy9060335

Du Jardin P (2012) The science of plant biostimulants–a bibliographic analysis, Ad hoc study report. European Commission

Du Jardin P (2015) Plant biostimulants: definition, concept, main categories and regulation. Sci Hortic 196:3–14

Fasciglione G, Casanovas EM, Quillehauquy V, Yommi AK, Goñi MG, Roura SI, Barassi CA (2015) *Azospirillum* inoculation effects on growth, product quality and storage life of lettuce plants grown under salt stress. Sci Hortic 195:154–162

Fawzy ZF (2012) Response of growth and yield of cucumber plants (Cucumis sativus L.) to different foliar applications of humic acid and bio-stimulators. Int Res J Appl Basic Sci 6:630–637

Fu ZQ, Dong X (2013) Systemic acquired resistance: turning local infection into global defence. Annu Rev Plant Biol 64:839–863

Gowtham HG, Singh SB, Shilpa N, Aiyaz M, Nataraj K, Udayashanka AC, Amruthesh KN, Murali M, Poczai P, Gafur A (2022) Insight into recent progress and perspectives in improvement of antioxidant machinery upon PGPR augmentation in plants under drought stress: a review. Antioxidants 11:1763. https://doi.org/10.3390/antiox11091763

Halpern M, Asher Bar-Taly A, Ofeky M, Minzy D, Mullerx T, Yermiyahu U (2015) Chapter two—the use of biostimulants for enhancing nutrient uptake. Adv Agron 130:141–174

Heidari M, Golpayegani A (2012) Effects of water stress and inoculation with plant growth promoting rhizobacteria (PGPR) on antioxidant status and photosynthetic pigments in basil (*Ocimum basilicum* L.). J Saudi Soc Agric Sci 11:57–61

Hernández-Herrera RM, Santacruz-Ruvalcaba F, Ruiz-López MA, Norrie J, Hernández-Carmona G (2014) Effect of liquid seaweed extracts on growth of tomato seedlings (*Solanum lycopersicum* L.). J Appl Phycol 26:619–628

Herrmann L, Lesueur D (2013) Challenges of formulation and quality of biofertilizers for successful inoculation. Appl Microbiol Biotechnol 97(20):8859–8873

Heuskin S, Verhehhen J, Haubruge E, Wathelet JP, Longnay G (2011) The use of semiochemical slow-release devices in integrated pest management strategies. Biotechnol Agron Soc Environ 15(3):459–470

Ikrina MA, Kolbin AM (2004) Regulators of plant growth and development, Stimulants, vol 1. Chimia, Moscow

Iriti M, Varoni EM (2015) Chitosan-induced antiviral activity and innate immunity in plants. Environ Sci Pollut Res 22:2935–2944

Jindo K, Olivares FL, Malcher DJDP, Sánchez-Monedero MA, Kempenaar C, Canellas LP (2020) From lab to field: role of humic substances under open-field and greenhouse conditions as biostimulant and biocontrol agent. Front Plant Sci 11:2020. https://doi.org/10.3389/fpls.2020.00426

Kałuzewicz A, Krzesinski W, Spizewski T, Zaworska A (2017) Effect of biostimulants on several physiological characteristics and chlorophyll content in broccoli under drought stress and re-watering. Not Bot Hortic Agrobo Cluj-Napoca 45:197–202
Kauffman GL, Kneivel DP, Watschke TL (2007) Effects of a biostimulant on the heat tolerance associated with photosynthetic capacity, membrane thermostability, and polyphenol production of perennial ryegrass. Crop Sci 47(1):261–267
Kunicki E, Grabowska AS, Ekara A, Wojciechowska R (2010) The effect of cultivar type, time of cultivation, and biostimulant treatment on the yield of spinach (*Spinaciao leracea* L.). Folia Hortic 22:9–13
Kunkel BN, Brooks DM (2002) Cross talk between signaling pathways in pathogen defense. Curr Opin Plant Biol 5(4):325–331
Kyriacou MC, Rouphael Y (2018) Towards a new definition of quality for fresh fruits and vegetables. Sci Hortic 234:463–469
La Torre A, Battaglia V, Caradonia F (2016) An overview of the current plant biostimulant legislations in different European Member States. J Sci Food Agric 96(3):727–734
Lee S, Yap M, Behringer G, Hung R, Bennett JW (2016) Volatile organic compounds emitted by Trichoderma species mediate plant growth. Fungal Biol Biotechnol 3:1–14
Looney NE (1993) Improving fruit size, appearance and other aspects of fruit crop quality with plant bioregulating chemicals. Acta Hortic 329:120–127
Mayak S, Tirosh T, Glick BR (2004) Plant growth-promoting bacteria confer resistance in tomato plants to salt stress. Plant Physiol Biochem 42:565–572
Mazhar AAM, Shedeed SI, Abdel-Aziz NG, Mahgoub MH (2012) Growth, flowering and chemical constituents of *Chrysanthemum indicum* L. plant in response to different levels of humic acid and salinity. J Appl Sci Res 8:3697–3706
Mejía-Teniente L, Durán-Flores BA, Torres-Pacheco I, González-Chavira MM, Rivera-Bustamante RF, Feregrino-Perez AA, Pérez-Ramírez I, Rocha-Guzmán NE, Reynoso-Camacho R, Guevara-González RG (2019) Hydrogen peroxide protects pepper (*Capsicum annuum* L.) against pepper golden mosaic geminivirus (PepGMV) infections. Physiol Mol Plant Pathol 106:23–29
Mendes R, Kruijt M, de Bruijn I, Dekkers E, van der Voort M, Schneider JH, Piceno YM, DeSantis TZ, Andersen GL, Bakker PA, Raaijmakers JM (2011) Deciphering the rhizosphere microbiome for disease-suppressive bacteria. Science 332(6033):1097–1100
Milic B, Tarlanovic J, Keserovic Z, Magazin N, Miodragovic M, Popara G (2018) Bioregulators can improve fruit size, yield and plant growth of northern highbush blueberry (*Vaccinium corymbosum* L.). Sci Hortic 235:214–220
Moghaddam ARL, Soleimani A (2012) Compensatory effects of humic acid on physiological characteristics of pistachio seedlings under salinity stress. Acta Hortic 940:252–255
Moraes J, Azevedo P (2016) Biostimulants: identification of regulatory challenges and proposals to make this agri-input viable in Brazil. Acta Hortic 1148:101–104
Morales-Payan JP (2015) Influence of foliar sprays of an amino acid formulation on fruit yield of "Edward" mango. Acta Hortic 1075:157–160
Mugnai S, Azzarello E, Pandolfi C, Salamagne S, Briand X, Mancuso S (2008) Enhancement of ammonium and potassium root influxes by the application of marine bioactive substances positively affects *Vitis vinifera* plant growth. J Appl Phycol 20:177–182
Naik K, Mishra S, Srichandan H, Singh PK, Choudhary A (2020) Microbial formulation and growth of cereals, pulses, oilseeds and vegetable crops. Sustain Environ Res 30:1–18
Pandey B, Bhardwaj V, Ramawat N (2022) The role of biostimulants in plant growth, development, and abiotic stress management: recent insights. In: Ramawat N, Bhardwaj V (eds) Biostimulants: exploring sources and applications: plant life and environment dynamics. Springer, Singapore. https://doi.org/10.1007/978-981-16-7080-0_9
Patkowska E, Mielniczuk E, Jamiołkowska A, Skwaryło-Bednarz B, Błażewicz-Woźniak M (2020) The influence of trichoderma harzianum rifai T-22 and other biostimulants on rhizosphere beneficial microorganisms of carrot. Agronomy 10(11):1637

Pellegrini M, Spera DM, Ercole C, Del Gallo M (2021) Allium cepa L. inoculation with a consortium of plant growth-promoting bacteria: Effects on plants, soil, and the autochthonous microbial community. Microorganisms 9(3):639
Remans R, Croonenborghs A, Torres Gutierrez R, Michiels J, Vanderleyden J (2007) Effects of plant growth-promoting rhizobacteria on nodulation of Phaseolus vulgaris L. are dependent on plant P nutrition. In: New perspectives and approaches in plant growth-promoting rhizobacteria research, pp 341–351
Rodriguez-Saona C, Blaauw D, Isaacs R (2012) Manipulation of natural enemies in agroecosystems: habitat and semiochemicals for sustainable insect pest control. In: Marcelo LL, Soloneski S (eds) Integrated pest management and pest control—current and future tactics. IntechOpen, Rijeka, pp 89–126
Romero AM, Vega D, Correa OS (2014) *Azospirillum brasilense* mitigates water stress imposed by a vascular disease by increasing xylem vessel area and stem hydraulic conductivity in tomato. Appl Soil Ecol 8:38–43
Rouphael Y, Colla G (2020) Biostimulants in agriculture. Front Plant Sci 11:40
Rouphael Y, Kyriacou MC, Petropoulos SA, De Pascale S, Colla G (2018) Improving vegetable quality in controlled environments. Sci Hortic 234:275–289
Rouphael Y, Jardin P, Brown P, Pascale S, Colla G (2020) Biostimulants for sustainable crop production. Burleigh Dodds Science
Sánchez-Montesinos B, Diánez F, Moreno-Gavira A, Gea FJ, Santos M (2019) Plant growth promotion and biocontrol of Pythium ultimum by saline tolerant Trichoderma isolates under salinity stress. Int J Environ Res Public Health 16(11):2053
Sible CN, Seebauer JR, Below FE (2021) Plant biostimulants: a categorical review, their implications for row crop production, and relation to soil health indicators. Agronomy 11:1297. https://doi.org/10.3390/agronomy11071297
Solano-Alvarez N, Valencia-Hernández JA, Rico-García E, Torres-Pacheco I, Ocampo-Velázquez RV, Escamilla-Silva EM, Romero-García AL, Alpuche-Solís ÁG, Guevara-González RG (2021) A novel isolate of Bacillus cereus promotes growth in tomato and inhibits *Clavibacter michiganensis* infection under greenhouse conditions. Plan Theory 10(3):506
Soppelsa S, Kelderer M, Testolin R, Zanotelli D, Andreotti C (2020) Effect of biostimulants on apple quality at harvest and after storage. Agronomy 10:12–14. https://doi.org/10.3390/agronomy10081214
Tarantino A, Lops F, Disciglio G, Lopriore G (2018) Effects of plant biostimulants on fruit set, growth, yield and fruit quality attributes of 'Orange rubis®' apricot (Prunus armeniaca L.) cultivar in two consecutive years. Sci Hortic 239:26–34
Vallad GE, Goodman RM (2004) Systemic acquired resistance and induced systemic resistance in conventional agriculture. Crop Sci 44(6):1920–1934
Van Wees SC, Van der Ent S, Pieterse CM (2008) Plant immune responses triggered by beneficial microbes. Curr Opin Plant Biol 11(4):443–448
Vargas-Hernandez M, Macias-Bobadilla I, Guevara-Gonzalez RG, Romero-Gomez SDJ, Rico-Garcia E, Ocampo-Velazquez RV, Alvarez-Arquieta LDL, Torres-Pacheco I (2017) Plant hormesis management with biostimulants of biotic origin in agriculture. Front Plant Sci 8:1762
Vázquez-Hernández MC, Parola-Contreras I, Montoya-Gómez LM, Torres-Pacheco I, Schwarz D, Guevara-González RG (2019) Eustressors: chemical and physical stress factors used to enhance vegetables production. Sci Hortic 250:223–229
Weber N, Schmitzer V, Jakopic J, Stampar F (2018) First fruit in season: seaweed extract and silicon advance organic strawberry (Fragaria × Ananassa Duch.) fruit formation and yield. Sci Hortic 242:103–109
Walters DR, Ratsep J, Havis ND (2013) Controlling crop diseases using induced resistance: challenges for the future. J Exp Bot 64(5):1263–1280
Walters DR, Newton AC, Lyon GD (2014) Induced resistance for plant defense: a sustainable approach to crop protection, 2nd edn. Wiley-Blackwell, Oxford

Wezel A, Marion C, Florian C, Jean-François V, Aurélie F, Joséphine P (2014) Agroecological practices for sustainable agriculture. A review. Agron Sustain Dev 34(1):1–20

Witzgall P (2001) Pheromones and other biological techniques for insect control in orchards and vineyards. Proceedings of the meeting in Hohenheim (Baden-Württemberg, Germany), 10–12 November, 1999. IOBC/WPRS Bull 24(2):114–122

Worrall D, Geoff HH, Jason PM, Marcin GPC, Taylor JE, Nigel DP, Michael RR (2012) Treating seeds with activators of plant defence generates long-lasting priming of resistance to pests and pathogens. New Phytol 193(3):770–778

Xu C, Leskovar DI (2015) Effects of *A. nodosum* seaweed extracts on spinach growth, physiology and nutrition value under drought stress. Sci Hortic 183:39–47

Yildirim E, Taylor AG, Spittler TD (2006) Ameliorative effects of biological treatments on growth of squash plants under salt stress. Sci Hortic 111:1–6

Zehra SA (2021) Evaluation of biological potential and phytochemical screening of some novel plants from Gilgit-Baltistan. Doctoral dissertation, Quaid-i-Azam University Islamabad

Chapter 9
The Role of Biochar in Improving Soil Quality and Horticulture Crop Performance

Deepshikha Thakur and Saiatluri Teja

Abstract Growing awareness of the detrimental effects of synthetic fertilizers and agrochemicals has spurred a shift toward organic inputs in agriculture. Biochar, generated through the pyrolysis process, emerges as a promising organic input, rich in carbon. The thermochemical breakdown of biomass at around 700 °C, irrespective of oxygen presence, produces biochar. Its incorporation into soils offers numerous benefits, including improvements in biological and physicochemical properties. Biochar enhances soil density, pore size distribution, color, pH, nutrient retention, cation exchange capacity, sorption capacity, moisture retention, aeration, and pollution remediation. Notably, it positively influences soil biological properties, impacting the microbial community's habitat, biomass, enzyme activity, abundance, and diversity. These advantages, encompassing the addition of essential cations, micronutrients, and increased water-holding capacity, position biochar as a potentially cost-effective and eco-friendly solution for horticulture crop production. Properly administered, biochar-treated soil demonstrates positive growth performance in horticultural plants, reinforcing its potential in sustainable agricultural practices.

Keywords Biochar · Soil · Microorganisms · Carbon · Micronutrients

9.1 Introduction

Due to growing consumer knowledge of the negative impacts of synthetic fertilizers and agrochemicals, there is a global demand for organic food. The use of organic farming inputs is a result of growing awareness. Biochar is one such organic input that shows promise for improving soil quality and crop performance in horticulture. As a kind of charcoal, biochar is a carbon-rich substance created through the pyrolysis process. At a temperature of approximately ≤700 °C, the biomass is subjected to

D. Thakur (✉) · S. Teja
Amity Institute of Organic Agriculture, Amity University Uttar Pradesh, Noida, Uttar Pradesh, India

M. Thakur, T. Belwal (eds.), *Advances in Postharvest and Analytical Technology of Horticulture Crops*, https://doi.org/10.1007/978-981-97-7247-6_9

a thermochemical breakdown in the absence or limited presence of oxygen. This results in a stable, carbon-rich material that can be used for various applications. The production of biochar sequesters carbon, which helps in mitigating climate change by reducing greenhouse gas emissions. Applying biochar to soils can also improve soil fertility, water retention, and nutrient availability, which makes it an important tool for environmental remediation and sustainable agriculture (Jien 2019). It is well recognized that biochar improves crop performance, productivity, and soil fertility. It also reduces greenhouse gas emissions by recycling biowaste in a sustainable manner. Beneficial properties of biochar such as alkaline pH, high cation exchange capacity (CEC), abundant surface functional groups, remarkable surface area, adequate porosity, excellent water holding capacity, and sufficient nutrient retention capacity can help repair the adverse effects of extreme weather events in the soil–plant system (Kumar et al. 2022).

The biochar applications and amendments have proved promising agronomic results in various horticulture crops (Sinha et al. 2023). Biochar application to soil or growing media at an appropriate rate improves seedling growth, affects some phytohormones, and increases photosynthetic pigments and photosynthesis (Zulfiqar et al. 2022). Biochar improves seed germination of horticultural crops (Zulfiqar et al. 2022). Biochar has also been reported to improve crop growth and yield and product quality. Biochar supplementation is an efficient and cost-effective management tool for improving crop productivity in terms of increasing yield and quality (Farooq et al. 2023).

9.2 Effect of Biochar on Physicochemical Properties of Soil

From both an ecocentric and an anthropocentric standpoint, soil is a very important resource and a very complicated ecosystem (Yang et al. 2020). As the primary natural resource that humans rely on for the production of food, feed, fiber, and renewable energy, it is an essential part of the Earth's ecosystem (Kalu et al. 2021). Unfortunately, soil has been and is currently being rapidly degraded at a global scale due to a range of invasive anthropic activities in intensive agriculture, with concomitant adverse effects on human and ecosystem health (Yang et al. 2020). Numerous studies have reported that applications of biochar into soil can potentially improve soil fertility by improving its physicochemical properties such as increasing organic C content (Wang et al. 2019a, b), reducing tensile strength and bulk density (Zhang et al. 2020), increasing water infiltration and retention (Chang et al. 2021), and increasing pH and CEC (Hussain et al. 2021). Biochar application in soil alters the physical properties of soils such as texture, structure, density, and pore size distribution with effects on soil aeration, plant growth, water retention capacity, and soil workability (Xiao et al. 2019; Horák et al. 2019). The addition of biochar to soil can have an impact on the chemical characteristics of the soil, including bulk density, temperature, and the retention and aggregation of water (Jien 2019). The addition of biochar to cropping soils has been shown to provide a range of agronomic benefits, especially in situations when soil fertility and production are declining (Singh et al. 2022). Under the following sections, the impacts of biochar amendment on various physicochemical properties of soil are evaluated and briefly addressed.

9.2.1 Soil Density

Biochar bulk density is much lower than soil, so biochar application can reduce the overall bulk density of mineral soil (Yang and Lu 2021), although it is also possible to increase bulk density. However, if the biochar applied has low mechanical strength and disintegrates rapidly into small particles and fills existing pore spaces in the soil, then the dry bulk density of the soil may increase (Razzaghi et al. 2020). Biochar influences soil porosity by either direct pore contribution, creation of accommodation pores, or improved aggregate stability (Diatta et al. 2020).

9.2.2 Soil Pore Size Distribution

Distribution and connectivity of Biochar pore size affect the overall porosity of the soil to either increase or decrease after biochar incorporation into the soil. Biochar porosity depends on the temperature of the pyrolysis (i.e., increases up to 750°C with higher temperatures) (Are 2019). Pore size ranges from <2 to >50 nm, with increasing pyrolysis temperature in smaller diameter fractions (Sharma et al. 2021).

9.2.3 Soil Color

Soils with high biochar content become darker in color (Glaser and Birk 2012). According to Islam et al. (2021), when biochar was added, the color of the dry soil changed. The Munsell value, which shows black (black) at the bottom and white (10) at the top, decreased from 5.5 to 4.8 at biochar concentrations of 10 g kg^{-1} to 3.6 at concentrations of 50 g kg^{-1}. The degree of darkening depends on (i) soil color before biochar addition, (ii) biochar color, (iii) soil biochar concentration, (iv) degree of mixing (related to biochar and soil particle sizes), (v) surface roughness, and (vi) changes in soil water retention that accompany biochar addition (Verheijen et al. 2010). Zhang et al. (2021a, b) have proved in their research that the addition of biochar to the terra preta soil has dramatically changed its color.

9.2.4 *Soil pH*

Biochar can be used as an amendment for acidic soils as it is known to decrease the soil acidity and in turn enhance the plant growth by increasing soil fertility (Shetty et al. 2021). Pyrolysis of plant biomass normally results in highly alkaline biochar (Lehmann and Joseph 2009; Yuan et al. 2011). However, alkalinity varies with respect to feedstock properties used for the biochar production. The greater the alkalinity of biochar, the greater is the reduction in the acidity of the soil (Shetty et al. 2021). In addition to the liming effect, high surface charge density, large surface area, internal porosity, and the presence of both polar and non-polar surface sites on biochar are critical for metal adsorption (Das and Ghosh 2022).

9.3 Effect of Biochar on Biological Properties of Soil

Understanding the impact of biochar on the biological properties of soil is crucial, as these properties play a vital role not only in crop growth but also in providing essential ecosystem services. Various studies, including those by Pietikäinen et al. (2000), Grossman et al. (2010), and Liang et al. (2017), have demonstrated that biochar enhances soil biological characteristics such as microbial habitat, diversity, abundance, biomass, and enzyme activity.

9.3.1 *Habitat for Soil Microorganisms*

Soil microorganisms are integral to soil ecosystems, contributing significantly to processes like organic matter mineralization and nutrient cycling. Approximately 90% of soil processes are mediated by microorganisms (Wu et al. 2019a, b). The addition of biochar to soil can impact microbial ecology and soil quality (Nyambo et al. 2022), leading to increased microbial abundance, as reported by Fu et al. (2019). Additionally, biochar can alter the communication between soil biota by adsorbing crucial signaling molecules.

9.3.2 *Microbial Biomass*

Biochar additions have been consistently linked to an increase in microbial biomass, with significant effects on microbial community composition and enzyme activities that influence element cycles, plant pathogens, and crop growth (Hardy et al. 2019). The porous nature of biochar provides a refuge for microbes, and an increase in microbial biomass is associated with enhanced nitrogen mineralization in the soil (Wang et al. 2019a, b).

9.3.3 Soil Enzyme Activity

Numerous studies have explored the short- and long-term impacts of biochar on soil enzyme activity. Jabborova et al. (2021) observed increased activities of urease, phosphatase, and dehydrogenase in soils treated with biochar compared with soils treated with mineral fertilizers alone. Feng et al. (2021) found that biochar application significantly raised alkaline phosphatase and dehydrogenase activities. The effects of biochar on enzyme activities vary, likely due to interactions between biochar and the target substrate (Li et al. 2020).

9.3.4 Microbial Abundance

Microbial abundance is influenced by the soil food web, which, in turn, depends on the quantity, quality, and distribution of organic matter. Biochar can increase microbial abundance by providing a habitat with surface and porosity that protects against predators (Pietikäinen et al. 2000; Song et al. 2022) and by the utilization of labile portions of pyrogenic C by microorganisms (Lehmann et al. 2006). Different types and application rates of biochar can impact microbial abundance (Wojewódzki et al. 2022; Khan et al. 2018).

9.3.5 Microbial Diversity

The use of biochar positively affects the microbial community in soil, contributing to increased microbial diversity (Wu et al. 2019a, b). Singh et al. (2022) found that bacterial community diversity and richness varied across environments, with neutral soils exhibiting the highest bacterial diversity. Imparto et al. (2016) suggested that low doses of gasification charcoal could increase microbial community structural diversity. In studies with biochar amendments, bacterial diversity increased with the addition rate of biochar (Xiang et al. 2023), and Ippolito (2014) demonstrated extreme changes in microbial community structure at a 10% application rate of hardwood biochar on calcareous soil.

9.4 Role of Biochar in the Growth, Quality, and Yield of Horticulture Crops

The application of biochar has shown promising improvements in crop performance, attributed to both direct and indirect effects. Directly, biochar, produced during pyrolysis, contains more nutrients than its biomass source. Indirectly, biochar application induces changes in the physical, chemical, and biological properties of the soil. Numerous studies, such as those by Jindo et al. (2014) and Zulfiqar et al. (2021), have reported positive impacts on fruit crops, indicating improved growth characteristics.

9.4.1 Germination and Early Growth

The impact of biochar application on seed germination varies based on factors such as quantity, feedstock, and pyrolysis temperature. Generally, positive effects are associated with lower biochar doses (Bai et al. 2022). Guo et al. (2021) observed improved root hydraulic conductance and leaf water potential in water-stressed tomatoes with miscanthus straw-based biochar. Positive effects on leaf relative water content were reported in pumpkin (Langeroodi et al. 2019) and tomato plants under deficit irrigation (Akhtar et al. 2014).

9.4.2 Nutrient Supply and Uptake

In both stressed and non-stressed situations, biochar enhances plant nutrition. Abd El-Mageed et al. (2021) reported increased N, P, K, and Ca^{2+} in drought-stressed faba beans with biochar. Yildirim et al. (2021) observed elevated mineral nutrient content in cabbage leaves and roots under deficit irrigation. Biochar affects soil nitrogen cycling, influencing N, P, K, and Mn uptake in Chinese cherry (Yang and Zhang 2022) and altering nutrient concentrations in tea plant leaves (Yan et al. 2021).

9.4.3 Root Growth Traits

Biochar positively influences root growth traits, as seen in studies with strawberry (Chiomento et al. 2021), sweet basil (Jabborova et al. 2021), and summer savory (Mehdizadeh et al. 2020). Positive effects on root length were observed in salt-stressed cowpea (Farooq et al. 2020) and apple seedlings (Wang et al. 2019a, b).

9.4.4 *Fruit Crops*

Biochar application has proven beneficial for various fruit crops. Jasmitha (2018) found improved germination and growth in mango seedlings with biochar addition. In citrus, biochar increased root system absorptive surface area, resulting in greater canopy volume and improved plant biomass (Zhang et al. 2021a, b). Wood sawdust biochar positively impacted the growth, production, and fruit quality of Grande Naine banana (Sharma et al. 2022). Zhang et al. (2020) reported improved nutritional quality of blueberry fruit with biochar and wood vinegar application.

9.4.5 *Vegetable Crops*

The effects of biochar on vegetable crops are well-documented. Coconut shell biochar and bamboo biochar are reported to significantly promote chili growth (Amaral et al. 2019). Biochar from pyrolyzed wood enhanced growth and yield in tomato plants (Tartaglia et al. 2020), while fenugreek responded positively to biochar and chitosan application in various soil conditions (Qaisi et al. 2022).

9.4.5.1 Phytohormones and Growth Regulation

Biochar can influence phytohormones in horticulture crops. Chinese cherry exhibited increased levels of zeatin, GA, and IAA with biochar application, impacting plant growth regulators (Yang and Zhang 2022). Changes in ABA levels were reported in pumpkin (Langeroodi et al. 2019) and tomato plants (Akhtar et al. 2014) with biochar application under deficit irrigation.

In summary, the application of biochar offers multifaceted benefits to horticulture crops, impacting germination, early growth, nutrient supply, root growth traits, and overall crop performance in diverse environments and conditions.

9.5 Conclusion

Horticultural intensification is required in an hour to meet the demands of the growing world population. The horticulture sector must contend with the difficulties of inadequate sustainability and climate change while expecting higher levels of horticultural output. Unfortunately, soil has been and is currently being rapidly degraded at a global scale due to a range of invasive anthropic activities in intensive agriculture, with concomitant adverse effects on human and ecosystem health. The growing public awareness of organic food has created new opportunities for the use of different organic inputs. One such organic input that shows promise for enhancing soil quality and horticultural crop performance is biochar. To fully understand the influence on horticultural crops, additional research is necessary in both field-oriented and molecular aspects.

References

Abd El-Mageed TA, Belal EE, Rady MOA, Abd El-Mageed SA, Mansour E, Awad MF, Semida WM (2021) Acidified biochar as a soil amendment to drought stressed (vicia faba l.) plants: influences on growth and productivity, nutrient status, and water use efficiency. J Agron 11:1290. https://doi.org/10.3390/agronomy11071290

Akhtar SS, Li G, Andersen MN, Liu F (2014) Biochar enhances yield and quality of tomato under reduced irrigation. Agric Water Manage 138:37–44

Amaral H, Parlindungan Y, Suarta (2019) The effects of compost and biochar on the growth and yield of red chili plants. J Phys Conf Ser 1402:033057. https://doi.org/10.1088/1742-6596/1402/3/033057

Are KS (2019) Biochar and soils physical health. In: Abrol V, Sharma P (eds) An imperative amendment for soil and the environment. IntechOpen, Rijeka, pp 21–33

Bai X, Zhang S, Shao J, Chen A, Jiang CA, Jinlin J, Chen A (2022) Exploring the negative effects of biochars on the germination, growth, and antioxidant system of rice and corn. J Environ Chem Eng 10(3):107398

Chang Y, Rossi L, Zotarelli L, Gao B, Shahid MA, Sarkhosh A (2021) Biochar improves soil physical characteristics and strengthens root architecture in Muscadine grape (*Vitis rotundifolia* L.). Chem Biol Technol Agric 8:1–11

Chiomento JLT, De Nardi FS, Filippi D, dos Santos Trentin T, Dornelles AG, Fornari M (2021) Morpho-horticultural performance of strawberry cultivated on substrate with arbuscular mycorrhizal fungi and biochar. Sci Hortic 282:110053

Das SK, Ghosh GK (2022) Soil hydro-physical properties affected by biomass-derived biochar and organic manure: a low-cost technology for managing acidic mountain sandy soils of north eastern region of India. Biomass Convers Biorefin 14:1–15

Diatta AA, Fike JH, Battaglia ML, Galbraith JM, Baig MB (2020) Effects of biochar on soil fertility and crop productivity in arid regions: a review. Arab J Geosci 13:1–17

Farooq M, Rehman A, Al-Alawi AK, Al-Busaidi WM, Lee DJ (2020) Integrated use of seed priming and biochar improves salt tolerance in cowpea. Sci Hortic 272:109507

Farooq U, Muhammad AQ, Furqan A (2023) Effects of biochar on poultry: a review. ASVS 5(1):69–73

Feng HL, Xu CS, He HH, Zeng Q, Chen N, Li XL, Ren TB, Ji XM, Liu GS (2021) Effect of biochar on soil enzyme activity & the bacterial community and its mechanism. Huan Jing ke Xue Huanj Kexue 42(1):422–432

Fu Q, Zhao H, Li H, Li T, Hou R, Liu D, Ji Y, Gao Y, Yu P (2019) Effects of biochar application during different periods on soil structures and water retention in seasonally frozen soil areas. Sci Total Environ 694:133732

Glaser B, Birk JJ (2012) State of the scientific knowledge on properties and genesis of Anthropogenic Dark Earths in Central Amazonia (terra preta de Índio). Geoch Cosmo Acta 82:39–51

Grossman JM, O'Neill BE, Tsai SM, Liang B, Neves E, Lehmann J, Thies JE (2010) Amazonian anthrosols support similar microbial communities that differ distinctly from those extant in adjacent, unmodified soils of the same mineralogy. Microb Ecol 60:192–205

Guo L, Yu H, Kharbach M, Zhang W, Wang J, Niu W (2021) Biochar improves soil-tomato plant, tomato production, and economic benefits under reduced nitrogen application in Northwestern China. Plants 10(4):759

Hardy B, Sleutel S, Dufey JE, Cornelis JT (2019) The long-term effect of biochar on soil microbial abundance, activity and community structure is overwritten by land management. Front Environ Sci 7:110

Horák J, Šimanský V, Igaz D (2019) Biochar and biochar with N fertilizer impact on soil physical properties in a silty loam Haplic Luvisol. J Ecol Eng 20(7):31–38

Hussain R, Ghosh KK, Ravi K (2021) Impact of biochar produced from hardwood of mesquite on the hydraulic and physical properties of compacted soils for potential application in engineered structures. Geoderma 385:114836

Imparto V, Hansen H, Santos SS, Nielsen T, Giagnoni L, Nielsen HH, Johansen A, Renella G, Anne W (2016) Gasification biochar has limited effects on functional and structural diversity of soil microbial communities in a temperate agroecosystem. Soil Biol Biochem 99:128–136

Ippolito J, Stromberger M, Lentz R, Dungan R (2014) Hardwood biochar influences calcareous soil physicochemical and microbiological status. J Environ Qual 43:681 10.2134/jeq2013.08.0324

Islam MU, Jiang F, Guo Z, Peng X (2021) Does biochar application improve soil aggregation? A meta-analysis. Soil Tillage Res 209:104926

Jabborova D, Ma H, Bellingrath-Kimura SD, Wirth S (2021) Impacts of biochar on basil (ocimum basilicum) growth, root morphological traits, plant biochemical and physiological properties and soil enzymatic activities. Sci Hortic 290:110518

Jasmitha BG, Anil S, Honnabyraiah MK, Swamy GSK, Patil SV (2018) Effect of enriched biochar on growth of mango seedlings in nursery. Int J Chem Stud:415–417

Jien SH (2019) Physical characteristics of biochars and their effects on soil physical properties. In: Biochar from biomass and waste. Elsevier, pp 21–35

Jindo K, Mizumoto H, Sawada Y, Sanchez-Monedero MA, Sonoki T (2014) Physical and chemical characterization of biochars derived from different agricultural residues. Biogeosciences 11:6613–6621

Kalu S, Simojoki A, Karhu K, Tammeorg P (2021) Long-term effects of softwood biochar on soil physical properties, greenhouse gas emissions and crop nutrient uptake in two contrasting boreal soils. Agric Ecosyst Environ 316:107454

Khan MA, Khan S, Ding X, Khan A, Alam M (2018) The effects of biochar and rice husk on adsorption and desorption of cadmium on to soils with different water conditions (upland and saturated). Chemosphere 193:1120–1126

Kumar A, Bhattacharya T, Mukherjee S, Sarkar B (2022) A perspective on biochar for repairing damages in the soil–plant system caused by climate change-driven extreme weather events. Biochar 4(1):22

Langeroodi ARS, Campiglia E, Mancinelli R, Radicetti E (2019) Can biochar improve pumpkin productivity and its physiological characteristics under reduced irrigation regimes? Sci Hortic 247:195–204

Lehmann J, Gaunt J, Rondon M (2006) Bio-char sequestration in terrestrial ecosystems–a review. Mitig Adapt Strat Glob Change 11(2):403–427

Lehmann J, Joseph S (2009) Biochar for environmental management: science and technology. Earthscan, London. Open J Soil Sci 23:121–143

Li X, Yao S, Bian Y, Jiang X, Song Y (2020) The combination of biochar and plant roots improves soil bacterial adaptation to PAH stress: insights from soil enzymes, microbiome, and metabolome. J Hazard Mater 400:123227

Liang J, Yang Z, Tang L, Zeng G, Yu M, Li X, Wu H, Qian Y, Li X, Luo Y (2017) Changes in heavy metal mobility and availability from contaminated wetland soil remediated with combined biochar-compost. Chemosphere 181:281–288

Mehdizadeh L, Moghaddam M, Lakzian A (2020) Amelioration of soil properties, growth and leaf mineral elements of summer savory under salt stress and biochar application in alkaline soil. Sci Hortic 267:109319

Nyambo P, Chiduza C, Araya T (2022) Effect of conservation agriculture on selected soil physical properties on a haplic cambisol in Alice, Eastern Cape, South Africa. Arch Agron Soil Sci 68(2):195–208

Pietikäinen J, Kiikkilä O, Fritze H (2000) Charcoal as a habitat for microbes and its effect on the microbial community of the underlying humus. Oikos 89(2):231–242

Qaisi A, Tawaha A, Rahman A, Imran, Al-Rifaee M (2022) Effects of chitosan and biochar-mended soil on growth, yield and yield components and mineral composition of fenugreek. Gesunde Pflanzen 75. https://doi.org/10.1007/s10343-022-00727-x

Razzaghi F, Obour PB, Arthur E (2020) Does biochar improve soil water retention? A systematic review and meta-analysis. Geoderma 361:114055

Sharma P, Abrol V, Sharma V, Chaddha S, Rao CS, Ganie AQ, Hefft DI, El-Sheikh MA, Mansoor S (2021) Effectiveness of biochar and compost on improving soil hydro-physical properties, crop yield and monetary returns in inceptisol subtropics. Saudi J Biol Sci 28(12):7539–7549

Sharma S, Rana VS, Rana N, Prasad H, Sharma U, Patiyal V (2022) Biochar from fruit crops waste and its potential impact on fruit crops. Sci Hortic 299:111052. https://doi.org/10.1016/j.scienta.2022.111052

Shetty R, Vidya CSN, Prakash NB, Lux A, Vaculik M (2021) Aluminum toxicity in plants and its possible mitigation in acid soils by biochar: a review. Sci Total Environ 765:142744

Singh H, Northup BK, Rice CW, Prasad PV (2022) Biochar applications influence soil physical and chemical properties, microbial diversity, and crop productivity: a meta-analysis. Biochar 4(1):8

Sinha P, Marchuk S, Harris P, Antille DL, McCabe BK (2023) Land application of biosolids-derived biochar in Australia: a review. Sustainability 15(14):10909

Song P, Ma W, Gao X, Ai S, Wang J, Liu W (2022) Remediation mechanism of cu, Zn, As, Cd, and Pb contaminated soil by biochar-supported nanoscale zero-valent iron and its impact on soil enzyme activity. J Clean Prod 378:134510

Tartaglia M, Arena S, Scaloni A, Mauro M, Mariapina R (2020) Biochar administration to san marzano tomato plants cultivated under low-input farming increases growth, fruit yield, and affects gene expression front. Plant Sci 11

Verheijen F, Jeffery S, Bastos AC, van der Velde M, Diafas I (2010) Biochar application to soils a critical scientific review of effects on soil properties. Processes and Functions JRC Scientific and Technical Reports

Wang G, Govinden R, Hafizah YC, Yan M, Dejie G, Gaidi R (2019a) Suppression of phytophthora blight of pepper by biochar amendment is associated with improved soil bacterial properties. Biol Fertil Soils 55:813–824

Wang Y, Zhong B, Shafi M, Ma J, Guo J, Wu J, Ye Z, Liu D, Jin H (2019b) Effects of biochar on growth, and heavy metals accumulation of moso bamboo (phyllostachy pubescens), soil physical properties, and heavy metals solubility in soil. Chemosphere 219:510–516

Wojewódzki P, Lemanowicz J, Debska B, Haddad SA (2022) Soil enzyme activity response under the amendment of different types of biochar. Agronomy 12(3):569

Wu Y, Cai P, Jing X, Niu X, Ji D, Ashry NM, Gao C, Huang Q (2019a) Soil biofilm formation enhances microbial community diversity and metabolic activity. Environ Int 132:105116

Wu B, Yang Q, Yao F, Chen S, He L, Hou K, Pi Z, Yin H, Fu J, Wang D, Li X (2019b) Evaluating the effect of biochar on mesophilic anaerobic digestion of waste activated sludge and microbial diversity. Bioresour Technol 294:122235

Xiang Y, Liu Y, Niazi NK, Bolan N, Zhao L, Zhang S, Xue J, Yao B, Li Y (2023 Oct) Biochar addition increased soil bacterial diversity and richness: large-scale evidence of field experiments. Sci Total Environ 1(893):164961. https://doi.org/10.1016/j.scitotenv.2023.164961. Epub 2023 Jun 17. PMID: 37336405

Xiao L, Yuan G, Feng L, Bi D, Wei J, Shen G, Liu Z (2019) Coupled effects of biochar use and farming practice on physical properties of a salt-affected soil with wheat–maize rotation. J Soils Sediments 20:3053–3061

Yan P, Shen C, Zou Z, Fu J, Li X, Zhang L (2021) Biochar stimulates tea growth by improving nutrients in acidic soil. Sci Hortic 283:110078

Yang CD, Lu SG (2021) Effects of five different biochars on aggregation, water retention and mechanical properties of paddy soil: a field experiment of three-season crops. Soil Tillage Res 205:104798

Yang W, Zhang L (2022) Biochar and cow manure organic fertilizer amendments improve the quality of composted green waste as a growth medium for the ornamental plant *Centaurea cyanus* l. Environ Sci Pollut Res 29:45474–45486. https://doi.org/10.1007/s11356-022-19144-8

Yang H, Ye S, Zeng Z, Zeng G, Tan X, Xiao R, Wang J, Song B, Du L, Qin M, Yang Y (2020) Utilization of biochar for resource recovery from water: a review. Chem Eng J125:502

Yildirim E, Ekinci M, Turan M (2021) Impact of biochar in mitigating the negative effect of drought stress on cabbage seedlings. J Soil Sci Plant Nutr 21:2297–2309. https://doi.org/10.1007/s42729-021-00522-z

Yuan J-H, Ren-Kou X, Zhang H (2011) The forms of alkalis in the biochar produced from crop residues at different temperatures. Bioresour Technol 102(3):3488–3497

Zhang C, Li X, Yan H, Ullah I, Zuo Z, Li L, Yu J (2020) Effects of irrigation quantity and biochar on soil physical properties, growth characteristics, yield and quality of greenhouse tomato. Agric Water Manag 241:106263

Zhang Y, Wang J, Feng Y (2021a) The effects of biochar addition on soil physicochemical properties: a review. Catena 202:105284

Zhang M, Zhang L, Riaz M, Xia H, Jiang C (2021b) Biochar amendment improved fruit quality and soil properties and microbial communities at different depths in citrus production. J Clean Prod 292:126062

Zulfiqar F, Wei X, Shaukat N, Chen J, Raza A, Younis A (2021) Effects of biochar and biochar–compost mix on growth, performance and physiological responses of potted Alpinia zerumbet. Sustainability 13(20):11226

Zulfiqar F, Moosa A, Nazir MM, Ferrante A, Ashraf M, Nafees M, Chen J, Darras A, Siddique KHM (2022) Biochar: an emerging recipe for designing sustainable horticulture under climate change scenarios. Front Plant Sci 13:1018646. https://doi.org/10.3389/fpls.2022.1018645

Part III
Advances in Analytical Technology for Horticulture Crops

Chapter 10
Nutrients and Secondary Metabolites Analysis of Horticulture Crops

Himani Agrawal, Robin Joshi, and Mahesh Gupta

Abstract Horticulture crops possess numerous functional properties with the potential to yield nutritional constituents or nutraceuticals, making them exemplary examples of edible plant harvests. Nutrients and secondary metabolites derived from various parts of horticulture crops exhibit several health-promoting properties. Nutrients, also referred to as primary metabolites, encompass carbohydrates, fatty acids, amino acids, and vitamins. These play crucial roles in various biological processes such as protein synthesis, hormone synthesis, respiration, photosynthesis, and growth and development. In contrast, secondary metabolites, including phenolic/polyphenolic compounds, alkaloids, terpenoids, and sulfur-containing compounds, do not directly contribute to the development and growth of plants. Instead, they serve as defenders against microorganisms, function as signal molecules under stress conditions, attract pollinators, and determine the color of different horticulture crops. Consequently, the beneficial effects of the diverse nutrients and secondary metabolites present in horticulture crops for their health-promoting properties continue to garner increasing attention. Furthermore, there is a growing interest in advancing techniques for the extraction, separation, and quantification of these valuable bioactive components from horticulture crops. This is to facilitate their use as value-added ingredients in the nutraceutical and food industries. This chapter provides comprehensive information on advanced analytical techniques for the analysis of nutrients and secondary metabolites in different horticulture crops.

Keywords Primary metabolites · Secondary metabolites · Analytical techniques · Horticulture crops · Fruits and vegetables

H. Agrawal (✉)
Institute of Bioorganic Chemistry, Polish Academy of Sciences, Poznan, Poland

R. Joshi · M. Gupta
CSIR-Institute of Himalayan Bioresource Technology, Palampur, Himachal Pradesh, India

M. Thakur, T. Belwal (eds.), *Advances in Postharvest and Analytical Technology of Horticulture Crops*, https://doi.org/10.1007/978-981-97-7247-6_10

10.1 Introduction

The field of nutrition research has seen significant advancements, reshaping our understanding of horticultural crops and their various products. The chemical composition of horticultural crops, encompassing both edible and non-edible biomass derived from plants, has gained attention for its metabolome capabilities, positioning it as a contender against traditional medicinal plants (Chen et al. 2022). Essential elements assimilated by the human body include water, carbohydrates, proteins, fats, fiber, vitamins, minerals, organic acids, pigments, and antioxidants, among others. Notably, fruits and vegetables are recognized as primary sources of phytochemicals, as well as other valuable components such as fatty acids and amino acids. Secondary metabolites, natural non-nutrient constituents in plants, are believed to harbor various health-promoting and/or toxic effects (Butnariu and Butu 2015; Vicente et al. 2014). Additionally, these compounds play a pivotal role in the growth, development, and defense mechanisms of plants (Reshi et al. 2023). Unlike primary metabolites, secondary metabolites are non-essential for both plants and humans. However, they may contribute to the protective effects of different fruits and vegetables and have been implicated in the prevention of various human diseases (Yeshi et al. 2022).

The prevalence of secondary metabolites is widespread in the plant kingdom, with fruits and vegetables standing out as rich sources, including polyphenols (e.g., flavonoids and phenolic acids), carotenoids, and polyacetylenes (Arfaoui 2021; Swallah et al. 2020). Fruits and vegetables with nutraceutical potential, abundant in secondary metabolites, encompass a diverse array such as endives, lettuces, spinach, asparagus, mustard, kale, turnips, Brussels sprouts, cauliflower, broccoli, carrot, bael, capsicum, chili, garlic, ginger, pomegranate, cranberry, Indian gooseberry or amla, vanilla, and citrus fruits like lemon and orange (Oz and Kafkas 2017; Liu 2013). This list continues to expand globally, driven by cutting-edge research in sectors like nutraceuticals and health foods, including medicinal foods.

The nutritional composition of horticulture crops is intricately complex and poses a considerable challenge in terms of measurement. Various factors significantly influence the levels of metabolites, including genetic, environmental, transportation, and storage conditions. Additionally, plant metabolites undergo alterations due to an array of growth factors such as light, humidity, temperature, soil type, fertilizers, heavy metals, UV radiation, and pesticides (Bolaji Umar et al. 2022; Chen et al. 2022; Xu 2022). These factors play a pivotal role in inducing diverse physical and chemical changes in fruits and vegetables, potentially leading to the loss of bioactive compounds inherent in fruits and vegetables (MacEvilly and Peltola 2003). However, recent advancements in analytical techniques for metabolite analysis have broadened the scope of research in horticulture crops.

Chromatographic techniques have made substantial contributions to the identification, separation, and characterization of bioactive constituents in horticulture crops. These techniques, coupled with various detection methods and biological assays, offer an efficient means of screening different horticulture crops for their nutrient and metabolite content. Furthermore, chromatographic techniques employing mass spectrometer detection systems have proven effective in rapidly and qualitatively determining novel components present in diverse horticulture crops' extracts (Hounsome et al. 2008).

In the past decade, numerous studies have been conducted to provide insights into the identification, biochemical characterization, and potential health benefits of various metabolites found in horticulture crops. This chapter represents an endeavor to review the knowledge pertaining to nutrients and secondary metabolites present in different horticulture crops, as well as the progress made in analytical techniques within this domain.

10.2 Primary Metabolites/Nutrients in Horticulture Crops

Horticulture crops yield a multitude of organic compounds as products of both primary and secondary metabolism. Primary metabolites encompass organic acids, fatty acids, amino acids, and carbohydrates. These compounds play pivotal roles in various biological processes, such as protein and hormone synthesis, growth and development, photosynthesis, and respiration. Interestingly, these primary metabolites are generated through shared biochemical pathways and are ubiquitous across diverse phylogenetic groups. In contrast, secondary metabolites, including alkaloids, phenolic acids, sterols, carotenoids, flavonoids, and glucosinolates, serve a range of functions. They act as defensive mechanisms against microorganisms, function as signaling molecules in response to stress conditions, attract seed-dispersing animals and pollinators, and contribute to the distinctive coloration of different horticulture crops (Crozier et al. 2006; Pott et al. 2019). Figure 10.1 illustrates the structural composition of some characteristic primary metabolites found in various fruits and vegetables crops.

10.2.1 Carbohydrate (CHO)

In fruits and vegetables, CHO manifests as various sugar monosaccharides, including rhamnose, galactose, arabinose, fructose, and glucose, as well as disaccharides such as trehalose, maltose, and sucrose. Additionally, they exist in the form of polysaccharides such as pectins, hemicellulose, cellulose, and starch, along with oligosaccharides such as fructo-oligosaccharides, stachyose, and raffinose and sugar alcohols such as xylitol, mannitol, and sorbitol (Brouns 2020). From a nutritional and physiological standpoint, CHO is classified as available (e.g., digestive starch and sucrose) and non-available (comprising non-starch polysaccharides, numerous oligosaccharides, and sugar alcohols). Enzymatically hydrolyzed into monosaccharides in the human gastrointestinal system, available CHO is readily absorbed into the bloodstream without requiring further digestion. On the contrary, unavailable

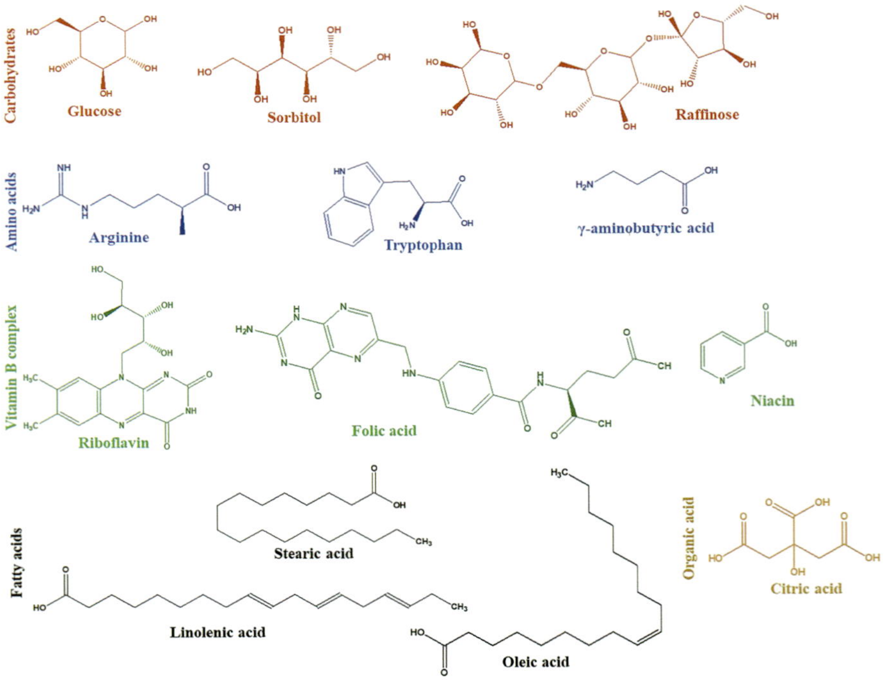

Fig. 10.1 Some representative primary metabolites present in different fruits and vegetables

CHO resists hydrolysis by endogenous enzymes in the gastrointestinal system, undergoing fermentation to varying degrees in the large intestine by microorganisms before absorption (Asp 1996; Gerschenson et al. 2022).

Dietary fiber, a category encompassing plant constituents such as lignin, resistant oligosaccharides, non-starch polysaccharides, and related components such as suberin, cutin, waxes, and resistant starch (Asp 1987; DeVries 2003), traverses the gastrointestinal tract, undergoing digestion and modification by microorganisms in the colon (Wang et al. 2020). Research has demonstrated that a high dietary fiber intake in daily nutrition reduces the risk of colitis, diverticular disease, and constipation (Stollman and Raskin 2004). Conversely, a low dietary fiber intake elevates the risk of developing several diseases, including cardiovascular disease, colon cancer, and obesity (Du et al. 2010).

10.2.2 Amino Acids

In animal and plant metabolism, amino acids serve as crucial intermediates, combining to form proteins. These proteins, in turn, play diverse roles as antibodies, hormones, and enzymes, providing structural materials for the human body.

Enzymatic breakdown of proteins into amino acids occurs predominantly in the small intestine, facilitating absorption. Dietary proteins stand out as the primary source of amino acids, with the human body capable of synthesizing ten amino acids (alanine, asparagine, aspartic acid, cysteine, glutamine, glutamic acid, glycine, proline, serine, and tyrosine). The remaining amino acids, categorized as essential amino acids, including histidine, isoleucine, leucine, lysine, methionine, phenylalanine, threonine, tryptophan, and valine, must be obtained from dietary sources (Dietzen 2018).

Meeting the body's requirements, fruits and vegetables encompass all essential amino acids, albeit in varying proportions (Mariotti and Gardner 2019). For instance, potatoes, watercress, Brussels sprouts, and asparagus boast high arginine content, while cauliflower, Brussels sprouts, and broccoli are rich in histidine. Phenylalanine is prevalent in tomatoes, spinach, parsley, carrots, and beetroot, whereas watercress, kale, radish, cauliflower, and cabbage are sources of methionine (Butnariu and Butu 2015).

Beyond their role in human metabolism, free amino acids significantly contribute to taste. Leucine and valine impart a bitter taste, alanine and glycine provide sweetness, and glutamate and aspartic acid contribute to umami (savory in Japanese) and sour tastes (Li-Chan and Cheung 2010).

10.2.3 Vitamin B Complex

Vitamin B, found in various fruits and vegetables, comprises a group of water-soluble vitamins, including thiamine (B1), riboflavin (B2), niacin (B3), pantothenic acid (B5), pyridoxine (B6), biotin (B7), folic acid (B9), and cobalamin (B12) (Zempleni et al. 2013). These compounds function as essential cofactors for enzymes in plants, playing crucial roles in processes such as cell division, flowering regulation, amino acid, and organic acid synthesis, as well as respiration and photosynthesis (Bowsher et al. 2008; Liu et al. 2022). In humans, B vitamins actively participate in amino acid, fatty acid, and carbohydrate metabolism, along with tissue respiration. Deficiency in B vitamins can lead to various disorders, including anemia, dermatitis, nausea, microcytic and peripheral neuropathies, glossitis, seborrhea, dementia, diarrhea, pellagra, angular stomatitis, and cheilosis (Schellack et al. 2019). Green leafy vegetables, such as lettuce, turnip, cauliflower, asparagus, spinach, and Brussels sprouts, serve as excellent sources of the B-complex vitamins (Butnariu and Butu 2015).

10.2.4 Fatty Acids

Fatty acids, acknowledged as the principal components of fats, consist of a hydrophilic carboxylate group attached to a long hydrocarbon chain. They play a crucial role in various physiological processes, including blood clotting, immune response, certain vitamin absorption, organ padding, and serving as structural components for cell membranes (Yashodhara et al. 2009). These fatty acids are categorized based on the number of double bonds they possess. Saturated fatty acids, characterized by the absence of double bonds, include palmitic, myristic, capric, and stearic acids. Monounsaturated fatty acids (MUFAs), featuring one double bond, encompass palmitoleic and oleic acids. Polyunsaturated fatty acids (PUFAs), containing two or more double bonds, consist of α-linolenic acid, eicosapentaenoic acid, arachidonic acid, and docosahexaenoic acid. Notably, α-linolenic and linoleic acids are deemed essential fatty acids as the body cannot synthesize them, necessitating their supply from dietary sources (Gunstone and Norris 2013). Linolenic acid is prevalent in various crops such as soybeans, pumpkin seeds, and wheat germ, while α-linolenic acid can be found in different vegetable oils such as safflower and sunflower (Jakab et al. 2002). Certain green leafy vegetables such as parsley, watercress, Chinese cabbage, and Brussels sprouts contain a notably higher proportion of PUFAs (Pereira et al. 2001). Research indicates that the intake of MUFAs contributes to decreased blood cholesterol levels and exhibits beneficial effects on type 2 diabetes and cardiovascular disease (Kuna and Achinna 2013).

10.2.5 Organic Acids

Organic acids constitute a category of organic compounds characterized by the presence of carboxylic groups. Organic acids release protons in solution, determining their distinct "acidic taste." Notable organic acids found in plants include ascorbic acid, malonic acid, citric acid, tartaric acid, quinic acid, fumaric acid, succinic acid, oxalic acid, malic acid, acetic acid, and other similar compounds (Bowsher et al. 2008). Predominantly, malic and citric acids are prevalent, while quinic, fumaric, and succinic acids are widely distributed.

Various fruits and vegetables, such as lamb's lettuce, lettuce, tomatoes, red and green peppers, lemons, oranges, peaches, grapes, and melons, have been studied, revealing the presence of citric acid, pyruvic acid, α-ketoglutaric acid, succinic acid, fumaric acids, malonic acid, malic acid, shikimic acid, quinic acid, glutamic acid, and tartaric acid (Flores et al. 2012). Additionally, plants contain pigments serving as natural pH indicators, impacting the color of fruits and vegetables. For instance, lettuce and cabbages contain anthocyanins responsible for color changes from red to blue as the pH increases (Davies 2009).

Vitamin C, also known as ascorbic acid, is a vital organic compound with robust antioxidant properties. It plays a crucial role in collagen synthesis, the production of steroid hormones and neurotransmitters, the conversion of cholesterol to bile acids, and the absorption of calcium and iron. Furthermore, vitamin C contributes to the prevention of degenerative diseases, including cardiovascular disorders, certain cancers, and cataracts (Santhi et al. 2020). Various vegetables such as beans, peas, sweet peppers, cauliflower, broccoli, cabbage, cress, spring onions, and spinach are rich sources of ascorbic acid (Hounsome et al. 2008).

10.3 Secondary Metabolites in Horticulture Crops

Secondary metabolites, as defined by Kumar et al. (2022), refer to organic components that do not directly participate in the growth and development of plants. The British Nutrition Foundation has established a classification system that categorizes plant secondary metabolites into four groups: phenolic and polyphenolic compounds, sulfur-containing compounds, alkaloids, and terpenoids (Divekar et al. 2022; Goldberg 2003). Figure 10.2 illustrates the structures of the major secondary metabolites found in various fruits and vegetables.

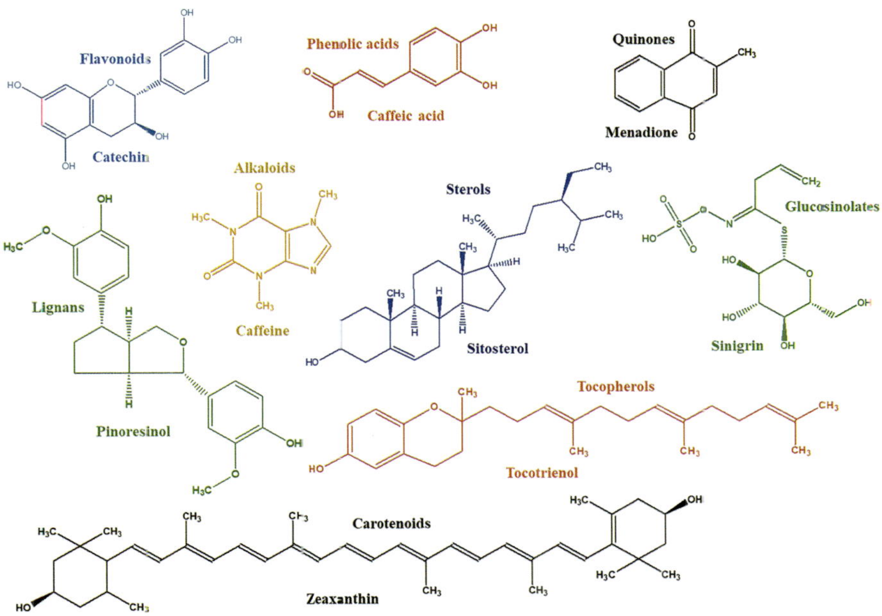

Fig. 10.2 Some representative secondary metabolites present in different fruits and vegetables

10.4 Phenolic and Polyphenolic Compounds

Phenolic and polyphenolic constituents play a crucial role in determining the nutritional and sensory quality of horticulture crops. These compounds are characterized by an aromatic ring bearing one or more hydroxyl groups. The structure of phenolic compounds can range from simple molecules to complex high-molecular masses. Based on the number of phenol rings, polyphenolic compounds are categorized as phenolic acids, flavonoids, tannins, lignans, and stilbenes (Shahidi and Ambigaipalan 2015).

Phenolic acids are present in plants in both free and bound forms, constituting one-third of dietary phenols. They consist of two subcategories: hydroxycinnamic acids and hydroxybenzoic acids. Hydroxycinnamic acids, such as sinapic, p-coumaric, ferulic, and caffeic, and hydroxybenzoic acids, including syringic, vanillic, protocatechuic, p-hydroxybenzoic, and gallic acids, are common representatives (Kumar and Goel 2019).

Flavonoids, a group of low molecular weight compounds, encompass flavonols (kaempferol, quercetin, and isorhamnetin), found in broccoli, endives, onions, and leeks; flavones (chrysoeriol, apigenin, and luteolin), present in celery, parsley, and thyme; flavanones; chalcones; isoflavones; flavanonols; and anthocyanidins (malvidin, cyanidin, and delphinidin), found in red lettuces, red cabbage, and radish (Yao et al. 2004). Citrus fruits are the primary source of flavanones, though they are also found in mint and tomatoes. Flavones, chalcones, and anthocyanidins, which act as water-soluble vacuolar plant pigments, determine the colors of various fruits and vegetables depending on pH (Panche et al. 2016). Flavonoids are frequently encountered phytochemicals, crucial due to their high redox potential, enabling them to serve as reducing agents, hydrogen donors, and singlet oxygen quenchers.

Tannins constitute the third essential group of phenolics, comprising relatively high molecular compounds categorized as condensed and hydrolyzable tannins. Condensed tannins, such as proanthocyanidins, are polymeric flavonoids. (−)-Epicatechin and (+)-catechin are extensively studied condensed tannins. Hydrolysable tannins are derivatives of gallic acid (3,4,5 trihydroxyl benzoic acid) (Smeriglio et al. 2017). Tannins are recognized as potential biological antioxidants, protein-precipitating agents, and metal ion chelators; however, their structure–activity relationship in biological systems is not clearly elucidated (Molino et al. 2023).

Lignans are generally found in nature in free form and are produced through the oxidative dimerization of two phenylpropane units. Due to their potential applications in various pharmacological effects and cancer therapy, interest in lignans and their synthetic derivatives is continually increasing (Saleem et al. 2005). Some plant lignans, such as pinoresinol, lariciresinol, matairesinol, and secoisolariciresinol, are absorbed into the human body after converting into "enterolignans" by intestinal microflora. Lignans exhibit numerous biological activities in the human body, including (anti)estrogenic and antioxidant effects, thereby potentially reducing the risk of certain cancers and cardiovascular diseases (Peterson et al. 2010). Foods rich in lignans include garlic, beans, Brussels sprouts, kale, and broccoli (Milder et al. 2005).

10.5 Alkaloids

Alkaloids constitute a category of naturally occurring compounds characterized by their basic nitrogen content. In addition to nitrogen, alkaloids may include elements such as hydrogen, carbon, sulfur, and oxygen, as well as trace amounts of phosphorus, bromine, and chlorine. The classification of alkaloids is based on shared molecular precursors, such as steroids (e.g., solanine), purine (e.g., caffeine), isoquinone (e.g., codeine, morphine), tropane (e.g., cocaine, atropine), and pyridine (e.g., nicotine, coniine) (Dey et al. 2020; Facchini 2001). Alkaloids display diverse pharmacological activities, including antihyperglycemic (e.g., piperine), antibacterial (e.g., chelerythrine), analgesic (e.g., morphine), antiarrhythmic (e.g., quinidine), vasodilatory (e.g., vinacamine), cholinomimetic (e.g., galantamine), anticancer (e.g., homoharringtonine), antiasthma (e.g., ephedrine), and antimalarial (e.g., quinine) effects (Bribi 2018).

Certain vegetable alkaloids, such as solanine, can form in unripe potatoes and tomatoes when exposed to light, potentially posing toxicity risks for humans. Solanine, a cholinesterase inhibitor, has been associated with gastrointestinal and neurological disorders (Dalvi and Bowie 1983). However, the toxicity of these vegetables has not been conclusively proven unless they constitute a significant proportion of the diet. Plant glucoalkaloids, such as solanine, chaconine, and tomatine, exhibit surfactant properties and are categorized as saponins. Generally found in potatoes, garlic, onions, asparagus, spinach, tomatoes, beans, and peas (Hounsome et al. 2008), dietary saponins have been identified to reduce blood cholesterol levels and inhibit the growth of cancer cells (Das et al. 2012; Francis et al. 2002). Some fruits and vegetables containing saponins possess an acrid and bitter taste due to the presence of these compounds (Drewnowski and Gomez-Carneros 2000). Notably, certain saponins, such as sapotoxin, can be toxic to humans, causing irritation of the digestive and respiratory tract membranes (Francis et al. 2002).

10.6 Terpenoids

Terpenoids, also referred to as isoprenoids, constitute a diverse and expansive family of naturally occurring chemical compounds. They are derived from 5-carbon isoprene units and isoprene polymers. Within plants, terpenoids assume various functional roles, acting as hormones, electron carriers, photosynthetic pigments, and integral components of membranes. The major dietary terpenoids include quinones, tocopherols, carotenoids, and sterols (Perveen 2018).

Quinones, a category of naturally occurring organic compounds, feature aromatic rings with two ketone substitutions. Phylloquinone, also known as vitamin K1, plays a crucial role in regulating blood clotting, bone formation, and repair in the human body. It is present in vegetables such as cabbage, asparagus, lettuce, spinach, kale, cauliflower, and broccoli (Hounsome et al. 2008; Tsugawa and Shiraki 2020). Quinones exhibit high reactivity and induce browning reactions in injured or cut fruits and vegetables (Moon et al. 2020; Singh et al. 2018).

Tocopherols, commonly known as vitamin E, are abundant in turnips, tomatoes, sweet potatoes, spinach, asparagus, lettuce, cauliflower, cabbage, Brussels sprouts, broccoli, and kale (Butnariu and Butu 2015; Eitenmiller and Lee 2004; Piironen et al. 1986). In humans, tocopherols serve as vital lipid-soluble chain-breaking antioxidants, protecting DNA, low-density lipoproteins, and PUFAs from free radical-induced oxidation. Additionally, tocopherols contribute to membrane structure stabilization, modulation of immune responses, and hemoglobin biosynthesis (Shahidi and De Camargo 2016).

Carotenoids, found in green leafy vegetables, carrots, tomatoes, pumpkins, and sweet potatoes, exist in the forms of α-carotenes, β-carotenes, xanthophylls, and lycopene, imparting vibrant yellow, red, and orange hues (Khoo et al. 2011; Sudhakar et al. 2016). Inadequate intake of carotenoids may lead to night blindness (xerophthalmia) and premature death, while a diet rich in carotenoids significantly decreases the risk of certain cancers, cataracts, and coronary heart diseases. Carotenoids also serve as biological antioxidants, playing a pivotal role in preventing oxidative damage to tissues and cells (Fiedor and Burda 2014).

Plant sterols, including stigmasterol, brassicasterol, β-sitosterol, and campesterol, are found in higher concentrations in spinach, cauliflower, broccoli, soy, and Brussels sprouts (Piironen et al. 2003). These plant sterols exhibit structural similarities to cholesterol and are beneficial for inhibiting cholesterol absorption in the human body. Additionally, they may possess various bioactivities such as antioxidant, anti-inflammatory, anti-atherosclerotic, and anticancer properties (Trautwein and Demonty 2007).

10.7 Determination of Different Metabolites in Horticulture Crops

10.7.1 Sample Preparation and Extraction

For the analysis of metabolites, solid plant materials undergo an initial grinding process under the influence of liquid nitrogen. This step is crucial, as proper grinding enhances the release of metabolites during the extraction process. In the case of liquid samples, freeze-drying serves as a concentration step, mitigating metabolite variances caused by variations in moisture content. Concentration steps are particularly advantageous in maximizing the yield of collected metabolites. For instance, lyophilization is employed for concentrating metabolites. The primary objective of the extraction process is to optimize the concentration and quantity of the target compounds. Consequently, extraction stands out as the most pivotal stage in metabolite analysis. Various techniques, including grinding, drying, lyophilization, or the mere soaking of fresh material, followed by solvent extraction, have been utilized for metabolite extraction from fruits and vegetables (Ignat et al. 2011).

10.7.2 Analytical Techniques for Metabolite Analysis

Two approaches, namely, destructive and non-destructive methods, can be employed to assess the nutritional quality of horticulture crops. Among these, the destructive approach, which has been endorsed by the European Commission, stands out as the most standardized for chemical analysis (Buchgraber and Karaali 2005), encompassing spectrophotometry and chromatography. Although destructive methods boast higher accuracy and detection limits, they are deemed time-consuming, expensive, and necessitate particularly skilled workers. Conversely, non-destructive techniques prove to be more economical and quicker. These techniques rely on nuclear magnetic resonance (NMR), electrochemical processes, and spectroscopy. However, when compared with destructive techniques, non-destructive methods are generally considered less precise and informative, identifying a limited number of chemical constituents (Hounsome et al. 2008).

Table 10.1 provides an overview of the analytical techniques applied for nutrient and secondary metabolite analysis in horticulture crops.

10.7.3 Non-Destructive Techniques

In non-destructive techniques, sensors predominantly rely on electromagnetic waves for detecting molecular interactions. These sensors operate across varying wavelengths, encompassing the ultraviolet range (10–400 nm), the visible range (400–700 nm), the infrared range (700–30,000 nm), X-rays (100 pm to 1 nm), microwaves (1–10 cm), and high-frequency electromagnetic waves (magnetic resonance, MHz to GHz). Additionally, sensors can be categorized based on their signal principles, including reflectance, transmittance, and absorbance. Further classification of sensors may be based on their application principles. For instance, nutrient components such as fiber, oil, and sugar content in diverse horticultural crops are determined using techniques within the visible region and near-infrared. Noteworthy methods include time-resolved diffuse reflectance spectroscopy and near-infrared transmittance (Beghi et al. 2018; El-Mesery et al. 2019; Kämper et al. 2020; Katayama et al. 1996; Xu et al. 2012).

Table 10.1 Analytical techniques applied for nutrients and secondary metabolite analysis in horticulture crops

Techniques	Metabolites	Horticulture crops	References
Non-destructive			
NIR spectroscopy	Sugar, carbohydrates, carotenoids, vitamins, organic acids, anthocyanin	Sweet potato, potato, tomato, watermelon, mango, cassava, passion fruit, jaboticaba fruit and pear	Abebe (2006), de Oliveira et al. (2014), Jie et al. (2004), Mariani et al. (2015), Pedro and Ferreira (2005), Rungpichayapichet et al. (2013), Tamburini et al. (2017), Xu et al. (2012), Zum Felde et al. (2013)
Raman spectroscopy	Fatty acids, carotenoids	Olives, tomato, carrot, orange, spinach, black grapes, potatoes, lettuce, yellow plums, yellow nectarines, apricots, yellow peaches, bananas, sweet potatoes	Bhosale et al. (2004), Darvin et al. (2007), El-Abassy et al. (2009), Muik et al. (2003), Qiu et al. (2019), Sebben et al. (2018)
NMR spectroscopy	Fatty acids, organic acids, carbohydrate sugars, amino acids, and vitamin	Pineapple, strawberry, orange, kiwi, pear, apricot, apple, peach, blueberry, citrus fruits, and tomato	Del Campo et al. (2006), Eisenmann et al. (2016), Mattoo et al. (2006), Sobolev et al. (2015), Villa-Ruano et al. (2019)
E-nose	Volatile compounds	Apple, mango, pineapple, hazelnut, peach, onion	Alasalvar et al. (2012), Lebrun et al. (2008), Russo et al. (2013), Saevels et al. (2004), Torri et al. (2010), Zhang et al. (2012)
Destructive techniques			
Liquid chromatography	Phenolics, tocopherols, organic acids, polysaccharides, carotenoids, flavonoids, anthocyanin	Apricot, pear, peach, papaya, apple, plum, cherry, onion, ginger, garlic, dark grapes, raspberry, black carrot, strawberry, blueberry, mango	Alasalvar et al. (2001), Bureau et al. (2009), Gu et al. (2019), Jakopic et al. (2012)

(continued)

Table 10.1 (continued)

Techniques	Metabolites	Horticulture crops	References
GC chromatography	Phenolics, flavonoids, vitamins, and organic acids	Papaya, cranberry, avocado	Canini et al. (2007), Hurtado-Fernández et al. (2014), Wang and Zuo (2011)
Capillary electrophoresis or capillary electrochromatography	Phenolics, carotenoids, organic acids	Berries, corn, spinach, tomato, carrot, apple, grape, mandarin, orange, pineapple	Ehala et al. (2005), Herrero-Martínez et al. (2006), Navarro-Pascual-Ahuir et al. (2017), Peng et al. (2005)

10.7.3.1 Near-Infrared Spectroscopy

Near-infrared (NIR) spectroscopy operates by analyzing the absorbance/transmittance spectrum of substances irradiated with wavelengths ranging from 700 to 25,000 nm. The radiation intensity, along with calibration equations, is then employed to calculate the concentrations of various components. The application of NIR spectroscopy is primarily limited to certain horticulture crops, such as small fruits and vegetables with thin peels. This technique finds use in the sorting and grading of small fruits and vegetables based on their sweetness in the food industry (Butz et al. 2005). Additionally, it has been employed to analyze carbohydrate content in potatoes (Jie et al. 2004), β-carotene and lycopene in tomato products (Pedro and Ferreira 2005), total sugar content in watermelons (Abebe 2006), β-carotene content in mango (Rungpichayapichet et al. 2013), pro-vitamin A component in cassava (Zum Felde et al. 2013), sucrose, glucose, fructose, malic acid, citric acids, and β-carotene contents in passion fruit (de Oliveira et al. 2014), total anthocyanin content in jaboticaba fruit (Mariani et al. 2015), and lycopene and β-carotene content in intact red-flesh watermelon (Tamburini et al. 2017).

10.7.3.2 Raman Spectroscopy

This spectroscopic technique is utilized to ascertain the intensity and wavelength of inelastically scattered light from molecules, known as Raman scattering, which exhibits a comparatively weak optical effect and involves the use of lasers for effective excitation. For Raman excitation, the optimal wavelength range is between 780 and 850 nm for rapid tissue analysis (Wahadoszamen et al. 2015). This nondestructive technique holds unlimited potential for the biochemical analysis of fatty acids and pigments in various fruits and vegetables and is also applied to determine carotenoids in vegetables. Raman spectroscopy has been employed to determine free fatty acids in olive oils and olives (El-Abassy et al. 2009; Muik et al. 2003; Qiu et al. 2019). Furthermore, the content of carotenoids in the pulp and peel of different fruits and vegetables, such as potatoes, tomatoes, lettuce, sweet potatoes, carrots, spinach, bananas, yellow peaches, yellow plums, yellow nectarines, apricots, orange, and black grapes, has been determined using resonance Raman spectroscopy (Bhosale et al. 2004; Darvin et al. 2007; Sebben et al. 2018).

10.7.3.3 NMR Spectroscopy

In contemporary times, the utilization of NMR spectroscopy has seen a consistent rise in applications for food analysis. Unlike Near-Infrared (NIR) spectroscopy, NMR does not necessitate a complex calibration method; instead, its measurement procedure is straightforward. Studies have indicated that this technique proves valuable in determining the composition and assessing the quality of various fruits and vegetables (Butz et al. 2005). In one particular investigation, a diverse array of fruits, including pineapple, strawberry, orange, kiwi, pear, apricot, and apple, underwent analysis for citric acid and malic acid content through ^{1}H NMR spectroscopy (Del Campo et al. 2006). Furthermore, the same spectroscopic technique was employed to ascertain the levels of sugars, amino acids, organic acids, and vitamins in different citrus cultivars (Villa-Ruano et al. 2019). Another study conducted by Mattoo et al. (2006) utilized NMR spectroscopy to scrutinize the metabolite profiles, specifically acids (citrate and malate) and sugars (glucose, fructose, and sucrose), in both transgenic and natural tomatoes.

10.7.3.4 Electronic Nose (e-Nose)

This sophisticated method comprises an electronic sensing device utilized for determining flavor, odors, and odorous volatile compounds. The device incorporates chemical gas sensors operating on various signal detection principles, encompassing magnetic properties, dielectric, optical, electrochemical activity, conductivity, electrical polarization, and heat generation. The chemical interaction between gas sensors and volatile compounds generates electrical signals, subsequently analyzed using multivariate pattern recognition methods. Data processing involves pattern recognition schedules and feature extraction methods, such as artificial neural networks, fuzzy logics, cluster analysis, functional discriminant analysis, partial least squares, and principal component analysis. The versatility of this technique extends to determining spoilage/freshness, detecting off-odors and taints, and overseeing the packaging methods and processing of diverse fruits and vegetables (Baldwin et al. 2011; Deisingh et al. 2004). Numerous studies have employed electronic noses (e-noses) to assess the postharvest quality of nectarines, bananas, peaches, and pears, track changes in the aroma profile of apples during storage, and identify spoilage in potatoes (Brezmes et al. 2001; Llobet et al. 1999).

10.7.4 Destructive Techniques

In destructive techniques, the destruction of plant material and the subsequent extraction and separation of constituents are necessary. The majority of chromatographic and mass spectrometric methods fall under this category.

10.7.4.1 Liquid Chromatography

This approach is commonly employed to ascertain the presence of various primary and secondary metabolites in foods. Liquid chromatography (LC) stands out as the preferred technique for separating and quantifying polyphenolic compounds found in diverse fruits and vegetables. Essential elements and conditions in LC techniques encompass the use of a column, detector, binary, and polar organic solvent systems. High-performance liquid chromatography (HPLC) typically utilizes refractive index (RI) detectors and UV detectors. Numerous chromatographic methods, including ion exchange chromatography, size exclusion chromatography, and reversed-phase (RP) chromatography, have been employed to analyze different metabolites in fruits and vegetables (Naczk and Shahidi 2006; Sakakibara et al. 2003).

The majority of HPLC techniques operate on an RP separation mode, utilizing a polar mobile phase and a non-polar stationary phase. Consequently, polar molecules elute more readily, while nonpolar molecules exhibit longer retention times. In a previous study, various vegetables such as parsley, lettuce, spinach, kale, red peppers, carrot, and tomato were scrutinized for lactucaxanthin, violaxanthin, neoxanthin, beta-carotene, lutein, lycopene, and carotenoids content using RP-HPLC and photodiode array (PDA) detection (Müller 1997).

RP-HPLC is commonly employed for determining fatty acids, organic acids, amino acids, sugars, carotenoids, phenolic acids, flavonoids, tocopherols, saponins, and non-starch polysaccharides in various fruits and vegetables (Lee 2000). Additionally, green pepper, tomato, broccoli, asparagus, celery, leek, and radish were analyzed for non-starch polysaccharides such as rhamnose, galactose, mannose, and arabinose using this technique (Villanueva-Suárez et al. 2003). In a study, the Golden Delicious variety of apples was examined for sugars and organic acids using HPLC with UV and RI detectors (Jakopic et al. 2012). Another investigation focused on diverse fruits and vegetables, including apricot, pear, peach, papaya, apple, plum, cherry, onion, ginger, garlic, dark grapes, raspberry, black carrot, strawberry, blueberry, and mango, determining various phenolic acids and flavonoids using HPLC-PDA (Gu et al. 2019).

Several potential anti-carcinogenic flavonoids, such as luteolin, quercetin, kaempferol, myricetin, and apigenin, have been identified in turnip, leek, endives, beans, broccoli, kale, and onions using LC (Hertog et al. 1992).

10.7.4.2 Gas Chromatography

Gas chromatography (GC) is primarily employed for the detection of volatile compounds in plants. In GC, the mobile phase consists of a carrier gas, typically an inert gas like unreactive nitrogen or helium. This mobile phase moves through a stationary phase, which is a microscopic layer of polymer or liquid on an inert solid support housed within a metal tubing or glass apparatus. Flame ionization and thermal conductivity detectors are commonly utilized in GC. GC is frequently employed for the analysis of mixtures containing compounds with relatively low molecular weights (less than 800 Da), such as essential oils, hydrocarbons, and fragrances. To enhance the volatility of compounds containing polar functional groups (e.g., −NH2, −OH, and −COOH), chemical derivatization can be employed (Lehotay and Hajšlová 2002). GC is widely employed for the analysis of fatty acids and their derivatives, encompassing flavonols, glucosinolates, sterols, phenolic acids, polysaccharides, and aroma compounds (Bilyk et al. 1984; Canini et al. 2007; Hurtado-Fernández et al. 2014; Normén et al. 1999; Shaw et al. 1989; Villanueva-Suárez et al. 2003; Wang and Zuo 2011).

10.7.4.3 Capillary Electrophoresis

Capillary electrophoresis (CE) is a suitable technique for the separation and quantification of charged and polar compounds with low to medium molecular masses. It is considered an alternative separation technique to HPLC due to its faster and more efficient compound separation capabilities (Caridi et al. 2007).

In a study, extracts from six different berries—red currant, black currant, strawberry, cranberry, cowberry, and bilberry—were analyzed for phenolic compounds using CE. The analytes of interest identified by CE included morin, quercetin, p-coumaric acid, ferulic acid, cinnamic acid, and trans-resveratrol (Ehala et al. 2005). Another study determined myricetin, chlorogenic acid, epicatechin, and phloridzin in apple juices and ciders using CE with electrochemical detection (Peng et al. 2005). Additionally, various vegetables such as corn, spinach, tomato, and carrot were assessed for their carotenoid content using capillary electrochromatography (CEC). Lutein, lycopene, and β-carotene were efficiently separated by CEC (Herrero-Martínez et al. 2006). The determination of the main organic acids in different fruits, including pineapple, orange, mandarin, grape, and apple, was carried out using capillary zone electrophoresis (Navarro-Pascual-Ahuir et al. 2017).

10.7.4.4 Mass Spectrometry

Mass spectrometry (MS) stands out as a powerful analytical tool capable of identifying chemical constituents based on their mass-to-charge ratio. Frequently employed in tandem with various separation techniques, such as those outlined earlier, MS analysis is enhanced by these complementary methods. Various ionization techniques, including matrix-assisted laser desorption/ionization, plasma desorption ionization, chemical ionization, electrospray ionization, electron impact ionization, and fast atom bombardment ionization, are commonly utilized for analyzing organic molecules (Herbert and Johnstone 2002).

In MS analysis, detectors such as ion cyclotron resonance, quadrupole ion trap, quadrupole, magnetic sector analyzer, and time-of-flight are commonly employed for precise mass detection (McLuckey and Wells 2001). Notably, researchers have utilized MS in conjunction with diverse chromatographic separation techniques to analyze the content of carotenoids in vegetables such as kale, Brussels sprouts, and tomatoes (Khachik et al. 1992); phenolics in onions (Price and Rhodes 1997); flavonoids in kale and soybean (Boué et al. 2003; Zhang et al. 2003); and oligosaccharides in red onion and Jerusalem artichoke (Laštovičková and Chmelík 2006).

Furthermore, MS coupled with tandem MS has been instrumental in determining various organic acids, including fumaric acid, succinic acid, pyruvic acid, citric acid, α-ketoglutaric acid, shikimic acid, malic acid, malonic acid, quinic acid, tartaric acid, and glutamic acid in different fruits and vegetables such as lamb's lettuce, lettuce, tomato, red and green pepper, lemon, orange, peach, grape, and melon (Flores et al. 2012).

10.8 Conclusion

Nutrients and secondary metabolites found in various horticultural crops are crucial phytochemicals that possess both functional and health-promoting properties. The utilization of advanced analytical techniques is essential for extracting, separating, identifying, and quantifying these compounds from horticultural crops, enabling a comprehensive understanding of the potential health benefits associated with various fruits and vegetables. This chapter furnishes information on the diverse nutrients and secondary metabolites present in horticultural crops, elucidating the advanced analytical techniques extensively employed for the isolation, separation, and identification of these phytochemicals. Furthermore, the chapter underscores the availability of several important technologies to achieve the necessary analytical objectives pertaining to the separation and quantification of different metabolites in horticulture crops.

References

Abebe AT (2006) Total sugar and maturity evaluation of intact watermelon using near infrared spectroscopy. J Near Infrared Spectrosc 14(1):67–70. https://doi.org/10.1255/jnirs.588

Alasalvar C, Grigor JM, Zhang D, Quantick PC, Shahidi F (2001) Comparison of volatiles, phenolics, sugars, antioxidant vitamins, and sensory quality of different colored carrot varieties. J Agric Food Chem 49(3):1410–1416. https://doi.org/10.1021/jf000595h

Alasalvar C, Pelvan E, Bahar B, Korel F, Ölmez H (2012) Flavour of natural and roasted Turkish hazelnut varieties (Corylus avellana L.) by descriptive sensory analysis, electronic nose and chemometrics. Int J Food Sci Technol 47(1):122–131. https://doi.org/10.1111/j.1365-2621.2011.02817.x

Arfaoui L (2021) Dietary plant polyphenols: effects of food processing on their content and bioavailability. Molecules 26(10):2959. https://doi.org/10.3390/molecules26102959

Asp NG (1987) Dietary fibre-definition, chemistry and analytical determination. Mol Asp Med 9(1):17–29. https://doi.org/10.1016/0098-2997(87)90014-8

Asp NG (1996) Dietary carbohydrates: classification by chemistry and physiology. Food Chem 57(1):9–14. https://doi.org/10.1016/0308-8146(96)00055-6

Baldwin EA, Bai J, Plotto A, Dea S (2011) Electronic noses and tongues: applications for the food and pharmaceutical industries. Sensors 11(5):4744–4766. https://doi.org/10.3390/s110504744

Beghi R, Giovenzana V, Tugnolo A, Guidetti R (2018) Application of visible/near infrared spectroscopy to quality control of fresh fruits and vegetables in large-scale mass distribution channels: a preliminary test on carrots and tomatoes. J Sci Food Agric 98(7):2729–2734. https://doi.org/10.1002/jsfa.8768

Bhosale P, Ermakov IV, Ermakova MR, Gellermann W, Bernstein PS (2004) Resonance Raman quantification of nutritionally important carotenoids in fruits, vegetables, and their juices in comparison to high-pressure liquid chromatography analysis. J Agric Food Chem 52(11):3281–3285. https://doi.org/10.1021/jf035345q

Bilyk A, Cooper PL, Sapers GM (1984) Varietal differences in distribution of Quercetin and Kaempferol in onion (Allium cepa L.) tissue. J Agric Food Chem 32(2):274–276. https://doi.org/10.1021/jf00122a024

Bolaji Umar O, Amudalat Ranti L, Shehu Abdulbaki A, Lukman Bola A, Khadijat Abdulhamid A, Ramat Biola M, Oluwagbenga Victor K (2022) Stresses in plants: biotic and abiotic. In: Current trends in wheat research. IntechOpen. https://doi.org/10.5772/intechopen.100501

Boué SM, Carter-Wientjes CH, Shih BY, Cleveland TE (2003) Identification of flavone aglycones and glycosides in soybean pods by liquid chromatography-tandem mass spectrometry. J Chromatogr A 991(1):61–68. https://doi.org/10.1016/S0021-9673(03)00209-7

Bowsher C, Steer M, Tobin A (2008) Plant biochemistry. Academic Press. https://doi.org/10.4324/9780203833483

Brezmes J, Llobet E, Vilanova X, Orts J, Saiz G, Correig X (2001) Correlation between electronic nose signals and fruit quality indicators on shelf-life measurements with pinklady apples. Sensors Actuators B Chem 80(1):41–50. https://doi.org/10.1016/S0925-4005(01)00867-X

Bribi N (2018) Pharmacological activity of alkaloids: a review. Asian J Bot 1:1–5

Brouns F (2020) Saccharide characteristics and their potential health effects in perspective. Front Nutr 7:75. https://doi.org/10.3389/fnut.2020.00075

Buchgraber M, Karaali A (2005) Compilation of standardised analytical methods for the analysis of active ingredients in functional foods (Issue Report EUR 21831 EN). http://www.irmm.jrc.be/activities/functional_foods_incl_antioxidative/Documents/eur21831en.pdf

Bureau S, Renard CMGC, Reich M, Ginies C, Audergon JM (2009) Change in anthocyanin concentrations in red apricot fruits during ripening. LWT 42(1):372–377. https://doi.org/10.1016/j.lwt.2008.03.010

Butnariu M, Butu A (2015) Chemical composition of vegetables and their products. In: Handbook of food chemistry. Springer, pp 627–692. https://doi.org/10.1007/978-3-642-36605-5_17

Butz P, Hofmann C, Tauscher B (2005) Recent developments in noninvasive techniques for fresh fruit and vegetable internal quality analysis. J Food Sci 70(9):R131–R141. https://doi.org/10.1111/j.1365-2621.2005.tb08328.x

Canini A, Alesiani D, D'Arcangelo G, Tagliatesta P (2007) Gas chromatography-mass spectrometry analysis of phenolic compounds from Carica papaya L. leaf. J Food Compos Anal 20(7):584–590. https://doi.org/10.1016/j.jfca.2007.03.009

Caridi D, Trenerry VC, Rochfort S, Duong S, Laugher D, Jones R (2007) Profiling and quantifying quercetin glucosides in onion (Allium cepa L.) varieties using capillary zone electrophoresis and high performance liquid chromatography. Food Chem 105(2):691–699. https://doi.org/10.1016/j.foodchem.2006.12.063

Chen T, Xing J, El-Mogy M, Wang Y (2022) Editorial: nutritional quality formation and maintenance of horticultural crops. Front Plant Sci 13:1005338. https://doi.org/10.3389/fpls.2022.1005338

Crozier A, Jaganath IB, Clifford MN (2006) Phenols, polyphenols and tannins: an overview. In: Plant secondary metabolites: occurrence, structure and role in the human diet. Wiley-Blackwell, Oxford, pp 1–24

Dalvi RR, Bowie WC (1983) Toxicology of solanine: an overview. Vet Hum Toxicol 25(1):13–15

Darvin ME, Gersonde I, Albrecht H, Sterry W, Lademann J (2007) Resonance Raman spectroscopy for the detection of carotenoids in foodstuffs. Influence of the nutrition on the antioxidative potential of the skin. Laser Phys Lett 4(6):452–456. https://doi.org/10.1002/lapl.200710004

Das TK, Banerjee D, Chakraborty D, Pakhira MC, Shrivastava B, Kuhad RC (2012) Saponin: role in animal system. Vet World 5(4):248–254. https://doi.org/10.5455/vetworld.2012.248-254

Davies K (2009) Annual plant reviews: plant pigments and their manipulation. Wiley

de Oliveira GA, de Castilhos F, Renard CMGC, Bureau S (2014) Comparison of NIR and MIR spectroscopic methods for determination of individual sugars, organic acids and carotenoids in passion fruit. Food Res Int 60:154–162. https://doi.org/10.1016/j.foodres.2013.10.051

Deisingh AK, Stone DC, Thompson M (2004) Applications of electronic noses and tongues in food analysis. Int J Food Sci Technol 39(6):587–604. https://doi.org/10.1111/j.1365-2621.2004.00821.x

Del Campo G, Berregi I, Caracena R, Santos JI (2006) Quantitative analysis of malic and citric acids in fruit juices using proton nuclear magnetic resonance spectroscopy. Anal Chim Acta 556(2):462–468. https://doi.org/10.1016/j.aca.2005.09.039

DeVries JW (2003) On defining dietary fibre. Proc Nutr Soc 62(1):37–43. https://doi.org/10.1079/pns2002234

Dey P, Kundu A, Kumar A, Gupta M, Lee BM, Bhakta T, Dash S, Kim HS (2020) Analysis of alkaloids (indole alkaloids, isoquinoline alkaloids, tropane alkaloids). In: Recent advances in natural products analysis. Elsevier, pp 505–567. https://doi.org/10.1016/B978-0-12-816455-6.00015-9

Dietzen DJ (2018) Amino acids, peptides, and proteins. In: Principles and applications of molecular diagnostics. Elsevier, pp 345–380. https://doi.org/10.1016/B978-0-12-816061-9.00013-8

Divekar PA, Narayana S, Divekar BA, Kumar R, Gadratagi BG, Ray A, Singh AK, Rani V, Singh V, Singh AK, Kumar A, Singh RP, Meena RS, Behera TK (2022) Plant secondary metabolites as defense tools against herbivores for sustainable crop protection. Int J Mol Sci 23(5):2690. https://doi.org/10.3390/ijms23052690

Drewnowski A, Gomez-Carneros C (2000) Bitter taste, phytonutrients, and the consumer: a review. Am J Clin Nutr 72(6):1424–1435. https://doi.org/10.1093/ajcn/72.6.1424

Du H, Van Der ADL, Boshuizen HC, Forouhi NG, Wareham NJ, Halkjær J, Tjønneland A, Overvad K, Jakobsen MU, Boeing H, Buijsse B, Masala G, Palli D, Sørensen TIA, Saris WHM, Feskens EJM (2010) Dietary fiber and subsequent changes in body weight and waist circumference in European men and women. Am J Clin Nutr 91(2):329–336. https://doi.org/10.3945/ajcn.2009.28191

Ehala S, Vaher M, Kaljurand M (2005) Characterization of phenolic profiles of Northern European berries by capillary electrophoresis and determination of their antioxidant activity. J Agric Food Chem 53(16):6484–6490. https://doi.org/10.1021/jf050397w

Eisenmann P, Ehlers M, Weinert CH, Tzvetkova P, Silber M, Rist MJ, Luy B, Muhle-Goll C (2016) Untargeted NMR spectroscopic analysis of the metabolic variety of new apple cultivars. Metabolites 6(3):29. https://doi.org/10.3390/metabo6030029

Eitenmiller RR, Lee J (2004) Vitamin E: food chemistry, composition, and analysis. CRC Press. https://doi.org/10.1201/9780203970140

El-Abassy RM, Donfack P, Materny A (2009) Rapid determination of free fatty acid in extra virgin olive oil by raman spectroscopy and multivariate analysis. J Am Oil Chem Soc 86(6):507–511. https://doi.org/10.1007/s11746-009-1389-0

El-Mesery HS, Mao H, Abomohra AEF (2019) Applications of non-destructive technologies for agricultural and food products quality inspection. Sensors 19(4):846. https://doi.org/10.3390/s19040846

Facchini PJ (2001) Alkaloid biosynthesis in plants: biochemistry, cell biology, molecular regulation, and metabolic engineering applications. Annu Rev Plant Biol 52:29–66. https://doi.org/10.1146/annurev.arplant.52.1.29

Fiedor J, Burda K (2014) Potential role of carotenoids as antioxidants in human health and disease. Nutrients 6(2):466–488. https://doi.org/10.3390/nu6020466

Flores P, Hellín P, Fenoll J (2012) Determination of organic acids in fruits and vegetables by liquid chromatography with tandem-mass spectrometry. Food Chem 132(2):1049–1054. https://doi.org/10.1016/j.foodchem.2011.10.064

Francis G, Kerem Z, Makkar HPS, Becker K (2002) The biological action of saponins in animal systems: a review. Br J Nutr 88(6):587–605. https://doi.org/10.1079/bjn2002725

Gerschenson LN, Rojas AM, Fissore EN (2022) Carbohydrates. In: Nutraceutical and functional food components. Academic Press, pp 49–126. https://doi.org/10.1016/B978-0-323-85052-0.00001-5

Goldberg G (2003) The report of a British Nutrition Foundation Task Force. Blackwell

Gu C, Howell K, Dunshea FR, Suleria HAR (2019) LC-ESI-QTOF/MS characterisation of phenolic acids and flavonoids in polyphenol-rich fruits and vegetables and their potential antioxidant activities. Antioxidants 8(9):405. https://doi.org/10.3390/antiox8090405

Gunstone FD, Norris FA (2013) Lipids in foods: chemistry, biochemistry and technology. Elsevier

Herbert CG, Johnstone RAW (2002) Mass spectrometry basics. In: Mass spectrometry basics. CRC Press. https://doi.org/10.1201/9781420040340

Herrero-Martínez JM, Eeltink S, Schoenmakers PJ, Kok WT, Ramis-Ramos G (2006) Determination of major carotenoids in vegetables by capillary electrochromatography. J Sep Sci 29(5):660–665. https://doi.org/10.1002/jssc.200500427

Hertog MGL, Hollman PCH, Hertog MGL, Katan MB (1992) Content of potentially anticarcinogenic flavonoids of 28 vegetables and 9 fruits commonly consumed in the Netherlands. J Agric Food Chem 40(12):2379–2383. https://doi.org/10.1021/jf00024a011

Hounsome N, Hounsome B, Tomos D, Edwards-Jones G (2008) Plant metabolites and nutritional quality of vegetables. J Food Sci 73(4):R48–R65. https://doi.org/10.1111/j.1750-3841.2008.00716.x

Hurtado-Fernández E, Pacchiarotta T, Mayboroda OA, Fernández-Gutiérrez A, Carrasco-Pancorbo A (2014) Quantitative characterization of important metabolites of avocado fruit by gas chromatography coupled to different detectors (APCI-TOF MS and FID). Food Res Int 62:801–811. https://doi.org/10.1016/j.foodres.2014.04.038

Ignat I, Volf I, Popa VI (2011) A critical review of methods for characterisation of polyphenolic compounds in fruits and vegetables. Food Chem 126(4):1821–1835. https://doi.org/10.1016/j.foodchem.2010.12.026

Jakab A, Nagy K, Héberger K, Vékey K, Forgács E (2002) Differentiation of vegetable oils by mass spectrometry combined with statistical analysis. Rapid Commun Mass Spectrom 16(24):2291–2297. https://doi.org/10.1002/rcm.862

Jakopic J, Slatnar A, Stampar F, Veberic R, Simoncic A (2012) Analysis of selected primary metabolites and phenolic profile of "golden delicious" apples from four production systems. Fruits 67(5):377–386. https://doi.org/10.1051/fruits/2012032

Jie YC, Miao Y, Zhang H, Matsunaga R (2004) Non-destructive determination of carbohydrate content in potatoes using near infrared spectroscopy. J Near Infrared Spectrosc 12(5):311–314. https://doi.org/10.1255/jnirs.439

Kämper W, Trueman SJ, Tahmasbian I, Bai SH (2020) Rapid determination of nutrient concentrations in hass avocado fruit by vis/NIR hyperspectral imaging of flesh or skin. Remote Sens 12(20):1–19. https://doi.org/10.3390/rs12203409

Katayama K, Komaki K, Tamiya S (1996) Prediction of starch, moisture, and sugar in sweetpotato by near infrared transmittance. HortScience 31(6):1003–1006. https://doi.org/10.21273/hortsci.31.6.1003

Khachik F, Goli MB, Beecher GR, Holden J, Lusby WR, Tenorio MD, Barrera MR (1992) Effect of food preparation on qualitative and quantitative distribution of major carotenoid constituents of tomatoes and several green vegetables. J Agric Food Chem 40(3):390–398

Khoo HE, Prasad KN, Kong KW, Jiang Y, Ismail A (2011) Carotenoids and their isomers: color pigments in fruits and vegetables. Molecules 16(2):1710–1738. https://doi.org/10.3390/molecules16021710

Kumar N, Goel N (2019) Phenolic acids: natural versatile molecules with promising therapeutic applications. Biotechnol Rep 24:e00370. https://doi.org/10.1016/j.btre.2019.e00370

Kumar S, Saini R, Suthar P, Kumar V, Sharma R (2022) Plant secondary metabolites: their food and therapeutic importance. In: Plant secondary metabolites: physico-chemical properties and therapeutic applications. Springer Nature, Singapore, pp 371–413. https://doi.org/10.1007/978-981-16-4779-6_12

Kuna A, Achinna P (2013) Mono unsaturated fatty acids for CVD and diabetes: a healthy choice. Int J Nutr Pharmacol Neurol Dis 3(3):236. https://doi.org/10.4103/2231-0738.114841

Laštovičková M, Chmelík J (2006) Simple and fast method for recognition of reducing and nonreducing neutral carbohydrates by matrix-assisted laser desorption/ionization time-of-flight mass spectrometry. J Agric Food Chem 54(14):5092–5097. https://doi.org/10.1021/jf052988s

Lebrun M, Plotto A, Goodner K, Ducamp MN, Baldwin E (2008) Discrimination of mango fruit maturity by volatiles using the electronic nose and gas chromatography. Postharvest Biol Technol 48(1):122–131. https://doi.org/10.1016/j.postharvbio.2007.09.010

Lee HS (2000) HPLC analysis of phenolic compounds. Food Anal 39(3):775–824

Lehotay SJ, Hajšlová J (2002) Application of gas chromatography in food analysis. Trends Anal Chem 21(9–10):686–697. https://doi.org/10.1016/S0165-9936(02)00805-1

Li-Chan ECY, Cheung IWY (2010) Flavor-active properties of amino acids, peptides, and proteins. In: Bioactive proteins and peptides as functional foods and nutraceuticals. Wiley, pp 341–358. https://doi.org/10.1002/9780813811048.ch23

Liu RH (2013) Health-promoting components of fruits and vegetables in the diet. Adv Nutr 4(3):384S–392S

Liu Z, Farkas P, Wang K, Kohli MO, Fitzpatrick TB (2022) B vitamin supply in plants and humans: the importance of vitamer homeostasis. Plant J 111(3):662–682. https://doi.org/10.1111/tpj.15859

Llobet E, Hines EL, Gardner JW, Franco S (1999) Non-destructive banana ripeness determination using a neural network-based electronic nose. Meas Sci Technol 10(6):538–548. https://doi.org/10.1088/0957-0233/10/6/320

MacEvilly C, Peltola K (2003) The effect of agronomy, storage, processing and cooking on bioactive substances in food. In: Plants: diet and health. Wiley, pp 226–239. https://doi.org/10.1002/9780470774465.ch12

Mariani NCT, De Almeida Teixeira GH, De Lima KMG, Morgenstern TB, Nardini V, Júnior LCC (2015) NIRS and iSPA-PLS for predicting total anthocyanin content in jaboticaba fruit. Food Chem 174:643–648. https://doi.org/10.1016/j.foodchem.2014.11.008

Mariotti F, Gardner CD (2019) Dietary protein and amino acids in vegetarian diets—a review. Nutrients 11(11):2661. https://doi.org/10.3390/nu11112661

Mattoo AK, Sobolev AP, Neelam A, Goyal RK, Handa AK, Segre AL (2006) Nuclear magnetic resonance spectroscopy-based metabolite profiling of transgenic tomato fruit engineered to accumulate spermidine and spermine reveals enhanced anabolic and nitrogen-carbon interactions. Plant Physiol 142(4):1759–1770. https://doi.org/10.1104/pp.106.084400

McLuckey SA, Wells JM (2001) Mass analysis at the advent of the 21st century. Chem Rev 101(2):571–606. https://doi.org/10.1021/cr990087a

Milder IEJ, Arts ICW, van de Putte B, Venema DP, Hollman PCH (2005) Lignan contents of Dutch plant foods: a database including lariciresinol, pinoresinol, secoisolariciresinol and matairesinol. Br J Nutr 93(3):393–402. https://doi.org/10.1079/bjn20051371

Molino S, Pilar Francino M, Ángel Rufián Henares J (2023) Why is it important to understand the nature and chemistry of tannins to exploit their potential as nutraceuticals? Food Res Int 173:113329. https://doi.org/10.1016/j.foodres.2023.113329

Moon KM, Kwon EB, Lee B, Kim CY (2020) Recent trends in controlling the enzymatic browning of fruit and vegetable products. Molecules 25(12):2754. https://doi.org/10.3390/molecules25122754

Muik B, Lendl B, Molina-Díaz A, Ayora-Cañada MJ (2003) Direct, reagent-free determination of free fatty acid content in olive oil and olives by Fourier transform Raman spectrometry. Anal Chim Acta 487(2):211–220. https://doi.org/10.1016/S0003-2670(03)00560-9

Müller H (1997) Determination of the carotenoid content in selected vegetables and fruit by HPLC and photodiode array detection. Eur Food Res Technol 204(2):88–94. https://doi.org/10.1007/s002170050042

Naczk M, Shahidi F (2006) Phenolics in cereals, fruits and vegetables: occurrence, extraction and analysis. J Pharm Biomed Anal 41(5):1523–1542. https://doi.org/10.1016/j.jpba.2006.04.002

Navarro-Pascual-Ahuir M, Lerma-García MJ, Simó-Alfonso EF, Herrero-Martínez JM (2017) Analysis of aliphatic organic acids in commercial fruit juices by capillary electrophoresis with indirect UV detection: application to differentiation of fruit juices. Food Anal Methods 10(12):3991–4002. https://doi.org/10.1007/s12161-017-0963-6

Normén L, Johnsson M, Andersson H, Van Gameren Y, Dutta P (1999) Plant sterols in vegetables and fruits commonly consumed in Sweden. Eur J Nutr 38(2):84–89. https://doi.org/10.1007/s003940050048

Oz AT, Kafkas E (2017) Phytochemicals in fruits and vegetables. In: Superfood and functional food—an overview of their processing and utilization. IntechOpen. https://doi.org/10.5772/66987

Panche AN, Diwan AD, Chandra SR (2016) Flavonoids: an overview. J Nutr Sci 5:e47. https://doi.org/10.1017/jns.2016.41

Pedro AMK, Ferreira MMC (2005) Nondestructive determination of solids and carotenoids in tomato products by near-infrared spectroscopy and multivariate calibration. Anal Chem 77(8):2505–2511. https://doi.org/10.1021/ac048651r

Peng Y, Liu F, Peng Y, Ye J (2005) Determination of polyphenols in apple juice and cider by capillary electrophoresis with electrochemical detection. Food Chem 92(1):169–175. https://doi.org/10.1016/j.foodchem.2004.08.001

Pereira C, Li D, Sinclair AJ (2001) The α-linolenic acid content of green vegetables commonly available in Australia. Int J Vitam Nutr Res 71(4):223–228. https://doi.org/10.1024/0300-9831.71.4.223

Perveen S (2018) Introductory chapter: terpenes and terpenoids. In: Terpenes and terpenoids. IntechOpen. https://doi.org/10.5772/intechopen.79683

Peterson J, Dwyer J, Adlercreutz H, Scalbert A, Jacques P, McCullough ML (2010) Dietary lignans: physiology and potential for cardiovascular disease risk reduction. Nutr Rev 68(10):571–603. https://doi.org/10.1111/j.1753-4887.2010.00319.x

Piironen V, Varo P, Koivistoinen P, Syvaöja EL, Salminen K (1986) Tocopherols and tocotrienols in finnish foods: vegetables, fruits, and berries. J Agric Food Chem 34(4):742–746. https://doi.org/10.1021/jf00070a038

Piironen V, Toivo J, Puupponen-Pimiä R, Lampi AM (2003) Plant sterols in vegetables, fruits and berries. J Sci Food Agric 83(4):330–337. https://doi.org/10.1002/jsfa.1316

Pott DM, Osorio S, Vallarino JG (2019) From central to specialized metabolism: an overview of some secondary compounds derived from the primary metabolism for their role in conferring nutritional and organoleptic characteristics to fruit. Front Plant Sci 10:835. https://doi.org/10.3389/fpls.2019.00835

Price KR, Rhodes MJC (1997) Analysis of the major flavonol glycosides present in four varieties of onion (Allium cepa) and changes in composition resulting from autolysis. J Sci Food Agric 74(3):331–339. https://doi.org/10.1002/(sici)1097-0010(199707)74:3<331::aid-jsfa806>3.3.co;2-3

Qiu J, Hou HY, Yang IS, Chen XB (2019) Raman spectroscopy analysis of free fatty acid in olive oil. Appl Sci 9(21):4510. https://doi.org/10.3390/app9214510

Reshi ZA, Ahmad W, Lukatkin AS, Javed SB (2023) From nature to lab: a review of secondary metabolite biosynthetic pathways, environmental influences, and in vitro approaches. Metabolites 13(8):895. https://doi.org/10.3390/metabo13080895

Rungpichayapichet P, Mahayothee B, Nagle M, Müller J (2013) Application of near-infrared spectroscopy to determine the ß-carotene content of mango. In: NIR 2013—16th international conference on near infrared spectroscopy, pp 162–166

Russo M, Di Sanzo R, Cefaly V, Carabetta S, Serra D, Fuda S (2013) Non-destructive flavour evaluation of red onion (Allium cepa L.) ecotypes: an electronic-nose-based approach. Food Chem 141(2):896–899. https://doi.org/10.1016/j.foodchem.2013.03.052

Saevels S, Lammertyn J, Berna AZ, Veraverbeke EA, Di Natale C, Nicolaï BM (2004) An electronic nose and a mass spectrometry-based electronic nose for assessing apple quality during shelf life. Postharvest Biol Technol 31(1):9–19. https://doi.org/10.1016/S0925-5214(03)00129-7

Sakakibara H, Honda Y, Nakagawa S, Ashida H, Kanazawa K (2003) Simultaneous determination of all polyphenols in vegetables, fruits, and teas. J Agric Food Chem 51(3):571–581. https://doi.org/10.1021/jf020926l

Saleem M, Hyoung JK, Ali MS, Yong SL (2005) An update on bioactive plant lignans. Nat Prod Rep 22(6):696–716. https://doi.org/10.1039/b514045p

Santhi VP, Poomaruthai M, Sarasu VP, Gurusamy K, Parthiban S, Geetha K, Indumathi K, Laveena KS, Umayal A (2020) Fruits: a potential source of vitamin c as essential human nutrition and immunity development: a review. Pharma Innov 9(9):132–144. https://doi.org/10.22271/tpi.2020.v9.i9b.5102

Schellack G, Harirari P, Schellack N (2019) Vitamin B-complex deficiency, supplementation and management. SA Pharm J 86(3):23–29

Sebben JA, da Silveira Espindola J, Ranzan L, Fernandes de Moura N, Trierweiler LF, Trierweiler JO (2018) Development of a quantitative approach using Raman spectroscopy for carotenoids determination in processed sweet potato. Food Chem 245:1224–1231. https://doi.org/10.1016/j.foodchem.2017.11.086

Shahidi F, Ambigaipalan P (2015) Phenolics and polyphenolics in foods, beverages and spices: antioxidant activity and health effects—a review. J Funct Foods 18:820–897. https://doi.org/10.1016/j.jff.2015.06.018

Shahidi F, De Camargo AC (2016) Tocopherols and tocotrienols in common and emerging dietary sources: occurrence, applications, and health benefits. Int J Mol Sci 17(10):1745. https://doi.org/10.3390/ijms17101745

Shaw GJ, Andrzejewski D, Roach JAG, Sphon JA (1989) Separation and identification of glucosinolates from Brassica vegetables using high-performance capillary gas chromatography (GC)-positive-ion chemical ionization mass spectrometry (PICIMS) and GC-PICIMS/MS. J Agric Food Chem 37(2):372–378. https://doi.org/10.1021/jf00086a022

Singh B, Suri K, Shevkani K, Kaur A, Kaur A, Singh N (2018) Enzymatic browning of fruit and vegetables: a review. In: Enzymes in food technology: improvements and innovations. Springer, pp 73–78. https://doi.org/10.1007/978-981-13-1933-4_4

Smeriglio A, Barreca D, Bellocco E, Trombetta D (2017) Proanthocyanidins and hydrolysable tannins: occurrence, dietary intake and pharmacological effects. Br J Pharmacol 174(11):1244–1262. https://doi.org/10.1111/bph.13630

Sobolev AP, Mannina L, Proietti N, Carradori S, Daglia M, Giusti AM, Antiochia R, Capitani D (2015) Untargeted NMR-based methodology in the study of fruit metabolites. Molecules 20(3):4088–4108. https://doi.org/10.3390/molecules20034088

Stollman N, Raskin JB (2004) Diverticular disease of the colon. Lancet 363(9409):631–639. https://doi.org/10.1016/S0140-6736(04)15597-9
Sudhakar P, Latha P, Reddy PV (2016) Phenotyping crop plants for physiological and biochemical traits. Academic Press. https://doi.org/10.1016/C2015-0-01450-2
Swallah MS, Sun H, Affoh R, Fu H, Yu H (2020) Antioxidant potential overviews of secondary metabolites (polyphenols) in fruits. Int J Food Sci 2020:9081686. https://doi.org/10.1155/2020/9081686
Tamburini E, Costa S, Rugiero I, Pedrini P, Marchetti MG (2017) Quantification of lycopene, β-carotene, and total soluble solids in intact red-flesh watermelon (Citrullus lanatus) using on-line near-infrared spectroscopy. Sensors 17(4):746. https://doi.org/10.3390/s17040746
Torri L, Sinelli N, Limbo S (2010) Shelf life evaluation of fresh-cut pineapple by using an electronic nose. Postharvest Biol Technol 56(3):239–245. https://doi.org/10.1016/j.postharvbio.2010.01.012
Trautwein EA, Demonty I (2007) Phytosterols: natural compounds with established and emerging health benefits. Oleagineux Corps Gras Lipides 14(5):259–266. https://doi.org/10.1684/ocl.2007.0145
Tsugawa N, Shiraki M (2020) Vitamin K nutrition and bone health. Nutrients 12(7):1–16. https://doi.org/10.3390/nu12071909
Vicente A, Ortiz C, Sozzi G, Manganaris G, Crisosto C (2014) Nutritional properties of fruits and vegetables. In: Postharvest handling a systems approach. Academic Press, pp 60–122
Villanueva-Suárez MJ, Redondo-Cuenca A, Rodríguez-Sevilla MD, De las Heras Martínez, M. (2003) Characterization of nonstarch polysaccharides content from different edible organs of some vegetables, determined by GC and HPLC: comparative study. J Agric Food Chem 51(20):5950–5955. https://doi.org/10.1021/jf021010h
Villa-Ruano N, Pérez-Hernández N, Zepeda-Vallejo LG, Quiroz-Acosta T, Mendieta-Moctezuma A, Montoya-García C, García-Nava ML, Becerra-Martínez E (2019) 1H-NMR based metabolomics profiling of citrus juices produced in Veracruz, México. Chem Biodivers 16(5):e1800479. https://doi.org/10.1002/cbdv.201800479
Wahadoszamen M, Rahaman A, Hoque NMR, Talukder AI, Abedin KM, Haider AFMY (2015) Laser Raman spectroscopy with different excitation sources and extension to surface enhanced raman spectroscopy. J Spectrosc 2015:895317. https://doi.org/10.1155/2015/895317
Wang C, Zuo Y (2011) Ultrasound-assisted hydrolysis and gas chromatography-mass spectrometric determination of phenolic compounds in cranberry products. Food Chem 128(2):562–568. https://doi.org/10.1016/j.foodchem.2011.03.066
Wang S, Xiao Y, Tian F, Zhao J, Zhang H, Zhai Q, Chen W (2020) Rational use of prebiotics for gut microbiota alterations: specific bacterial phylotypes and related mechanisms. J Funct Foods 66:103838. https://doi.org/10.1016/j.jff.2020.103838
Xu X (2022) Major challenges facing the commercial horticulture. Front Hortic 1:980159. https://doi.org/10.3389/fhort.2022.980159
Xu H, Qi B, Sun T, Fu X, Ying Y (2012) Variable selection in visible and near-infrared spectra: application to on-line determination of sugar content in pears. J Food Eng 109(1):142–147. https://doi.org/10.1016/j.jfoodeng.2011.09.022
Yao LH, Jiang YM, Shi J, Tomás-Barberán FA, Datta N, Singanusong R, Chen SS (2004) Flavonoids in food and their health benefits. Plant Foods Hum Nutr 59:113–122
Yashodhara BM, Umakanth S, Pappachan JM, Bhat SK, Kamath R, Choo BH (2009) Omega-3 fatty acids: a comprehensive review of their role in health and disease. Postgrad Med J 85(1000):84–90. https://doi.org/10.1136/pgmj.2008.073338
Yeshi K, Crayn D, Ritmejerytė E, Wangchuk P (2022) Plant secondary metabolites produced in response to abiotic stresses has potential application in pharmaceutical product development. Molecules 27(1):313. https://doi.org/10.3390/molecules27010313
Zempleni J, Suttie JW, Gregory JF III, Stover PJ (2013) Handbook of vitamins. CRC Press

Zhang J, Satterfield MB, Brodbelt JS, Britz SJ, Clevidence B, Novotny JA (2003) Structural characterization and detection of kale flavonoids by electrospray ionization mass spectrometry. Anal Chem 75(23):6401–6407. https://doi.org/10.1021/ac034795e

Zhang SUM, Ye B, Li ZSJ (2012) Non-destructive detection of peach (Prunus persica) fruit volatiles using an electronic nose. J Fruit Sci 29:809–813

Zum Felde T, Alamu EO, Porras E, Maroya N, Maziya-Dixon B (2013) Screening for pro-vitamin A components in Cassava (Manihot esculenta) using Nirs for supporting bioforyication. In: 16th International Conference on Near Infrared Spectroscopy (ICNIRS2013), 34280 LG Grande Motte, France

Chapter 11
Detection of Pesticides and Other Contaminants in Horticultural Crops

Vigya Mishra, Vishal Chugh, Shubham Gangwar, Priya Awasthi, and Akhilesh Kumar Singh

Abstract The presence of pesticide residues and other contaminants poses a significant risk to both human health and the environment, particularly in horticultural crops. Thus, there is a pressing need to develop effective detection methods to ensure the safety and quality of food. Given that fruits and vegetables are often consumed raw, the potential health implications of these residues are especially concerning. Consequently, it is imperative to regularly employ various analytical techniques to scrutinize fruits and vegetables for pesticide residues and other pollutants. Understanding maximum residue levels (MRLs) is crucial for determining contamination levels in produce.

This chapter outlines current detection technologies utilized for examining pollutants in horticultural crops, focusing on pesticide residue detection. Techniques such as mass spectrometry coupled with chromatography (including liquid and gas chromatography) have been instrumental in identifying and quantifying pesticide residues with remarkable sensitivity and selectivity. However, these methods necessitate expensive equipment, skilled personnel, and time-intensive sample preparation.

To enhance detection accuracy and streamline analysis, high-selectivity and sensitivity methods such as enzyme inhibition, immunoassays, and DNA hybridization-based biosensors are employed. Integration of different detection technologies with advanced data analysis techniques such as chemometrics and machine learning algorithms facilitates improved detection accuracy and expedites the analysis of large datasets. Consequently, these technologies significantly reduce the time and

V. Mishra (✉) · S. Gangwar · P. Awasthi
Department of Post-Harvest Technology, College of Horticulture, Banda University of Agriculture and Technology, Banda, India

V. Chugh
Department of Basic and Social Sciences, College of Horticulture, Banda University of Agriculture and Technology, Banda, India

A. K. Singh
Department of Entomology, College of Agriculture, Rani Laxmi Bai Central Agriculture University, Jhansi, India

M. Thakur, T. Belwal (eds.), *Advances in Postharvest and Analytical Technology of Horticulture Crops*, https://doi.org/10.1007/978-981-97-7247-6_11

cost associated with standard laboratory analysis, enabling quick screening of a large number of samples. In recent years, there has been notable progress in the development and application of detection technologies for investigating pollutants, particularly pesticide residues in horticultural crops. By combining traditional analytical methods with rapid, nondestructive techniques, biosensors, and advanced data analysis tools, the identification of pesticide residues and other contaminants has become more accurate and efficient. These advancements not only safeguard food safety and quality but also promote environmentally friendly gardening practices and protect public health.

Keywords Horticultural crops · Pesticides · Toxicity · Maximum residue limits · Analytical GC-MS · QuEChERS · LC-MS/MS · Heavy metals · Mycotoxins

11.1 Introduction

In contemporary times, consumers exhibit a heightened awareness regarding the presence of organic residues and contaminants within the food supply chain, reflecting apprehensions regarding potential toxicological ramifications. The proliferation of contaminants within the environment has markedly escalated since the advent of the industrial revolution and the onset of economic globalization, attributable to a plethora of anthropogenic factors. Consequently, global attention toward the multifaceted and evolving issues pertaining to food security has intensified, particularly due to their intrinsic nexus with human health (Rai 2018; Säumel et al. 2012; Tóth et al. 2016).

Horticultural crops, encompassing fruits, vegetables, and spices, assume a pivotal role in human nutrition and fortification of immunity, owing to their abundant reservoir of functional compounds such as carotenoids, trace minerals, vitamins, phenolics, bioactive compounds, and fiber (Shashirekha et al. 2015). However, these commodities may serve as reservoirs for toxic residual pesticides, heavy metals, or mycotoxins stemming from the utilization of pesticides in the agricultural production continuum (Lee et al. 2009; Wang et al. 2013). Given their customary consumption in raw form subsequent to washing and devoid of further processing, they emerge as primary conduits for pesticide residue and other contaminants assimilated by humans. The plausible human intake of deleterious compounds via pesticide residues within agricultural produce might surpass their exposure to such substances through water and atmospheric sources (Juraske et al. 2007; Szpyrk et al. 2015). Consequently, it is imperative to vigilantly monitor the presence of residual pesticides in fruits and vegetables and evaluate their potential threat to human health.

Pesticide residues, in conjunction with other contaminants such as mycotoxins and heavy metals, possess a propensity to accumulate readily in food matrices throughout cultivation and postharvest operations, encompassing washing, storage, processing, packaging, and distribution. These chemical contaminants or their metabolites have been identified as carcinogenic, teratogenic, and immunotoxic, precipitating a broad spectrum of consequences ranging from acute fatalities to chronic maladies, inclusive of tumors, in both human and animal populations (Isohata et al. 1986; Cabras et al. 1997; Jimenez et al. 2003). The utilization of pesticides plays an indispensable role in augmenting agricultural productivity by safeguarding crops against insect pests, diseases, and weed infestations. Nevertheless, the excessive application of these chemical agents and their by-products engenders enduring repercussions on both crops and soil, engendering substantial challenges within the ecosystem and public health domain (Lee et al. 2009; Wang et al. 2013; Chowdhury et al. 2012; Fan et al. 2015; Li et al. 2015). Consequently, prolonged human exposure to hazardous pesticide levels can precipitate an array of ailments, exerting a deleterious impact on human well-being.

The global utilization of pesticides has undergone a notable 30% escalation over the preceding 23 years (FAO 2022). Predominantly concentrated between 2000 and 2012, this surge witnessed a subsequent stabilization until 2017, followed by a marginal decline. The Americas have emerged as the principal contributors, trailed by Asia, Europe, Africa, and Oceania. The Americas' proportion of worldwide pesticide usage has ascended from 44% to 51%, whereas Asia and Europe have experienced a 4–5% downturn, reaching 25% and 18%, respectively. In the year 2020, the United States emerged as the foremost pesticide user, accounting for 0.41 million tons or 15% of the global total, followed closely by Brazil (0.38 million tons) and China (0.27 million tons). Pesticide application rates in 2020 were most pronounced in the Americas, followed by Oceania, Europe, Asia, and Africa. Saint Lucia, with 20 kg/ha, Maldives, with 17 kg/ha, and Oman, with 16 kg/ha, emerged as the top three nations in terms of pesticide application rate in 2020 (FAO 2022).

Over the past decade, the global proliferation of pesticide usage has surged, propelled by the burgeoning world populace and rapid urbanization. The residues stemming from these pesticides, whether in the form of active constituents, metabolites, or degradation by-products, harbor the potential to adversely impact human physiology. Thus, it becomes paramount to enlighten consumers regarding the plausible risks entailed in pesticide utilization (Fan et al. 2014). Emerging research underscores the diverse array of pesticides, each wielding distinct mechanisms of action, which pose tangible risks, manifesting in cognitive impairments, depressive tendencies, diabetic conditions, and respiratory ailments such as rhinitis upon prolonged exposure (Li et al. 2015). Delving into the analysis of residues not only facilitates gauging the magnitude and endurance of chemical contamination within the environment but also aids in pinpointing and ameliorating potential health hazards.

Predominantly found in fruits and vegetables, particularly prevalent in citrus fruits, grapes, and potatoes (Szpyrk et al. 2015), pesticides necessitate prompt scrutiny in these dietary staples, given their propensity to elevate health risks. Evaluation of pesticide contamination in fruits and vegetables mandates a comprehensive grasp of maximum residue limits (MRLs), delineating the highest legally permissible concentrations of pesticide residues in food commodities post-application (Szpyrk et al. 2015). Defined by the Codex Alimentarius Commission, MRLs represent the maximum allowable concentration of pesticide residue (measured in milligrams per kilogram) deemed fit for use in or on food items and animal feeds. Anchored upon evidence derived from sound agricultural practices, foods adhering to these stipulated MRLs are anticipated to be toxically acceptable. Although MRLs themselves do not serve as toxicity indicators, they serve as pivotal benchmarks in trade, sanctioned by both national and international entities such as the Codex Alimentarius, for monitoring hazardous residues in the global food exchange.

Furthermore, concerns of heavy metal pollution reverberate globally, casting a shadow of significant health hazards upon humanity. The genesis of this concern within soil–crop systems is often ascribed to rapid urbanization, shifts in land usage, and industrial proliferation, particularly discernible in densely populated developing nations such as India and China (UN-HABITAT 2004). In a broad sense, heavy metals encompass metallic elements surpassing tolerable density thresholds, encompassing entities such as mercury (Hg), cadmium (Cd), chromium (Cr), nickel (Ni), lead (Pb), copper (Cu), among others (Jabeen et al. 2020; Sankhla et al. 2016). The consumption of commonplace vegetables tainted with heavy metals has been implicated in potential health hazards among humans, encompassing risks of carcinogenesis and renal dysfunction (Table 11.1).

Mycotoxins, resultant from the metabolic activities of filamentous fungi, wield a profound influence on the health of both human and animal populations, thereby precipitating substantial economic ramifications (Moretti et al. 2017). Prominent among these mycotoxins found in diverse foodstuffs are Aflatoxin A, zearalenone, fumonisins, ergot alkaloids, enniatins, patulin, Alternaria toxins, and trichothecenes (TCs) such as deoxynivalenol (DON), T-2, and HT-2 toxins (Agriopoulou 2016; Barac 2018). Fungi belonging to the *Aspergillus*, *Fusarium*, *Penicillium*, *Claviceps*, and *Alternaria* genera serve as primary producers of these mycotoxins (Agriopoulou et al. 2020). Owing to climatic conditions, suboptimal agricultural practices, and inadequate storage facilities, toxigenic fungi are more prevalent in food and feed originating from developing nations. Nonetheless, mycotoxin-contaminated products can disseminate globally via international trade routes (Al-Jaal et al. 2019).

Table 11.1 Studies reported for heavy metal contamination in some popular horticultural crops (Source: Manwani et al. 2022)

Sr no.	Horticultural crops with heavy metals	References
1.	Spinach and tomato (Pb and Cd)	Mohod (2015)
3.	Cauliflower, Chinese cabbage, spinach, lettuce, napa cabbage, Amaranthus, coriander, cowpea, common bean (Pb, Cd, Cu, Zn, and As)	Zhou et al. (2016)
4.	Tomato, bottle gourd, brinjal, Amaranthus, pepper (Cr, Ni, Cu, Zn, As, Cd, and Pb)	Islam and Hoque (2014)
5.	Garden cress, cucumber, spinach (As, Cd, Cr, Pb, and Zn)	Shakya and Khwaounjoo (2013)
6.	Turmeric, nutmeg, fenugreek, thyme, ginger, cardamom, nigella (Pb, Co, and Cd)	Al-Eed et al. (2002); Sharma et al. (2012)
7.	Mango, melon, spinach, banana (Pb, Cu, Zn, Co, Ni, and Cd)	Elbagermi et al. (2012)

Horticultural commodities, encompassing fresh and dried fruits, vegetables, beverages such as wine and grape juice, nuts, coffee, cocoa, spices, and herbs, are susceptible to mycotoxin contamination at all stages of the food supply chain (Sanzani et al. 2016; Gonçalves et al. 2018; Welke 2019; Pascari et al. 2018; Kluczkovski 2019; Bessaire et al. 2019; Huertas-Pérez et al. 2016; Gambacorta et al. 2019). Among these, Aflatoxins, notably Aflatoxin AFB1, are deemed the most perilous, distinguished for their carcinogenic, mutagenic, teratogenic, and immunosuppressive properties (Agriopoulou et al. 2016). AFB1 has earned classification as a Group 1 carcinogen (carcinogenic to humans) by the International Agency for Research on Cancer (IARC) (Anonymous 1993). Owing to the profound public health implications and consequential economic reverberations, a plethora of analytical methodologies have been devised to ascertain the presence of chemical contaminants in both fresh and processed horticultural produce. This chapter endeavors to furnish an up-to-date discourse on advanced techniques for the detection of pesticide residues and other contaminants within horticultural crops.

11.2 Analytical Methods for Pesticide Residues Detection

The escalation in pesticide application across global horticultural domains over recent decades has evoked profound apprehensions regarding its potential long-term ramifications on human health. The imperative to rigorously control and monitor pesticide residues in fruits and vegetables arises as a paramount concern to ensure the well-being of consumers (Wahab et al. 2022). Thus, the availability of accurate, sensitive, and reliable analytical methodologies becomes indispensable in regulating pesticide levels within food items effectively.

A myriad of techniques, spanning from conventional extraction methodologies to sophisticated detection processes, have been devised for the extraction and identification of pesticides present in various fruits and vegetables. Typically, a sequential two-step approach is adopted in the assessment of pesticide residues within these agricultural commodities: initially, the targeted analytes are extracted and purified from the matrix, followed by the identification and quantification of these compounds (Prodhan et al. 2017).

Despite the existence of numerous conventional extraction and detection methods, the absence of universally accepted standards for pesticide extraction under laboratory conditions remains conspicuous (Stocka et al. 2011). This lacuna underscores the necessity for further refinement and standardization in extraction protocols to ensure consistency and reliability across analytical procedures.

11.3 Pretreatment and Extraction Methods

Pesticide residues are typically extracted from samples using solvents to isolate them from the rest of the material. This extraction process follows a systematic procedure designed to release the targeted analyte from the matrix. However, the absence of widely recognized standards for pesticide extraction in laboratory settings remains a notable issue.

Following extraction, a cleanup process, consisting of one or multiple steps, is employed within the analytical procedure to physically or chemically eliminate most of the potential interfering co-extracts subsequent to the extraction of the desired analyte from the matrices (see Fig. 11.1). It is imperative that the extraction method is meticulously devised to ensure the quantitative and efficient removal of pesticides from the matrix, employing clean, readily available, and cost-effective equipment.

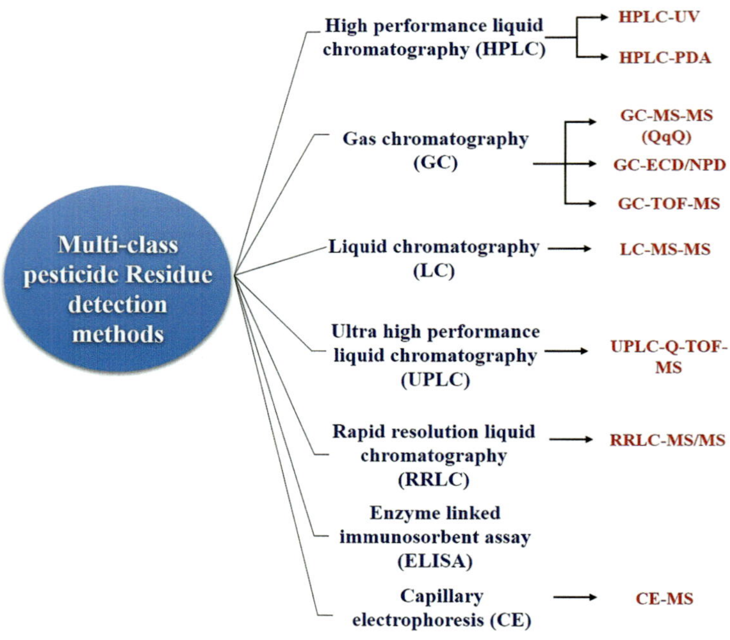

Fig. 11.1 Various detection for pesticidal residues in horticultural crops. Abbreviations: HPLC-UV: HPLC with ultraviolet detector; HPLC-PDA: HPLC with photodiode array detector; GC-MS/MS (QqQ): gas chromatography-tandem mass spectrometry (triple quadrupole); GC-ECD/NPD: gas chromatography with electron capture detector/nitrogen phosphorus detector; GC-TOF-MS: gas chromatography-time-of-flight mass spectrometry; LC-MS/MS: liquid chromatography with tandem mass spectrometry; UPLC-Q-TOF-MS: ultra-high performance liquid chromatography with quadrupole time-of-flight mass spectrometry; RRLC-MS/MS: rapid resolution liquid chromatography tandem mass spectrometry; CE-MS: capillary electrophoresis-mass spectrometry

Various pretreatment and extraction techniques can be utilized to determine pesticide residues in horticultural crops, including but not limited to liquid–liquid extraction (LLE), solid phase extraction (SPE), solid phase microextraction (SPME), matrix solid phase dispersion (MSPD), the quick, easy, cheap, effective, rugged, and safe method (QuEChERS), as well as other alternatives such as gel permeation chromatography (GPC), liquid–liquid microextraction (LLME), and capillary electrophoresis (CE) (Narenderan et al. 2020).

Pesticide residue analysis can be conducted on a variety of substrates, including solids such as soil, meat, and green plant materials (utilizing sorbents), as well as liquids such as water, fruit juices, and body fluids (through partitioning). The selection of the extraction solvent takes into account both the substrate and the type of pesticide. Acetonitrile, ethyl acetate, dichloromethane, methanol, and toluene are among the most commonly employed solvents for assessing pesticides in fruits and vegetables, as indicated by studies. However, it is imperative that the solvent does not undergo chemical modifications or reactions with the pesticide, and it should possess low solubility for co-extracts while demonstrating good solubility for pesticides (Boyapati 2021).

The initial stage of the extraction process involves the preparation of subsamples. The starting material comprises samples weighing between 0.5 and 2 kg, which are initially cleaned and then homogenized using a mixer. Subsequently, for further extraction, homogenized subsamples weighing between 0.5 and 100 g are selected. Acetonitrile, ethyl acetate, dichloromethane, methanol, and toluene are among the solvents commonly utilized for the analysis of pesticide residue in horticultural crops (Narenderan et al. 2020). Occasionally, solvent mixes may be employed to enhance the recovery process. Furthermore, to neutralize the matrices, substances such as sodium hydroxide and acetic acid are used, which contributes to a higher recovery rate (Andrade et al. 2015).

11.3.1 Liquid–Liquid Extraction (LLE)

Liquid–liquid extraction (LLE), also referred to as solvent extraction and partitioning, is a method utilized to separate substances based on their solubility in specific immiscible liquids (Berk 2018). Among the earliest and most reliable techniques for pesticide extraction, LLE is renowned for its flexibility and compatibility with various instruments. Ethyl acetate, recognized as one of the most commonly employed solvents of medium polarity, is frequently utilized for LLE in conjunction with other extraction solvents such as hexane, acetonitrile, chloroform, and 1,2-dichloromethane (Cho et al. 2013).

In a study by Wang et al. (2012), the LLE approach was employed to identify seven neonicotinoid insecticides using a mixture of chloroform and dichloromethane. The method exhibited a recovery range of 76–123% and achieved limits of detection (LODs) ranging from 0.002 to 0.005 mg/kg, with a correlation coefficient (R^2) ≥0.99. Grimalt et al. (2010) utilized another LLE approach to simultaneously detect 11 pesticides from various classes, employing a methanol–water solution for extraction followed by filtration through nylon syringe filters before analysis.

Goulart et al. (2012) optimized and confirmed the liquid–liquid extraction with low-temperature partitioning to assess aldicarb, carbofuran, and carbaryl in chocolate milk beverages and grape juice using HPLC coupled with a UV–Vis detector. Bedassa et al. (2017) and Amir et al. (2022) investigated the quantification of atrazine, ametryn, terbutryn, carbaryl, and chlorothalonil in beer, wine, and Ethiopian honey wine using LLE in combination with HPLC equipped with a UV–Vis detector, demonstrating the efficacy of the suggested method for sample preparation.

Araoud et al. (2007) evaluated a multiresidue approach for the measurement of numerous pesticides in four fruits and vegetables using gas chromatography/mass spectrometry, employing LLE and solid phase extraction (SPE) to remove pesticides. The majority of pesticide recoveries from all four crops exceeded 60%, with estimated limits of detection and quantification ranging between 0.01 and 0.1 mg/kg and 0.02 and 0.3 mg/kg, respectively.

Similarly, Cho et al. (2008) reported the extraction and quantification of iprodione, chlorpyrifos-methyl, EPN, and endosulfan (with its metabolites) from kiwi fruit using the LLE method in conjunction with GC–MS, yielding recoveries ranging from 75.6% to 127.1%.

Despite its efficacy, LLE procedures have notable drawbacks, including the utilization of hazardous solvents, prolonged analysis times, challenges in automation, and reduced effectiveness against polar molecules. Liquid–liquid extraction is a complex and time-consuming process, as it extracts all molecules from the matrix without selectivity to specific analytes, resulting in significant matrix interactions.

11.3.2 Solid-Phase Extraction (SPE)

Solid phase extraction (SPE) emerges as the most frequently employed method due to its simplicity, rapidity, and ability to handle a large number of samples while ensuring high recovery rates. Various SPE cartridges, such as Envicarb, C18, and florisil columns, are commonly utilized for the pretreatment and detection of pesticide residues in horticultural crops (Liu et al. 2010; Balinova et al. 2007). Common sorbents such as graphitized carbon black (GCB) and primary secondary amine (PSA) are predominantly used for residual estimation (Bakırcı et al. 2014). Occasionally, to enhance sensitivity, sorbents such as PSA-GCB-C18 may be combined during the cleanup process. Multiwalled carbon nanotubes (MWCNTs), known for their efficiency, are employed alongside aforementioned sorbents in pesticide extraction. Fan et al. (2014) utilized MWCNTs as an SPE sorbent to extract 36 pesticide residues from cauliflower and spinach, achieving LODs in the range of 0.1–5 g/kg with recovery rates between 57% and 108%.

A recent innovation involves the use of magnetic nanoparticles (MNPs) in SPE, garnering attention due to their exceptional selectivity and capacity to yield high analyte recovery rates, particularly from samples with substantial volume (Wierucka and Biziuk 2014; Wahab et al. 2021). In a study by Eticha (2020), a solid phase extraction adsorbent based on multiwalled carbon nanotubes was developed to remove organophosphorus pesticides from fruit juices.

For the quantitative determination of 88 pesticide residues in berry fruits, Yang et al. (2011) employed solid phase extraction (SPE) utilizing Envi-Carb cartridges coupled with NH2-LC cartridges, achieving remarkable recoveries ranging from 63% to 137% at fortification levels of 0.05–0.5 mg/kg. Similarly, Zhou et al. (2023) compared dispersive solid phase extraction with multi-pesticide residues analysis in jasmine flower and its aromatic tea, demonstrating recoveries ranging from 64% to 108% for 25 pesticides.

Organic solvents, either alone or in combination, are commonly employed for the extraction and elution of pesticide residues, necessitating the selection of an appropriate solvent based on the molecular properties (ionic and nonionic) of the pesticide under investigation. A wide array of solvents, including acetonitrile, methanol, ethyl acetate, dichloromethane, acetone, acetic acid, hexane, toluene, petroleum ether, cyclohexane, and diethyl ether, have been utilized for this purpose. Despite offering enhanced separation and recovery from complex matrices, SPE approaches can be time-consuming and may result in low recovery due to sorbent–analyte interactions and cartridge clogging with suspended debris from the samples.

11.3.3 Solid Phase Microextraction (SPME)

Solid phase microextraction (SPME), introduced by Arthur and Pawliszyn in (1990), stands out as a straightforward sample preparation technique offering several notable advantages. This method is solvent-free, rapid, portable, and applicable to the analysis of both volatile and nonvolatile pesticides. SPME has proven effective in combination with both gas chromatography (GC) and liquid chromatography (LC). While interfacing with LC has posed challenges, coupling with GC has been relatively straightforward and requires minimal modification of existing equipment (Abdulra'uf et al. 2012).

At the core of SPME lies the partitioning of analytes between the sample matrix and a phase immobilized on the SPME fiber. In a study by Abdulra'uf and Tan (2015), a method was outlined for determining multiclass pesticide residues in fruits and vegetables using headspace-solid phase microextraction coupled with gas chromatography–mass spectrometry (GC–MS). Similarly, Kin and Huat (2010) utilized headspace solid phase microextraction with GC-electron capture detector (ECD) to detect pesticide residues in strawberry and cucumber samples. Guillet et al. (2009) employed an analytical approach combining focused microwave-assisted extraction (FMAE) and solid phase microextraction (SPME) to identify 25 pesticides used in tomato cultivation.

For the detection of organophosphorus pesticides in fruit (pears) and fruit juice samples, Simplício and Boas (1999) developed an SPME method combined with GC-flame photometric detector in phosphorous mode. The method demonstrated limits of detection below 2 μg/kg for all pesticides in fruit and fruit juice matrices, with relative standard deviations for analyses performed in triplicate on fortified samples not exceeding 8.7%.

Despite its advantages, SPME does have notable limitations, including challenging sample carryover, fragility of the fiber, and a relatively short extraction lifetime. These shortcomings should be considered when employing SPME for pesticide residue analysis.

11.3.4 Matrix Solid Phase Dispersion (MSPD)

The MSPD method, initially described by Barker et al. (1989), serves as a technique for extracting solid and semisolid materials. This method integrates extraction and cleanup into a single phase, streamlining the process while minimizing sample loss and solvent usage. However, implementing the MSPD approach requires careful adjustment of several variables, including sample size, dispersant type and quantity, and elution composition.

In a study by Guan et al. (2011), nine organophosphorus pesticides (OPPs), including methamidophos, monocrotophos, and malathion, were isolated from eight different fruit and vegetable samples using MSPD. The proposed method, combining MSPD with rapid resolution liquid chromatography-tandem mass spectrometry (RRLC-MS/MS), exhibited recovery rates with a relative standard deviation (RSD) ranging from 2.0% to 11.8% over a range of 71.2–102.8%, with LODs below 0.2 g/kg.

Dos Santos et al. (2019) developed the ultrasound-assisted matrix solid phase dispersion (UA-MSPD) method, employing sand as a solid support, for extracting various pesticide classes from fruits and vegetables. Recovery rates between 55 and 140% with an RSD under 20% were achieved, with limits of quantification (LOQs) ranging from 0.005 to 0.5 mg/kg. In strawberry samples, levels of thiamethoxam, captan, chlorpyrifos, dimethoate, and pyrimethanil were found to range from 0.01 to 0.06 mg/kg. Similarly, in a sample of tomato, acephate and tebuconazole were detected at quantities of 0.45 and 0.30 mg/kg, respectively.

Recently, Yuan et al. (2023) detected nine pesticide residues in various fruits and vegetables using an MSPD method with *n*-octadecylamine functionalized mesoporous carbon nanocomposite (ODA-MPC) as the sorbent. The study reported relative standard deviations smaller than 5% and recoveries between 85.0 and 95.8% ($n = 3$).

11.3.5 Quick, Easy, Cheap, Effective, Rugged, and Safe Method (QuEChERS)

In 2003, a pioneering method combining acetonitrile extraction and dispersive-solid phase extraction (d-SPE) cleaning was introduced, named QuEChERS, signifying "Quick, Easy, Cheap, Effective, Rugged, and Safe." This technique gained widespread acceptance owing to its microscale extraction process, which is simpler, faster, and employs organic solvents, diverging from traditional methods. Acetonitrile, renowned for its high recovery rates, is commonly employed in the extraction process. Initially, liquid–liquid extraction (LLE) and solid phase extraction (SPE) were favored for extraction. However, the QuEChERS method's adoption has surged in the past decade due to its microscale extraction procedure (Lehotay et al. 2010).

The extraction of organic compounds from diverse matrices, such as food, biological, and environmental samples, typically demands significant time and resources. However, the QuEChERS method streamlines this process by employing fewer reagents in smaller quantities, minimizing analysis steps, and yielding higher recovery rates. Setting the pH value at approximately 5, the QuEChERS technique has facilitated efficient extraction of pesticide residues from various matrices.

In a study by Jallow et al. (2017), the presence of pesticide residues in popular fruits and vegetables consumed in Kuwait was investigated. Employing QuEChERS multi-residue extraction followed by GC/MS or liquid chromatography-tandem mass spectrometry (LC/MS), 150 samples were tested for 34 pesticides. Results indicated that 21% of the samples contained pesticide residues exceeding maximum residue limits (MRL), while 79% either had no residues or residues below the MRL.

More recently, Deveci et al. (2023) utilized liquid chromatography-tandem mass spectrometry (LC-MS/MS) to analyze 260 pesticides in fruit samples from Turkey's Tokat area. Mean recovery values for the pesticides ranged from 70.3% to 119%, with a relative standard deviation of 20%. Linearities (R2), limits of detection, limits of quantification, and mean recovery values fell within specific ranges. Notably, residues of permethrin in cherries, diflubenzuron in apples, and dimethoate in pears, apples, and peaches exceeded EU-MRL. Health risk assessments indicated the acute and chronic toxicity potential of omethoate and dimethoate, highlighting the necessity of reducing pesticide usage in the research area to mitigate health hazards.

11.3.6 Other Methods

In laboratory settings, gel permeation chromatography (GPC), also known as size exclusion chromatography, occasionally serves as an alternative method for pesticide residue extraction. Originating in the 1970s, GPC emerged as a robust cleanup technique for pesticides. This process relies on molecular size separation, wherein larger molecules are eluted first, followed by smaller ones. Recognized for its efficacy in multi-residue pesticide detection, GPC is often recommended for cleaning up extracts from biological samples. For instance, in a study examining the concentration of 100 pesticides in 240 samples of fruits and vegetables, researchers utilized GPC, employing ethyl acetate–cyclohexane as the eluent (Knežević and Serdar 2009). However, GPC comes with its drawbacks, including the requirement for specialized equipment, high cost, and involvement of multiple steps.

Dispersive liquid–liquid microextraction (DLLME) represents another microextraction technique employed for pesticide detection. As reported by Shamsipur et al. (2016), this method is utilized for detecting pesticides in various samples such as fruit juice, water, milk, and honey, often coupled with solid phase extraction (SPE). The detection threshold ranges between 0.5 and 1.0 ng/kg.

11.4 Pesticide Residue Detection Strategies

Developing effective techniques for detecting pesticides in real samples poses challenges due to matrix interference. In recent years, gas chromatography (GC) and liquid chromatography (LC), along with advanced methods such as gas chromatography–mass spectrometry (GC–MS), ultra-high-performance liquid chromatography-tandem mass spectrometry (UHPLC-MS/MS), and liquid chromatography–mass spectrometry (LC–MS), have emerged as the most favored strategies for detecting and quantifying pesticides in horticultural crops. Additionally, capillary electrophoresis (CE) and enzyme-linked immunosorbent assay (ELISA) are among the alternative techniques employed. Figure 11.1 summarizes various detection methods utilized to assess the presence of pesticides in horticultural crops.

11.4.1 Gas Chromatography (GC)

Gas chromatography (GC) has proven to be an excellent separation technique in various studies involving volatile, rapidly vaporized insecticides. Pesticides have been effectively detected using GC in combination with a range of sensitive detectors. Examples of such detectors include electron capturing detectors (ECD), flame photometric detectors (FPD), nitrogen phosphorus detectors (NPD), and mass selective detectors (MSD) (Łozowicka et al. 2017).

In a particular investigation, for instance, the flame photometric detector (FPD) was employed to identify 11 organophosphorus pesticide residues in samples of mustard kale and cabbage (Sapahin et al. 2019).

A study monitoring 381 individual pesticides in grapes by GC/MS-MS found mass spectrometer detectors (MS and tandem MS) to be equally favored solutions for pesticide analysis (Balinova et al. 2006). Similarly, gas chromatography with electron capture detector (GC-ECD) was developed to simultaneously determine specific electronegative pesticides in fruits and vegetables with high water content (Tankiewicz 2019). The study reported limits of detection (LODs) in the range of 5.6–15 μg/kg and limits of quantification (LOQs) in the range of 17–45 μg/kg for all pesticides under investigation. Within the studied concentration range up to 10 mg/kg, good reproducibility and stability were demonstrated, with relative standard deviations (RSDs) under 10%.

Organochlorine and organophosphorus pesticides were identified in vegetable samples by Abbasalizadeh et al. (2023) using GC-FID following dispersive solid phase extraction based on CNO/MOF nanocomposites and liquid microextraction. Precision was assessed using repeated intra-day and inter-day ($n = 5$) studies, which exhibited RSD values less than 6.9%.

Additionally, Kalojeh et al. (2023) employed GC-FID to identify several pesticides in fruit juices, achieving detection and quantification limits of 0.22–0.72 and 0.73–2.40 ng/mL, respectively. The depicted calibration curves were reported to be linear in the range of 0.73–1000 μg/L for all analytes.

To enhance method sensitivity, mass detection techniques such as ion trap (IT), quadrupole, triple quadrupole (QqQ), and time of flight mass analyzer (TOF) have been utilized (Tao et al. 2009; de Oliveira et al. 2012; Wu 2017; Cervera et al. 2012).

While gas chromatography (GC) has traditionally been the preferred method for separating volatile insecticides, its usage has declined in the past decade due to the rising prevalence of polar pesticides. These pesticides, characterized by high toxicity and low persistence, are unsuitable for GC detection methods due to their volatility and low thermal stability.

Recently, Tankiewicz and Berg (2022) employed the QuEChERS method coupled with GC-MS/MS to estimate 31 pesticides, including acaricides, fungicides, and herbicides, in various fruit and vegetable samples. Lower limits of quantification (LOQs) were determined to be between 0.63 μg/kg for fenazaquin and 39 μg/kg for imazalil, with a coefficient of variation (CV) under 6%. Recoveries ranged from 82% (etofenprox) to 112% (captan) at two distinct spiking levels.

In a similar study, Kottadiyil et al. (2023) detected and quantified 52 pesticides in samples of commonly consumed vegetables and fruits gathered from three different regions of Gujarat state, India. Bifenthrin, fenvalerate, and cypermethrin were among the most frequently detected pesticides, with the majority of samples found to be below the maximum residue limit (MRL).

Deveci et al. (2023) detected 363 pesticides in leafy vegetables (dill, rocket, and parsley) using QuEChERS coupled with LC-MS/MS and GC-MS/MS. Approximately 35% of the samples had no measurable residues recorded, while 130 green leafy vegetables included 43 residues from 24 distinct chemical classes. Pendimethalin, diuron, and pymetrozine were among the most frequently discovered pesticides, with some samples exceeding EU maximum residue levels (EU MRLs).

11.4.2 Liquid Chromatography (LC)

Liquid chromatography-based methods for analyzing pesticide residues often incorporate detectors such as UV, photodiode array (PDA), diode array detector (DAD), and mass spectrometry (MS). The most common stationary phase for chromatographic separation is octadecyl (C18), and gradient mode is often used for multiresidue analysis to shorten runtime.

Abd Al-Rahman et al. (2012) developed a method using HPLC-DAD analysis to estimate the degradation rate of the acaricide fenpyroximate in apple, citrus, and grape. Similarly, Kandeh et al. (2022) detected eight particular pesticides in food and vegetable samples using poly(vinyl alcohol)/Stevia/MIL-88A@Au nanoparticles composite nanofibers.

Boontongto and Burakham (2021) employed magnetic dual-template molecularly imprinted polymer (DMIP) sorbent extraction with HPLC-PDA detection to quantify organophosphorus pesticides in fruits and vegetables.

Wahab et al. (2022) identified 60 different pesticides in vegetables and fruits using time-of-flight mass spectrometry (TOF-MS) and ultra-high performance liquid chromatography (UHPLC). Martins et al. (2017) used tandem mass spectrometry (UHPLC-MS/MS) and UHPLC to identify 21 pesticides in sweet pepper and tomato samples.

Liquid chromatography coupled with mass detection, particularly tandem mass spectrometry (MS/MS), has been employed to overcome challenges in providing structural information for pesticide residues in fruits and vegetables. Reverse phase columns (C-8, C-12, and C-18), organic phases (acetonitrile and methanol), and buffers (formic acid, ammonium formate, acetic acid, and ammonium acetate) are often utilized in these studies (Christia et al. 2015; Jallow et al. 2017; Stachniuk et al. 2017; Aslantas et al. 2023).

Electrospray ionization (ESI) is frequently used as an ionization source in MS detection due to its ability to ionize both polar and nonpolar analytes (Dzuman et al. 2015). Mass analyzers such as the triple quadrupole (QqQ) and Q-Trap are commonly employed for qualitative and quantitative analysis (Rong et al. 2017; Crnogorac et al. 2008). Additionally, single quadrupoles have been utilized for analyzing pesticide residues using multi-residue techniques (Bakırcı and Hışıl 2012).

Various studies have utilized LC-MS/MS for analyzing pesticide residues, including the estimation of organophosphorus pesticides in fruit and vegetables by Guan et al. (2011) and the simultaneous measurement of penflufen and its metabolite in vegetables and grains by Tian et al. (2016).

Ultra-performance liquid chromatography (UPLC) has emerged as a new method for analyzing pesticides in horticultural crops due to its chromatographic efficiency and sensitivity, as reported in recent studies (Sivaperumal et al. 2015; de Castro et al. 2023; Chen et al. 2023; Wang et al. 2022).

11.4.3 Other Detection Methods

CE (capillary electrophoresis) and ELISA (enzyme-linked immunosorbent assay) are considered fast and cost-effective techniques for pesticide analysis (Table 11.2).

ELISA relies on antigen–antibody interaction and offers high sensitivity and selectivity for specific pesticides. For instance, Yang et al. (2008) used ELISA to detect carbofuran in various matrices, achieving high average recoveries. Similarly,

Table 11.2 Compilation of selected studies investigating different methodologies for the extraction, pretreatment, and detection of pesticide residues in horticultural crops

Sr. no.	Name of the extraction method	Matrix used	Group of pesticide	Recovery (%)	Detection method employed	Reference
1.	Liquid–liquid extraction (LLE)	Apples, grapes, pears, cherry, tamarillos, cucumbers, tomatoes, papaya, and broccoli	Dithiocarbamate fungicide (DMDs, EBDs, and PBDs)	97–101	LC-MS/MS	Crnogorac et al. (2008)
		Green beans	Multiclass pesticides	90–110	LC-MS/MS	Afify et al. (2012)
		Bean sprouts	Fungicide and plant growth regulator	80.4–96.3	LC–ESI-MS/MS	Cho et al. (2013)
		Orange peel and flesh, banana peel and flesh, strawberry, tomato, grape fruit, cucumber, and pepper	Multiclass pesticides	70–110	UPLC-Q-TOF-MS	Grimalt et al. (2010)
		Grape	Carbamates	>90	HPLC-UV	Goulart et al. (2012)
		Cauliflower, onions, tomatoes, potatoes, spinach, cucumber, kiwis, apples, grapes, pineapple, peaches, and strawberries	Organochlorines (Aldrin) and organophosphorus (Malathion)	65–100	HPLC-UV	Batool (2022)
2.	Solid phase extraction (SPE)	Peaches and lettuce	Multiclass pesticides	73–117	GC–MS and GC–ECD	Balinova et al. (2007)
		Rape, cauliflower, and leek	Organophosphorus pesticides (trichlorfon and monocrotophos)	88.5–94.2	HPLC-UV	Wang et al. (2014)
		Berry fruits, raspberry, strawberry, blueberry, and grape	Multiclass pesticides	63–137	HPLC-PDA	Yu and Yang (2017)
		Brinjal, cauliflower, cabbage, okra, potato, onion, apple, banana, grape, guava, mango, pomegranate, and orange	Multiclass pesticides	74–111	UHPLC/TOF-MS	Sivaperumal et al. (2015)
		Spinach and cauliflower	Multiclass pesticides	57 108	LC MS/MS	Fan et al. (2014)
		Raspberry, strawberry, blueberry, and grape	Multiclass pesticides	63–137	GC–MS	Yang et al. (2011)
		Cucumber, tomato	Multiclass pesticides	91.8–99.1	GC–MS	Shirani et al. (2023)
		Tomato, apple, and pear	Triazole fungicide	81.0–119	GC-FID	Sun et al. (2022)

(continued)

Table 11.2 (continued)

Sr. no.	Name of the extraction method	Matrix used	Group of pesticide	Recovery (%)	Detection method employed	Reference
3.	Solid phase microextraction (SPME)	Apple, tomato, cucumber, and cabbage	Multiclass pesticides	73–118	GC–MS	Abdulra'uf and Tan (2015)
		Peach, cucumber, and cabbage	Pyrethroid	85–103.5	GC-μECD (GC-microelectron capture detector)	Zhang et al. (2017)
		Strawberry and cucumber	Organophosphorus (diazinon, malathion, chlorpyrifos, quinalphos, and profenofos) and organochlorine (chlorothalonil, α-endosulfan, and β-endosulfan)	77.3–93.3	GC-ECD	Kin and Huat (2010)
		Pears	Organophosphorus pesticides (diazinon, fenitrothion, fenthion, quinalphos, triazophos, phosalon, and pyrazophos)	75.9–102.6	GC-FPD (GC with flame photometric detector in phosphorous mode)	Simplı́cio and Boas (1999)
		Strawberry juice, lemon juice, orange juice, tomato juice, and honey	Abamectin and amitraz	87.4–110.0	HPLC–UV	Hasani et al. (2022)
4.	Matrix solid phase dispersion (MSPD)	Apple, grape, strawberry, tomato, cabbage, spinach, rape, and banana	Organophosphorus pesticides	71.2–102.8	LC-MS/MS	Guan et al. (2011)
		Collard greens, tomato, strawberry, lettuce, and green pepper	Multiclass pesticides	55–140	GC–MS and LC-MS/MS	Dos Santos et al. (2019)
		Apple, peach, pear, and plum	Multiclass pesticides	72–113	UHPLC-MS/MS	Kemmerich et al. (2020)
		Lettuce, cucumber, carrot, tomato, pepper, shallot, Chinese flowering cabbage, and cabbage	Multiclass pesticides	71.0–110.8	UHPLC-MS/MS	Liu et al. (2021)
		Eggplant, capsicum, apple gourd, cauliflower, and sponge gourd	Multiclass pesticides	88.5–116.9	HPLC-UV	Mujahid et al. (2022)

5.	Quick, easy, cheap, effective, rugged, and safe method (QuEChERS)	Apple, tomatoes, carrot, oranges, and olives	Multiclass pesticides	70–120	GC-TOF-MS	Cervera et al. (2012)
		Tomato, eggplant, bell pepper, zucchini, cucumber, cabbage, carrot, potato, watermelon, apple, strawberry, and grapes	Multiclass pesticides	85–106	GC–MS and LC-MS/MS	Jallow et al. (2017)
		Strawberries, plums, carrots, green peppers, milk, molasses, alfalfa oats, corn silage, dry pet food, soybean, almonds, and foliage	Multiclass pesticides	70–120	HPLC/MS/MS	Lehotay et al. (2010)
		Potato, celery, cauliflower, cabbage, garlic, broccoli, peas, leek, carrot, ginger, lettuce, plum, baby apple, peach, strawberry, pear, and passion fruits	Organophosphate pesticide	76.8–110.30	GC–MS and LC-MS/MS	Narenderan et al. (2019)
		Apples, carrots, pears, red beets, courgettes (zucchinis), cauliflowers, broccoli, kohlrabi, tomatoes, cucumbers, pumpkins, parsley, celery, leeks, peppers, rhubarbs, radishes, and white radishes	Multiclass pesticides	82–112	GC-MS/MS	Tankiewicz and Berg (2022)
		Pepper	Multiclass pesticides	74–100.6	LC-MS/MS and GC-MS/MS	Yıldırım and Çiftçi (2022)
		Corn, grapes, and alfalfa	Multiclass pesticides	70–120	LC-MS/MS	García-Vara et al. (2023)
6.	Gel permeation chromatography (GPC)	Kidney beans and cucumber	Organophosphate pesticide (trichlorfon, omethoate, and monocrotophos)	78.8–106.3	Biomimetic immunoassay-capillary electrophoresis	Li et al. (2017)
		Watermelons, muskmelons, melons, cucumbers, carrots, lettuces, tomatoes, and cabbages	Multiclass pesticides	81–136	GC–MS	Hadian et al. (2019)
		Cucumber, pear, orange, apple, pepper, lettuce, and tomato	Multiclass pesticides	80.9–112.6	GC–MS	Wang et al. (2019)

Navarro et al. (2013) employed duplex ELISA to detect OPPs (organophosphorus pesticides) in tangerine juice samples, demonstrating high sensitivity and recoveries.

However, ELISA's primary limitation is the instability of antibodies and inadequate blocking of immobilized antigens, which can lead to inaccurate results.

CE achieves excellent separation efficiency in a short time with minimal reagent and sample volume requirements. For example, Yue et al. (2022) developed an indirect competitive ELISA and a competitive lateral-flow immunochromatographic strip for detecting methyl parathion and triazophos in fruits and vegetables. Liu et al. (2023) reported the detection of methomyl in cabbage and tomato samples using ELISA and immunochromatographic assay strips. Li et al. (2017) utilized CE with a biomimetic immunoassay (BICE) to measure trichlorfon residue in vegetable matrices, achieving high recovery rates and low LODs. CE combined with tandem mass spectrometry was employed by Daniel et al. (2015) to detect halosulfuron-methyl residue in tomato and sugarcane juice samples, demonstrating a low detection limit. Carbonell-Rozas et al. (2023) reviewed CE-based methods for determining neonicotinoid residues in complex matrices, highlighting the versatility of CE in pesticide residue analysis.

Overall, a wide range of pretreatment, extraction and detection techniques are utilized for pesticide residue analysis in fruits and vegetables, including CE and ELISA, which offer specific advantages such as high sensitivity, selectivity, and efficiency.

11.5 Mycotoxins

Mycotoxins, produced by certain molds, are toxic substances that can contaminate various food crops, particularly in warm and humid conditions, posing risks to human and animal health if ingested or inhaled (Bryden 2007). These harmful by-products of fungal metabolism have the potential to infiltrate the entire food chain, affecting agricultural and horticultural crops from cultivation to consumption (Rossi et al. 2020; Malir et al. 2023). Their presence fluctuates annually, creating challenges for food safety across all stages of production, including planting, growth, harvesting, processing, and distribution. The Food and Agriculture Organization of the United Nations reports that approximately a quarter of the global food supply is contaminated with varying levels of mycotoxins, highlighting the magnitude of this global concern (Zhou and Tang 2020). The ongoing investigation into mycotoxins reflects their concealed threat to human health, prompting extensive research efforts to understand their toxicity and associated risks (Yang et al. 2020; Alshannaq and Yu 2017). Table 11.3 outlines the major health hazards associated with different classes of mycotoxins.

Table 11.3 Classes of major mycotoxins and their physiological effects

Class of mycotoxin	Detrimental effects	Reference
Aflatoxins	Carcinogenic, acute hepatitis, impaired immune system	IARC (1993); Lye et al. (1995)
Citrinin	Nephrotoxic	Reddy et al. (1991)
Citreoviridin	Neurotoxicity	Bennett and Klich (2003)
Cyclopiazonic acid	Carcinogenic, liver lesions, kodua poisoning, diarrhea, depression hyperesthesia, lactation loss, kidney lesions, weight loss	D'Mello and Macdonald (1997)
Fumonisins	Carcinogenic, hepatotoxic, causative agent in leukoencephalomalacia in horses	Kellerman et al. (1990)
Moniliformin	Carcinogenic weight loss, intestinal hemorrhage, Keshan disease	Meerdink (2002)
Ochratoxins	Carcinogenic, nephrotoxic, hepatotoxic, teratogenic	Maff (1993)
Patulin	Lung and brain hemorrhaging.	McKinley et al. (1982)
Sterigmatocystin	Carcinogenic, diarrhea, liver lesions, lactation loss, kidney lesions	EMAN (2003)
Trichothecenes	Immuno-depressants, gastrointestinal hemorrhaging	D'Mello et al. (1999)
Zearalenone	Estrogenic activity	D'Mello et al. (1999)

11.5.1 Detection Approaches for Mycotoxins Testing

Due to the potential harm of mycotoxins even at low concentrations, accurate and sensitive analysis requires the implementation of appropriate and accredited detection and quantification methods. Alongside conventional mycotoxin extraction techniques using organic solvents, various approaches have been employed for the isolation, separation, and extraction of mycotoxins. These include methods such as quick, easy, cheap, effective, rough, and safe (QuEChERS), liquid–liquid extraction (LLE), solid–liquid extraction (SLE), accelerated solvent extraction (ASE), supercritical fluid extraction (SFE), microwave-assisted extraction (MAE), vortex-assisted low-density solvent–microextraction (VALDS–ME), solid phase extraction (SPE), BSA (bovine serum albumin)-based sample clean-up columns, aptamer-affinity columns (AACs), molecularly imprinted polymers (MIPs), and immunoaffinity columns (IACs) (Al-Jaal et al. 2019; Asema and Parveen 2018; Alshannaq and Yu 2017).

Over time, a range of analytical techniques has been applied for mycotoxin analysis, including thin-layer chromatography (TLC), high-performance liquid chromatography (HPLC) coupled with various detectors (e.g., fluorescence, diode array, UV), liquid chromatography coupled with mass spectrometry (LC–MS), liquid chromatography-tandem mass spectrometry (LC-MS/MS), and gas chromatography-tandem mass spectrometry (GC-MS/MS), with chromatographic methods predominating (Alshannaq and Yu 2017). LC–MS is commonly used for analyzing most extracted samples, and in recent years, there has been a growing interest in the development of LC-MS/MS for simultaneous identification of multiple mycotoxins (Al-Jaal et al. 2019; Bryden 2007). Immunoassay-based approaches such as enzyme-linked immunosorbent assay (ELISA) and lateral-flow devices (LFDs) are particularly useful for rapid mycotoxin analysis, while biosensors provide another valuable tool for mycotoxin detection (Busman 2018; Chung and Lam 2012; Cody et al. 2005; Cunha et al. 2009).

Emerging technologies such as proteomic and genomic techniques, molecular methods, electronic noses, and hyperspectral imaging (HSI) offer novel approaches for detecting and analyzing mycotoxins in foods (D'mello and Macdonald 1997; D'mello et al. 1999; DeBerardis et al. 2018; Demayo et al. 1982). It is crucial to accurately characterize critical procedures such as extraction, purification, and chromatographic separation to minimize matrix effects (Dong et al. 2019). In immunoassay-based methods, samples containing color compounds that have not been adequately pretreated may decrease sensitivity and lead to overestimation of mycotoxin levels due to interference from matrix effects (EMAN 2003). Additionally, matrix component co-elution in LC–MS analysis can either suppress or amplify chromatographic signals, emphasizing the importance of thorough method validation for each matrix/mycotoxin combination (Ezeilo et al. 2020; Dong et al. 2019).

11.5.2 Gas Chromatography

Gas chromatography (GC) is widely utilized for detecting and quantifying mycotoxins in food samples. However, many mycotoxins are nonvolatile and require chemical derivatization prior to GC analysis (Scott 1995). Ghorbal et al. (2023) investigated the effectiveness of indigenous essential oils as a biocontrol agent for ensuring the safety of Algerian date palm fruits, with detection conducted via gas chromatography–mass spectrometry. Similarly, Cunha et al. (2009) employed gas chromatography–mass spectrometry to accurately measure trace levels (10 g/L) of patulin in apple products and quince jam. The study reported recoveries of patulin from samples spiked at levels of 8–50 g/kg ranging between 71% and 89%, with limits of detection and quantification of 0.4 and 1.6 g/kg, respectively. Given its high sensitivity, selectivity, and capability to handle complex matrices, gas chromatography plays a crucial role in pesticide residue measurement, ensuring compliance with regulatory standards and safeguarding both consumers and the environment through food safety and environmental monitoring efforts.

11.5.3 High-Performance Liquid Chromatography (HPLC)

High-performance liquid chromatography (HPLC) is a widely utilized method for segregating, characterizing, and quantifying chemical compounds within mixtures. It operates by propelling a liquid mobile phase through a column containing a stationary phase under high pressure. The various components of the mixture interact differently with the stationary phase, leading to their separation through distinct partitioning or adsorption mechanisms. Efficacy of HPLC in separation hinges on factors such as polarity, size, charge, and affinity between the analyte molecules and the stationary phase. Various types of detectors can be employed in HPLC to monitor the eluting compounds. Commonly used detectors encompass UV–Vis detectors, which gauge the absorption of ultraviolet or visible light by the analytes, and diode array detectors, which furnish spectral information across a spectrum of wavelengths (Chung and Lam 2012). Additionally, other detectors such as fluorescence detectors, refractive index detectors, and mass spectrometers can be utilized based on the specific needs of the analysis.

These methods commonly employ gradient elution, where the organic solvent concentration is increased linearly. Conversely, isocratic elution is seldom utilized (Bagheri et al. 2016). HPLC analytical techniques are particularly effective in separating high-polarity and nonvolatile analytes recovered from samples. In certain scenarios, HPLC can be coupled with various detectors. For instance, UV detectors are typically employed to analyze pyrethroid residues in fruits and vegetables, while MS and tandem MS detectors are commonly used for detecting triadimefon, imidacloprid, diazinon, and malathion in fruit juices such as pineapple, apple, raspberry, orange, and cherry (Timofeeva et al. 2017).

In a different study, a UHPLC-MS/MS method was applied to detect seven mycotoxins (ochratoxin A, patulin, trichothecenes, alternariol, alternariol monomethyl ether, tenuazonic acid, and tentoxin) in various fruit (strawberry) and vegetable (cherry tomato and two types of leafy vegetables: lettuce and pak choi) samples (Dong et al. 2019). All mycotoxins exhibited linear concentration ranges with an R^2 value exceeding 0.999. The respective ranges for the limit of detection (LOD) and limit of quantification (LOQ) were 0.05–3.0 μg/kg and 0.2–10.0 μg/kg. With intra-day and inter-day precision in the ranges of 3.0–6.2% and 4.2–6.1%, respectively, average recoveries ranged from 81.1% to 116%. Similarly, a liquid chromatography-tandem mass spectrometry method was developed using a QuEChERS approach for the extraction of mycotoxins (DeBerardis et al. 2018) for the determination of four Alternaria toxins, namely alternariol, alternariol monomethyl ether, tentoxin, and tenuazonic acid in tomato-based and fruit-based products. High linearity ($R^2 > 0.997$) and precision (RSD% < 10) were reported for all analytes. Tenuazonic acid exhibited good recovery (98.8–108.9%) in both tomato-based and fruit-based products.

11.5.4 Ambient Ionization Mass Spectrometry

Ambient ionization represents a novel advancement in mass spectrometry (MS) ion sources, enabling the rapid ionization of small molecules under ambient conditions. Two prominent techniques, desorption electrospray ionization (DESI) (Takàs et al. 2004; Cooks et al. 2006) and direct analysis in real time (DART) (Cody et al. 2005), have been developed and commercialized based on different principles.

In DESI, analytes are extracted from sample surfaces, and secondary ions are propelled toward the mass analyzer by a fast-moving charged solvent stream angled to the sample surface. On the other hand, DART-MS allows for the analysis of samples at atmospheric pressure, essentially in an open laboratory environment. It is capable of analyzing substances desorbed from surfaces into the environment or deposited/adsorbed on surfaces. Therefore, DART-MS is particularly well-suited for studying the constituents of plant materials and monitoring pesticides on vegetables.

DART-MS relies on a gas-phase ionization process, where ionizing species are generated by a corona discharge in a helium atmosphere. Excited helium atoms subsequently initiate gas-phase processes, leading to the production of reagent ions from atmospheric water vapor or solvent vapor near the surface under study, triggering a chemical ionization process (Gross 2014).

Ambient ionization MS for mycotoxins has been the focus of several reviews (Hajslova et al. 2011, Busman 2018, Li et al. 2013, Zhang et al. 2021). Maragos et al. (2015) utilized both DART-MS and liquid chromatography (LC) to identify patulin on the surface of fruit leathers. They observed that while patulin deposited on the peel or leather surface could be detected, DART-MS lacked the sensitivity required for quantification from the surface of homemade apple leathers. Similarly, Song et al. (2015) employed DART in conjunction with time-of-flight MS to evaluate the potential value of grape seeds from various sources, laying the groundwork for the effective utilization of grape seeds.

11.6 Detection of Heavy Metals

The presence of heavy metals in fruits and vegetables has emerged as a significant concern due to the potential health risks associated with their consumption. Heavy metals are toxic substances that can accumulate in the body over time, leading to various health issues such as organ damage and neurological diseases (Oyugi et al. 2021). The adverse effects of different heavy metals on human health are detailed in Table 11.4. To safeguard public health and ensure food safety, it is crucial to detect heavy metals and pesticide residues in fruits.

According to Reilly (1980), heavy metals are metallic elements characterized by high atomic weights and densities that are substantially greater (at least five times) than those of water. Accurate and precise procedures are required to detect and

Table 11.4 Detrimental effects of heavy metals on human health (Source: Jyothi 2020)

Heavy metals	Effect on health
Chromium (Cr)	Lung disorders (bronchitis, cancer), renal and reproductive system
Lead (Pb)	Serious effect on mental health (Alzheimer's disease), nervous system
Arsenic (As)	Highly affects dermal region (cancer), brain and cardiac problems
Mercury (Hg)	Sclerosis, blindness, Minamata disease, deafness, gastric problems, renal disorders
Cadmium (Cd)	Osteo-related problems, prostate cancer, lung diseases, renal issues

quantify the presence of these pollutants in fruit samples. This detection process typically involves a combination of sample preparation, extraction, and analytical techniques (Tsagkaris et al. 2021).

Techniques such as atomic absorption spectroscopy (AAS), inductively coupled plasma-optical emission spectroscopy (ICP-OES), or inductively coupled plasma-mass spectrometry (ICP-MS) are commonly employed in the analysis of heavy metals. These methods offer high sensitivity and accuracy, enabling the identification and measurement of heavy metal levels in fruit samples.

A comparison of fruit quality parameters and heavy metal concentrations in strawberries grown using organic and conventional methods revealed the presence of heavy metals. Organic and conventional strawberry samples were compared by contrasting the levels of detected heavy metals (Pb, Cd, Cu, and Zn) (Sevinç and Mehmet 2023). For instance, the lead concentration in various fruits and vegetables, including lettuce, spinach, broccoli, and endive, collected from four different regions of the United States, ranged from 0.01 to 0.71 mg/kg (Kleinman 1968). Moreover, lead concentrations in tomatoes grown in pure sludge reached 0.4 mg/kg (dry weight), posing significant risks to human health (Demayo et al. 1982).

11.6.1 Atomic Absorption Spectroscopy

A study was conducted to assess the levels of harmful heavy metals, including lead, cadmium, and copper, in various fruits and vegetables, such as guava, tomato, onion, okra, sorrel, garden eggs, potato, cucumber, carrot, and spinach. Among these, spinach exhibited the highest levels of the metals, with concentrations of 0.2250 ppm for lead and 0.0325 ppm for cadmium. Notably, the concentration of lead was significantly higher ($P < 0.05$) in unwashed samples (ranging from 0.1121 to 0.5010 ppm) compared to washed samples (ranging from 0.0132 to 0.2250 ppm). Similarly, cadmium concentration ranged from 0.0023 to 0.721 ppm in unwashed samples and from 0.0010 to 0.0325 ppm in washed samples. The metal concentrations exhibited significant variations ($P < 0.05$) both between and within different crop samples, indicating the influence of crop types on metal concentration. According to Igwegbe et al. (2013), leafy vegetables generally contained higher ($P < 0.05$) metal contents compared to fruits.

In another investigation, AAS was employed to determine the content of heavy metals, including Cu, Pb, Cd, Hg, As, and Zn, in fruits and vegetables sourced from specific marketplaces (Ezeilo et al. 2020). Given the established norms by the WHO and other international regulators, proper management of heavy metal concentrations in foods is imperative (Rotimi et al. 2018). As an illustration, samples of various jam flavors were collected and analyzed for heavy metal presence, such as Fe, Cu, Zn, and Pb, using AAS. Iron was detected in the highest proportion, with a concentration of 2.3462 ppm. Zinc exhibited a minimum value of 0.0697 ppm in the mixed fruit flavor and a maximum value of 0.3823 ppm for lead in the strawberry flavor. Copper levels were below the detectable range in three varieties, with the strawberry flavor showing a copper concentration of 0.0941 ppm (Asema and Parveen 2018).

Additionally, using AAS and GC, levels of heavy metals and pesticide residues in commonly consumed Nigerian food crops, including Amaranthus leaf, fluted pumpkin, waterleaf, okra, carrot, cucumber, and watermelon, were evaluated. The detected heavy metals included cadmium (Cd), chromium (Cr), arsenic (As), cobalt (Co), lead (Pb), iron (Fe), manganese (Mn), nickel (Ni), and mercury (Hg), while pesticide residues comprised Carbofuran, Aldrin, Chlorpyrifos, g-chlordane, Dichloro-Biphenyl, DDT, Dichlorvos, DDD, Endosulfan, Heptachlor, Isopropylamine, Lindane, HCB, t-nonachlor, and profenofos. The quantities of heavy metals and pesticides varied among the food crops investigated, with most falling below the specified maximum allowed limits, although there were exceptions (Omeje et al. 2021).

11.6.2 ICP-MS

ICP-MS, or inductively coupled plasma-mass spectrometry, is an analytical technique utilized to determine isotopic ratios and elemental compositions across various materials. It integrates mass spectrometry with inductively coupled plasma, resulting in a powerful analytical method. Typically, liquid samples are nebulized or solid samples are laser-ablated to form an aerosol, which is then introduced into the ICP for analysis. An inert gas, typically argon, is used to transport the aerosol into the high-temperature plasma. Within the plasma, the sample undergoes atomization, excitation, and ionization reactions. Subsequently, the ions are extracted from the sample, ionized within the ICP, and introduced into the mass spectrometer for analysis. The mass spectrometer separates and detects the ions based on their mass-to-charge ratio (*m/z*). It consists of three primary components: an ion lens, a mass analyzer, and a detector. The ion lens focuses and accelerates the ions upon entry into the mass analyzer, while the mass analyzer separates the ions according to their *m/z* ratio, enabling the resolution of different isotopes and elements. Commonly employed mass analyzers in ICP-MS include quadrupole, magnetic sector, and time-of-flight (TOF) analyzers.

ICP-MS has been utilized in various studies to screen for heavy metals in different food samples. For example, Zhang and Rui (2007) used ICP-MS to assess heavy metal concentrations in the peel and pulp of Fuji apples, revealing significantly higher levels of heavy metals in the fruit peel compared to the pulp. Similarly, Ul-Haq et al. (2021) employed ICP to detect harmful trace metals in common vegetables, with findings indicating potential health risks associated with cadmium intake, particularly for children. Another study utilized ICP-MS to measure heavy metal levels in leafy vegetables grown in Saudi Arabia, concluding that the vegetables were safe for consumption based on oral reference doses and target hazard quotient values.

In summary, ICP-MS is a valuable tool for analyzing heavy metal concentrations in various food samples, providing crucial information for assessing food safety and potential health risks associated with heavy metal exposure.

11.7 Conclusions

The presence of pesticides, mycotoxins, heavy metals, and other pollutants in food can pose significant health risks to consumers, even at low concentrations. Therefore, continuous monitoring of these contaminants in food is essential to safeguard public health. However, identifying all pesticides directly presents challenges due to limitations in detection sensitivity and interference effects within the food matrix. Thus, it is imperative to refine existing techniques and develop advanced methods to achieve more precise analytical results.

The complexity of legislation, residual limits, and evolving lists of target compounds and matrices complicates the selection of methods and equipment by food laboratories for pesticide residue detection. Striking a balance between cost, accuracy, selectivity, and sensitivity becomes crucial when choosing the optimal sample preparation technique. Developing sample preparation protocols capable of identifying multi-residues in food is vital, given the necessity to analyze several physically and chemically distinct compounds. Despite advancements in chromatographic separation and detection, effective cleaning remains essential for reliable results.

The International Food Standards Organization underscores the importance of effective, eco-friendly, and time-efficient solutions for sample treatment and pesticide residue evaluation in fatty food matrices. Future sample preparation technologies should prioritize reliability, availability, cost-effectiveness, and environmental sustainability. High resolution, selectivity, and speed are vital for meeting increasingly stringent regulatory limits, necessitating the development of new approaches and instruments. Ensuring a quick analytical turnaround time is critical for efficient contaminant residue detection in food matrices. Laboratories require sensitive mass spectrometers with lower cost per analysis, higher throughput, and shorter sample preparation times to meet regulatory limits. These instruments should be compatible with commonly used analytical procedures to minimize errors. Analytical chemistry advancements aim to simplify and reduce sample preparation steps, as well as decrease the size and amount of organic solvents used.

Future chromatographic–mass spectrometric techniques will focus on detecting pesticides at the required low levels. Despite challenges in assembling multi-class multi-residue systems due to the diverse nature and properties of pesticides, there is a growing trend in this direction. In conclusion, detecting pesticides and contaminants in horticultural crops is crucial for ensuring food safety, environmental protection, and promoting sustainable agricultural practices. A multifaceted approach involving innovative technologies, sustainable farming practices, and public awareness initiatives is essential. Prioritizing sustainable detection methods contributes to a safer food supply chain, environmental preservation, and the long-term sustainability of the horticultural industry, promoting a healthier and more sustainable future for agriculture and the well-being of both people and the planet.

References

Abbasalizadeh A, Ghalkhani M, Khosrowshahi EM, Mazani A, Hosseini A, Sohouli E, Ahmadi F (2023) Determination of selected pesticides by GC-FID after CNO/MOF nanocomposites-based dispersive solid phase extraction coupled with liquid microextraction. Diam Relat Mater 137:110087

Abd Al-Rahman SH, Almaz MM, Osama IA (2012) Determination of degradation rate of acaricide fenpyroximate in apple, citrus, and grape by HPLC-DAD. Food Anal Methods 5:306–311

Abdulra'uf LB, Tan GH (2015) Chemometric approach to the optimization of HS-SPME/GC-MS for the determination of multiclass pesticide residues in fruits and vegetables. Food Chem 177:267–273

Abdulra'uf LB, Hammed WA, Tan GH (2012) SPME fibres for the analysis of pesticide residues in fruits and vegetables: a review. Crit Rev Anal Chem 42(2):152–161

Afify AEMM, Attallah ER, El-Gammal HA (2012) A modified multi-residue method for analysis of 150 pesticide residues in green beans using liquid chromatography-tandem mass spectrometry. Adv Food Sci 34(1):24–35

Agriopoulou S (2016) Enniatins: An emerging food safety issue. EC Nutr 3:1142–1146

Agriopoulou S, Koliadima A, Karaiskakis G et al (2016) Kinetic study of aflatoxins' degradation in the presence of ozone. Food Control 61:221–226

Agriopoulou S, Stamatelopoulou E, Varzakas T (2020) Advances in occurrence, importance, and mycotoxin control strategies: prevention and detoxification in foods. Foods 9(2):137

Al-Eed MA, Assubaie FN, El-Garawany MM et al (2002) Determination of heavy metal levels in common spices. J Appl Sci 17:87–98

Al-Jaal B, Salama S, Al-Qasmi N et al (2019) Mycotoxin contamination of food and feed in the Gulf Cooperation Council countries and its detection. Toxicon 171:43–50

Alshannaq A, Yu JH (2017) Occurrence, toxicity, and analysis of major mycotoxins in food. Int J Environ Res Public Health 14(6):632

Amir M, Zafar A, Ahmad R, Ahmad W et al (2022) Quality control standardization, contaminant detection and in vitro antioxidant activity of *Prunus domestica* Linn. Fruit. Plants (Basel) 11:706

Andrade G, Monteiro S, Francisco J et al (2015) Liquid chromatography-electrospray ionization tandem mass spectrometry and dynamic multiple reaction monitoring method for determining multiple pesticide residues in tomato. Food Chem 175:57–65

Anonymous (1993) IARC Monographs on the Evaluation of Carcinogenic Risks to Humans. International Agency for Research on Cancer, p 56

Araoud M, Douki W, Rhim A et al (2007) Multiresidue analysis of pesticides in fruits and vegetables by gas chromatography-mass spectrometry. J Environ Sci Health Part B 42(2):179–187

Arthur CL, Pawliszyn J (1990) Solid phase microextraction with thermal desorption using fused silica optical fibres. Anal Chem 62(19):2145–2148

Asema SUK, Parveen N (2018) Study of heavy metal content by AAS in a variety of flavours of jam samples and its physicochemical characterization. Int J Sci Res Sci Eng Technol 4:1259–1261

Aslantas S, Golge O, González-Curbelo MÁ et al (2023) Determination of 355 pesticides in lemon and lemon juice by LC-MS/MS and GC-MS/MS. Foods 12(9):1812

Bagheri H, Yamini Y, Safari M et al (2016) Simultaneous determination of pyrethroids residues in fruit and vegetable samples via supercritical fluid extraction coupled with magnetic solid phase extraction followed by HPLC-UV. J Supercrit Fluids 107:571–580

Bakırcı GT, Hışıl Y (2012) Fast and simple extraction of pesticide residues in selected fruits and vegetables using tetrafluoroethane and toluene followed by ultrahigh-performance liquid chromatography/tandem mass spectrometry. Food Chem 135(3):1901–1913

Bakırcı GT, Acay DBY, Bakırcı F et al (2014) Pesticide residues in fruits and vegetables from the Aegean region, Turkey. Food Chem 160:379–392

Balinova AM, Mladenova RI, Shtereva DD (2006) Effects of processing on pesticide residues in peaches intended for baby food. Food Addit Contam 23:895–901

Balinova A, Mladenova R, Shtereva D (2007) Solid-phase extraction on sorbents of different retention mechanisms followed by determination by gas chromatography-mass spectrometric and gas chromatography-electron capture detection of pesticide residues in crops. J Chromatogr A 1150(1–2):136–144

Barac A (2018) Mycotoxins and human disease. In: Presterl I (ed) Clinically relevant Mycoses. Springer, Cham, pp 213–225

Barker SA, Long AR, Short CR (1989) Isolation of drug residues from tissues by solid phase dispersion. J Chromatogr A 475(2):353–361

Batool Z (2022) Estimation of toxic concentrations of agrochemicals (aldrin, malathion) in food products by using HPLC and liquid- liquid extraction (LLE) method. Int Res J Mod Eng Technol Sci 4(9):2643–2656

Bedassa T, Megersa N, Gure A (2017) Salting-out assisted liquid-liquid extraction for the determination of multiresidue pesticides in alcoholic beverages by high performance liquid chromatography. Sci J Anal Chem 5:38–45

Bennett JW, Klich M (2003) Mycotoxins. Clin Microbiol Rev 16(3):497–516

Berk Z (2018) Chapter 9-centrifugation. In: Food process engineering and technology. Academic Press, San Diego, CA, pp 243–259

Bessaire T, Perrin I, Tarres A, Bebius A et al (2019) Mycotoxins in green coffee: occurrence and risk assessment. Food Control 96:59–67

Boontongto T, Burakham R (2021) Eco-friendly fabrication of a magnetic dual-template molecularly imprinted polymer for the selective enrichment of organophosphorus pesticides for fruits and vegetables. Anal Chim Acta 1186:339128

Boyapati T (2021) Review on methods for analysis/estimation of/estimation of pesticide residue in food. Int J Creat Res Thoughts 8:2320–2882

Bryden WL (2007) Mycotoxins in the food chain: human health implications. Asia Pac J Clin Nutr 16(1):95–101

Busman M (2018) Quantitation of mycotoxins using direct analysis in real time mass spectrometry (DART-MS). J AOAC Int 101(3):643–646

Cabras P, Angioni A, Garau VL et al (1997) Persistence of insecticide residues in olives and olive oil. J Agric Food Chem 45:2244–2247

Carbonell-Rozas L, Lara FJ, García-Campaña AM (2023) Analytical methods based on liquid chromatography and capillary electrophoresis to determine neonicotinoid residues in complex matrices. A comprehensive review. Crit Rev Anal Chem:1–29. https://doi.org/10.1080/10408347.2023.2186700

Cervera MI, Portolés T, Pitarch E et al (2012) Application of gas chromatography time-of-flight mass spectrometry for target and non-target analysis of pesticide residues in fruits and vegetables. J Chromatogr A 1244:168–177
Chen Y, Gao D, Wu Y et al (2023) Determination of the dissipation dynamics and terminal residue of bupirimate and its metabolites in cucumber by QuEChERS-based UPLC-MS/MS. ACS Omega 8:23975–23981
Cho SK, Abd El-Aty AM, Jeon HR et al (2008) Comparison of different extraction methods for the simultaneous determination of pesticide residues in kiwi fruit using gas chromatography-mass spectrometry. Biomed Chromatogr 22(7):727–735
Cho SK, Abd El-Aty AM, Park KH et al (2013) Simple multiresidue extraction method for the determination of fungicides and plant growth regulator in bean sprouts using low temperature partitioning and tandem mass spectrometry. Food Chem 136(3-4):1414–1420
Chowdhury MAZ, Banik S et al (2012) Organophosphorus and carbamate pesticide residues detected in water samples collected from paddy and vegetable fields of the Savar and Dhamrai Upazilas in Bangladesh. Int J Environ Res Public Health 9:3318–3329
Christia C, Bizani E, Christophoridis C et al (2015) Pesticide residues in fruit samples: comparison of different QuEChERS methods using liquid chromatography-tandem mass spectrometry. Environ Sci Pollut Res 22:13167–13178
Chung SWC, Lam CH (2012) Development and validation of a method for determination of residues of 15 pyrethroids and two metabolites of dithiocarbamates in foods by ultra-performance liquid chromatography-tandem mass spectrometry. Anal Bioanal Chem 403:885–896
Cody RB, Laramée JA, Durst HD (2005) Versatile new ion source for the analysis of materials in open air under ambient conditions. Anal Chem 77:2297–2302
Cooks RG, Ouyang Z, Takats Z, Wiseman JM (2006) Ambient mass spectrometry. Science 311(5767):1566–1570
Crnogorac G, Schmauder S, Schwack W (2008) Trace analysis of dithiocarbamate fungicide residues on fruits and vegetables by hydrophilic interaction liquid chromatography/tandem mass spectrometry. Rapid Commun Mass Spectrom 22(16):2539–2546
Cunha SC, Faria MA, Fernandes JO (2009) Determination of patulin in apple and quince products by GC-MS using 13C5-7 patulin as internal standard. Food Chem 115(1):52–359
D'Mello JPF, Macdonald AMC (1997) Mycotoxins. Anim Feed Sci Technol 69:155–166
D'Mello JPF, Placinta CM, Macdonald AMC (1999) Fungal toxins as disease elicitors. In: Rose J (ed) Environmental toxicology: current developments. Gordon and Breach, London, pp 253–289
Daniel D, Dos Santos VB, Vidal DTR et al (2015) Determination of halosulfuron-methyl herbicide in sugarcane juice and tomato by capillary electrophoresis-tandem mass spectrometry. Food Chem 175:82–84
de Castro IM, Bastos LHP, dos Anjos MR et al (2023) Multi-residue method of pesticides by UPLC-MS/MS in bivalve mollusks samples as a tool for food quality and safety. Food Sci Technol 43:e116722
de Oliveira MLG, Madureira FD, Aurélio F et al (2012) A multi-residue method for the determination of pesticides in high water content matrices by gas chromatography-single quadrupole mass spectrometry with electron ionisation (EI-GC/MS). Food Addit Contam Part A Chem Anal Control Expo Risk Assess 29(4):657–664
DeBerardis S, Paola D, Montevecchi G et al (2018) Determination of four Alternaria alternata mycotoxins by QuEChERS approach coupled with liquid chromatography-tandem mass spectrometry in tomato-based and fruit-based products. Food Res Int 106:677–685
Demayo A, Taylor MC, Taylor KW et al (1982) Toxic effects of lead and lead compounds on human health, aquatic life, wildlife, plants and live stocks. Crit Rev Environ Cont 12(4):257–305
Deveci B, Golge O, Kabak B (2023) Quantification of 363 pesticides in leafy vegetables (Dill, Rocket and Parsley) in the Turkey Market by Using QuEChERS with LC-MS/MS and GC-MS/MS. Foods 12(5):1034

Dong H, Xian Y, Xiao K, Wu Y et al (2019) Development and comparison of single-step solid phase extraction and QuEChERS clean-up for the analysis of 7 mycotoxins in fruits and vegetables during storage by UHPLC-MS/MS. Food Chem 274:471–479

Dos Santos EO, Gonzales JO, Ores JC, Marube LC et al (2019) Sand as a solid support in ultrasound-assisted MSPD: A simple, green and low-cost method for multiresidue pesticide determination in fruits and vegetables. Food Chem 297:124926

Duan Y, Guan N, Li PP et al (2016) Monitoring and dietary exposure assessment of pesticide residues in cowpea (*Vigna unguiculata* L Walp) in Hainan, China. Food Control 59:250–255

Dzuman Z, Zachariasova M, Veprikova Z, Godula M (2015) Multi-analyte high performance liquid chromatography coupled to high resolution tandem mass spectrometry method for control of pesticide residues, mycotoxins, and pyrrolizidine alkaloids. Anal Chim Acta 863:29–40

Elbagermi MA, Edwards HG, Alajtal AI (2012) Monitoring of heavy metal content in fruits and vegetables collected from production and market sites in the Misurata area of Libya. Int Sch Res Net:827645. https://doi.org/10.5402/2012/827645

EMAN (2003) European Mycotoxin Awareness Network co-ordinated by Leatherhead Food Research Association (UK)

Eticha SA (2020) Review: sample preparation methods for the pesticide residue analysis in food samples. Int J Pharm Chem 6:65

Ezeilo CA, Okonkwo SI, Chibuzor C et al (2020) Determination of heavy metals in some fruits and vegetables from selected market's in Anambra state. ACTA Sci Nutr Heal 4:163–171

Fan S, Zhao P, Yu C, Pan C (2014) Simultaneous determination of 36 pesticide residues in spinach and cauliflower by LC-MS/MS using multi-walled carbon nanotubes-based dispersive solid-phase clean-up. Food Addit Contam Part A Chem Anal Control Expo Risk Assess 31(1):73–82

Fan L, Niu H et al (2015) Factors affecting farmers' behaviour in pesticide use: insights from a field study in northern China. Sci Total Environ 557:360–368

FAO (2022) World Food and Agriculture - Statistical Yearbook. FAO, Rome

Gambacorta L, El Darra N, Fakhoury R et al (2019) Incidence and levels of Alternaria mycotoxins in spices and herbs produced worldwide and commercialized in Lebanon. Food Control 106:106724

García-Vara M, Postigo C, Palma P (2023) Development of QuEChERS-based multiresidue analytical methods to determine pesticides in corn, grapes and alfalfa. Food Chem 405:134870

Ghorbal I, Salah SR, Abdallah MF (2023) Investigating the efficacy of indigenous essential oils as a biocontrol agent for Safe Consumption of Algerian Date Palm (*Phoenix dactylifera*) Fruits. In: 44th Mycotoxin-Workshop, The Society for Mycotoxin Research

Gonçalves BL, Coppa CC, Neeff DV et al (2018) Mycotoxins in fruits and fruit-based products: occurrence and methods for decontamination. Toxin Rev 38:263–272

Goulart SM, Alves RD, De Paula WX et al (2012) Determination of carbamates in beverages by liquid-liquid extraction with low temperature partitioning and liquid chromatography. J Braz Chem Soc 23:1154–1165

Grimalt S, Sancho JV, Pozo ÓJ (2010) Quantification, confirmation and screening capability of UHPLC coupled to triple quadrupole and hybrid quadrupole time-of-flight mass spectrometry in pesticide residue analysis. J Mass Spectrom 45(4):421–436

Gross JH (2014) Direct analysis in real time-a critical review on DART-MS. Anal Bioanal Chem 406:63–80

Guan Sx, Yu Zg, Yu, Hn (2011) Multi-walled carbon nanotubes as matrix solid-phase dispersion extraction adsorbent for simultaneous analysis of residues of nine organophosphorus pesticides in fruit and vegetables by rapid resolution LC–MS–MS. Chromatographia 73:33–41. https://doi.org/10.1007/s10337-010-1840-2*

Guillet V, Fave C, Montury M (2009) Microwave/SPME method to quantify pesticide residues in tomato fruits. J Environ Sci Health B 44(5):415–422

Hadian Z, Eslamizad S, Yazdanpanah H (2019) Pesticide residues analysis in Iranian fruits and vegetables by Gas Chromatography-Mass Spectrometry. Iran J Pharm Res 18(1):275

Hajslova J, Cajka T, Vaclavik L (2011) Challenging applications offered by direct analysis in real time (DART) in food-quality and safety analysis. Trends Anal Chem 30:204–218

Hasani F, Raoof JB, Ghani M (2022) In situ electrodeposition of Cu-BDC metal-organic framework on pencil graphite substrate for solid-phase microextraction of some pesticides. Microchim Acta 189(11):432

Huertas-Pérez JF, Arroyo-Manzanares N, García-Campaña A et al (2016) Solid phase extraction as sample treatment for the determination of ochratoxin a in foods a review. Crit Rev Food Sci Nutr 57:3405–3420

IARC (1993) International agency for research on cancer, vol 56. IARC, Geneva, p 489

Igwegbe AO, Agukwe CH, Negbenebor CA (2013) A survey of heavy metal (lead, cadmium and copper) contents of selected fruit and vegetable crops from Borno State of Nigeria. Int J Eng Sci 2(1):01–05

Islam MS, Hoque MF (2014) Concentrations of heavy metals in vegetables around the industrial area of Dhaka city, Bangladesh and health risk assessment. Int Food Res J 21(6):2121

Isohata E, Toyoda M, Saito Y (1986) Studies on the chemical analysis of mycotoxin (XIII). Determination of aflatoxin B1 in edible oil. Eisei Shikenjo Hokoku 104:138–142. PMID: 3105630

Jabeen F, Aslam A, Salman M (2020) Heavy metal contamination in vegetables and soil irrigated with sewage water and associated health risks assessment. J Agric Environ Sci 22(1):23–31

Jallow MF, Awadh DG, Albaho MS, Devi VY, Ahmad N (2017) Monitoring of pesticide residues in commonly used fruits and vegetables in Kuwait. Int J Environ Res Public Health 14(8):833

Jimenez MS, Velarte R, Castillo JR (2003) On-line emulsions of olive oil samples and ICP-MS multi-elemental determination. J Anal Atom Spectrom 18:1154–1162

Juraske R, Antón A et al (2007) Human intake fractions of pesticides via greenhouse tomato consumption: Comparing model estimates with measurements for Captan. Chemosphere 67:1102–1107

Jyothi NR (2020) Heavy metal sources and their effects on human health. In: Heavy metals-their environmental impacts and mitigation. IntechOpen, London. https://doi.org/10.5772/intechopen.95370

Kalojeh MN, Sabet RH, Farajzadeh MA (2023) Application of ZIF-67 coated by N-doped carbon dots in dispersive solid phase extraction of several pesticides from fruit juices and its combination with dispersive liquid-liquid microextraction followed by GC-FID. J Food Compos 121:105372

Kandeh SH, Amini S, Ebrahimzadeh H (2022) PVA/Stevia/MIL-88A@ AuNPs composite nanofibers as a novel sorbent for simultaneous extraction of eight agricultural pesticides in food and vegetable samples followed by HPLC-UV analysis. Food Chem 386:132734

Kellerman TS, Marasas WFO, Thiel PG et al (1990) Leukoencephalomalacia in two horses induced by oral dosing of fumonisin B1. J Onderstepoort Vet Res 57:269–275

Kemmerich M, Demarco M, Bernardi G et al (2020) Balls-in-tube matrix solid phase dispersion (BiT-MSPD): An innovative and simplified technique for multiresidue determination of pesticides in fruit samples. J Chromatogr A 1612:460640

Kin CM, Huat TG (2010) Headspace solid-phase microextraction for the evaluation of pesticide residue contents in cucumber and strawberry after washing treatment. Food Chem 123(3):760-764

Kleinman A (1968) Investigation of lead residues on growing fruits and vegetables. Pestic Monit J 1(4):8–10

Kluczkovski AM (2019) Fungal and mycotoxin problems in the nut industry. Curr Opin Food Sci 29:56–63

Kneževìc Z, Serdar M (2009) Screening of fresh fruit and vegetables for pesticide residues on Croatian market. Food Cont 20:419–422

Kottadiyil D, Mehta T, Thasale R, Sivaperumal P (2023) Determination and dietary risk assessment of 52 pesticide residues in vegetable and fruit samples by GC-MS/MS and UHPLC-QTOF/MS from Gujarat, India. J Food Compos 115:104957

Lee HJ, Choe WJ, Lee JY et al (2009) Monitoring of ergosterol biosynthesis inhibitor (EBI) pesticide residues in commercial agricultural products and risk assessment. J Korean Soc Food Sci Nutr 38:1779–1784

Lehotay SJ, Mastovska K, Lightfield AR, Gates RA (2010) Multi-analyst, multi-matrix performance of the QuEChERS approach for pesticide residues in foods and feeds using HPLC/MS/MS analysis with different calibration techniques. J AOAC Int 93(2):355–367

Li LP, Feng BS, Yang JW et al (2013) Applications of ambient mass spectrometry in high-throughput screening. Analyst 138(11):3097–3103

Li ZX, Nia JY et al (2015) Risk assessment and ranking of pesticide residues in Chinese pears. J Integr Agric 14:2328–2339

Li J, Lu J, Qiao X, Xu Z (2017) A study on biomimetic immunoassay-capillary electrophoresis method based on molecularly imprinted polymer for determination of trace trichlorfon residue in vegetables. Food Chem 221:1285–1290

Liu S, Zheng Z, Wei F et al (2010) Simultaneous determination of seven neonicotinoid pesticide residues in food by ultraperformance liquid chromatography tandem mass spectrometry. J Agric Food Chem 58(6):3271–3278

Liu J, Ji C, Liu X et al (2021) Fe3O4 nanoparticles as matrix solid-phase dispersion extraction adsorbents for the analysis of thirty pesticides in vegetables by ultrahigh-performance liquid chromatography-tandem mass spectrometry. J Chromatogr B 1165:122532

Liu J, Guo L, Wu A, Song S et al (2023) Immunochromatographic assay for the analysis of methomyl in cabbage and tomato. Food Chem 409:135273

Łozowicka B, Rutkowska E, Jankowska M (2017) Influence of QuEChERS modifications on recovery and matrix effect during the multi-residue pesticide analysis in soil by GC/MS/MS and GC/ECD/NPD. Environ Sci Pollut Res 24:7124–7138

Lye MS, Ghazali AA, Mohan A et al (1995) An outbreak of acute hepatic encephalopathy due to severe aflatoxicosis in Malaysia. J Trop Med Hyg 53:68–72

Maff U (1993) Aluminium in Food: 39th Report of the Steering Group on Chemical Aspects of Food Surveillance, pp 1–52

Malir F, Pickova D, Toman J et al (2023) Hazard characterisation for significant mycotoxins in food. Mycotoxin Res 39(2):81–93

Manwani S, Vanisree CR, Jaiman V et al (2022) Heavy metal contamination in vegetables and their toxic effects on human health. In: Sustainable crop production: recent advances. IntechOpen, London, p 181

Maragos CM, Busman M, Ma L et al (2015) Quantification of patulin in fruit leathers by ultra-high-performance liquid chromatography-photodiode array (UPLC-PDA). Food Addit Contam Part A Chem Anal Control Expo Risk Assess 32(7):1164–1174

Martins ML, Kemmerich M, Prestes OD et al (2017) Evaluation of an alternative fluorinated sorbent for dispersive solid-phase extraction clean-up of the quick, easy, cheap, effective, rugged, and safe method for pesticide residues analysis. J Chromatogr A 1514:36–43

McKinley ER,Carlton WW, Boon GD (1982) Patulin mycotoxicosis in the rat: Toxicology,pathology and clinical pathology. Food Chem Toxicol 20:289-300.

Meerdink GL (2002) Mycotoxins. Clin Tech Equine Pract 1:89–93

Mohod CV (2015) A review on the concentration of the heavy metals in vegetable samples like spinach and tomato grown near the area of Amba Nalla of Amravati City. Int J Innov Res Technol Sci Eng 4(5):2788–2792

Moretti A, Logrieco AF, Susca A (2017) Mycotoxins: an underhand food problem. Methods Mol Protoc 1542:154–196

Mujahid M, Latif S, Ahmed M et al (2022) Modified matrix solid phase dispersion-HPLC method for determination of pesticide residue in vegetables and their impact on human health: a risk assessment. Front Chem 10:1084350

Narenderan ST, Meyyanathan SN, Babu BJFRI (2020) Review of pesticide residue analysis in fruits and vegetables. Pre-treatment, extraction and detection techniques. Food Res Int 133:109141

Narenderan ST, Meyyanathan SN, Karri VVSR, Babu B, Chintamaneni P (2019) Multivariate response surface methodology assisted modified QuEChERS extraction method for the evaluation of organophosphate pesticides in fruits and vegetables cultivated in Nilgiris. South India Food chemistry 300:125188

Navarro P, Pérez AJ, Gabaldón JA et al (2013) Detection of chemical residues in tangerine juices by a duplex immunoassay. Talanta 116:33–38

Omeje KO, Ezema BO, Okonkwo F et al (2021) Quantification of heavy metals and pesticide residues in widely consumed nigerian food crops using atomic absorption spectroscopy (AAS) and gas chromatography (GC). Toxins 13(12):870

Oyugi AM, Kibet JK, Adongo JO (2021) A review of the health implications of heavy metals and pesticide residues on khat users. Bull Natl Res Cent 45:1–22

Pascari X, Ramos AJ, Marín S et al (2018) Mycotoxins and beer impact of beer production process on mycotoxin contamination: a review. Food Res Int 103:121–129

Prodhan MDH, Alam SN, Uddin MJ (2017) Analytical methods in measuring pesticides in foods. In: Pesticide residue in foods: sources, management, and control. Springer, Heidelberg, pp 135–145

Rai PK (2018) Heavy metal phyto-technologies from Ramsar wetland plants: green approach. Chem Ecol 34(8):786–796

Reddy CS, Berndt WO, Sharma RP, Salunkhe DK (eds) (1991) Mycotoxins and phyloalexins. CRC Press, Boca Raton, FL. Open J Med Microbiol 5(3):237

Reilly C (1980) Metal contamination of food. Applied Science Publishers Ltd., London

Rong L, Wu X, Xu J et al (2017) Simultaneous determination of three pesticides and their metabolites in unprocessed foods using ultraperformance liquid chromatography-tandem mass spectrometry. Food Addit Contam 35(2):273–281

Rossi F, Gallo A, Bertuzzi T (2020) Emerging mycotoxins in the food chain. Med J Nutr Metab 13:7–27

Rotimi OT, Ogunyebi AL, Fingesi TS (2018) Evaluation of heavy metals level and proximate analysis of common fruits from selected markets in Lagos, Nigeria. J Appl Sci Environ Manag 22(5):791–795

Säumel I, Kotsyuk I, Hölscher M, Lenkereit C, Weber F, Kowarik I (2012) How healthy is urban horticulture in high traffic areas? Trace metal concentrations in vegetable crops from plantings within inner city neighbourhoods in Berlin, Germany. Environ Pollut 165:124–132

Sankhla MS, Kumari M, Nandan M et al (2016) Heavy metals contamination in water and their hazardous effect on human health-a review. Int J Curr Microbiol Appl Sci 5(10):759–766

Sanzani SM, Reverberi M, Geisen R (2016) Mycotoxins in harvested fruits and vegetables: Insights in producing fungi, biological role, conducive conditions, and tools to manage postharvest contamination. Postharvest Boil Technol 122:95–105

Sapahin HA, Makahleh A, Saad B (2019) Determination of organophosphorus pesticide residues in vegetables using solid phase micro-extraction coupled with gas chromatography–flame photometric detector. Arab J Chem 12:1934–1944

Scott PM (1995) Mycotoxin methodology. Food Addit Contam 12:395–403

Sevinç S, Mehmet C (2023) Comparison of fruit quality criteria and heavy metal contents of strawberries grown in organic and conventional agriculture. Appl Sci 13:7919

Shakya PR, Khwaounjoo NM (2013) Heavy metal contamination in green leafy vegetables collected from different market sites of Kathmandu and their associated health risks. Sci World J 11(11):37–42

Shamsipur M, Yazdanfar N, Ghambarian M (2016) Combination of solid-phase extraction with dispersive liquid–liquid microextraction followed by GC–MS for determination of pesticide residues from water, milk, honey and fruit juice. Food Chem 204:289–297

Sharma N, More B, Bhandari D, Wavhal SD (2012) Analysis of heavy metals content in spices collected from local market of Mumbai by using atomic absorption spectrometer. Glob J Res Anal 3:56–57

Shashirekha MN, Mallikarjuna SE, Rajarathnam S (2015) Status of bioactive compounds in foods with focus on fruits and vegetables. Crit Rev Food Sci Nutr 55:1324–1339

Shirani M, Aslani A, Ansari F et al (2023) Zirconium oxide/titanium oxide nanorod decorated nickel foam as an efficient sorbent in syringe filter based solid-phase extraction of pesticides in some vegetables. Microchem J. https://doi.org/10.1016/j.microc.2023.108507

Simplício AL, Boas LV (1999) Validation of a solid-phase microextraction method for the determination of organophosphorus pesticides in fruits and fruit juice. J Chromatogr A 833(1):35–42

Sivaperumal P, Anand P, Riddhi L (2015) Rapid determination of pesticide residues in fruits and vegetables, using ultra-high-performance liquid chromatography/time-of-flight mass spectrometry. Food Chem 168:356–365

Song YQ, Liao J, Dong JX, Chen L (2015) Rapidly differentiating grape seeds from different sources based on characteristic fingerprints using direct analysis in real time coupled with time-of-flight mass spectrometry combined with chemometrics. J Sep Sci 38(17):3084–3089

Stachniuk A, Szmagara A, Czeczko R et al (2017) LC-MS/MS determination of pesticide residues in fruits and vegetables. J Environ Sci Health B 52(7):446–457

Stocka J, Tankiewicz M, Biziuk M et al (2011) Green aspects of techniques for the determination of currently used pesticides in environmental samples. Int J Mol Sci 12:7785–7805

Sun H, Feng J, Feng J et al (2022) Carbon aerogels derived from waste paper for pipette-tip solid-phase extraction of triazole fungicides in tomato, apple and pear. Food Chem. https://doi.org/10.1016/j.foodchem.2022.133633

Szpyrk E, Kurdziel A, Matyaszek A et al (2015) Evaluation of pesticide residues in fruits and vegetables from the region of south-eastern Poland. Food Control 48:137–142

Takàs Z, Wiseman JM, Gologan B (2004) Mass spectrometry sampling under ambient conditions with desorption electrospray ionization. Science 306:471–473

Tankiewicz M (2019) Determination of selected priority pesticides in high water fruits and vegetables by modified QuEChERS and GC-ECD with GC-MS/MS confirmation. Molecules. https://doi.org/10.3390/molecules24030417

Tankiewicz M, Berg A (2022) Improvement of the QuEChERS method coupled with GC–MS/MS for the determination of pesticide residues in fresh fruit and vegetables. Microchem J. https://doi.org/10.1016/j.microc.2022.107794

Tao CJ, Hu JY, Li JZ et al (2009) Multi-residue determination of pesticides in vegetables by gas chromatography/ion trap mass spectrometry. Bull Environ Contam Toxicol 82:11–115

Timofeeva I, Shishov A, Kanashina D et al (2017) On-line in-syringe sugaring-out liquid-liquid extraction coupled with HPLC-MS/MS for the determination of pesticides in fruit and berry juices. Talanta 167:761–767

Tian F, Liu X, Wu Y, Xu J, Dong F, Wu X, Zheng Y (2016) Simultaneous determination of penflufen and one metabolite in vegetables and cereals using a modified quick, easy, cheap, effective, rugged, and safe method and liquid chromatography coupled to tandem mass spectrometry. Food Chem 213:410–416

Tóth G, Hermann T, Da Silva MR, Montanarella LJEI (2016) Heavy metals in agricultural soils of the European Union with implications for food safety. Environ Int 88:299–309

Tsagkaris AS, Pulkrabova J, Hajslova J (2021) Optical screening methods for pesticide residue detection in food matrices: advances and emerging analytical trends. Foods 10(1):88

Ul-Haq I, Ahmed E, Sharif A et al (2021) Optimization of ultrasound-assisted extraction of essential and non-essential/toxic trace metals in vegetables and their determination by FAAS and ICP-OES: An evaluation of human health risk. Food Anal Methods 14(11):2262–2275

UN-Habitat Self-Assessment Report (2004) The present self-assessment report is a contribution from UN-Habitat to the planned in-depth evaluation by OIOS of the Human Settlements Programme for the period 2002–2003 English 19-May-04

Wahab S, Ahmad I, Irfan S et al (2021) Hydrogel: an encouraging nanocarrier system for the delivery of herbal bioactive compounds. Curr Nanosci 17:797–807

Wahab S, Muzammil K, Nasir N et al (2022) Advancement and new trends in analysis of pesticide residues in food: a comprehensive review. Plants. https://doi.org/10.3390/plants11091106

Wang P, Yang X, Wang J (2012) Multi-residue method for determination of seven neonicotinoid insecticides in grains using dispersive solid-phase extraction and dispersive liquid–liquid micro-extraction by high performance liquid chromatography. Food Chem 134(3):1691–1698

Wang S, Wang Z, Zhang Y et al (2013) Pesticide residues in market foods in Shaanxi province of China in 2010. Food Chem 138:2016–2025
Wang X, Tang Q, Wang Q et al (2014) Study of a molecularly imprinted solid-phase extraction coupled with high-performance liquid chromatography for simultaneous determination of trace trichlorfon and monocrotophos residues in vegetables. J Sci Food Agric 94(7):1409–1415
Wang J, Duan HL, Ma SY et al (2019) Solidification of a switchable solvent-based QuEChERS method for detection of 16 pesticides in some fruits and vegetables. J Agric Food Chem 67(28):8045–8052
Wang M, Wang J, Wang K et al (2022) Magnetic mesoporous material derived from MIL-88B modified by l-alanine as modified QuEChERS adsorbent for the determination of 6 pesticide residues in 4 vegetables by UPLC-MS/MS. Food Chem. https://doi.org/10.1016/j.foodchem.2022.132325
Welke JE (2019) Fungal and mycotoxin problems in grape juice and wine industries. Curr Opin Food Sci 29:7–13
Wierucka M, Biziuk M (2014) Application of magnetic nanoparticles for magnetic solid-phase extraction in preparing biological, environmental and food samples. TrAC Trends Anal Chem 59:50–58
Wu CC (2017) Multiresidue method for the determination of pesticides in Oolong tea using QuEChERS by gas chromatography-triple quadrupole tandem mass spectrometry. Food Chem 229:580–587
Yang J, Wang H, Jiang Y et al (2008) Development of an enzyme-linked immuno-sorbent assay (ELISA) method for carbofuran residues. Molecules 13(4):871–881
Yang X, Zhang H, Liu Y et al (2011) Multiresidue method for determination of 88 pesticides in berry fruits using solid-phase extraction and gas chromatography–mass spectrometry: Determination of 88 pesticides in berries using SPE and GC–MS. Food Chem 127(2):855–865
Yang Y, Li G, Wu D et al (2020) Recent advances on toxicity and determination methods of mycotoxins in foodstuffs. Trends Food Sci Technol 96:233–252
Yıldırım İ, Çiftçi U (2022) Monitoring of pesticide residues in peppers from Çanakkale (Turkey) public market using QuEChERS method and LC–MS/MS and GC–MS/MS detection. Environ Monit Assess 194(8):570
Yu X, Yang H (2017) Pyrethroid residue determination in organic and conventional vegetables using liquid-solid extraction coupled with magnetic solid phase extraction based on polystyrene-coated magnetic nanoparticles. Food Chem 217:303–310
Yuan C, You Y, Wang Y et al (2023) Matrix solid-phase extraction of nine pesticide residues from vegetables based on mesoporous carbon via an alternative sample preparation approach. J Biobaased Mater Bioenergy 17(2):133–141
Yue Y, Chen J, Zhang M et al (2022) Determination of organophosphorus pesticides in vegetables and fruit by an indirect competitive enzyme-linked immunosorbent assay (IC-ELISA) and a lateral-flow immunochromatographic (LFIC) strip assay. Anal Lett 11:1701–1718
Zhang H, Rui YK (2007) Application of ICP-MS to detecting heavy metals in apple. Spectrosc Spectr Anal 27(8):1632
Zhang S, Yang Q, Yang X et al (2017) A zeolitic imidazolate framework based nanoporous carbon as a novel fiber coating for solid phase microextraction of pyrethroid pesticides. Talanta 166:46–53
Zhang X, Ren X, Chingin K (2021) Applications of direct analysis in real time mass spectrometry in food analysis: a review. Rapid Commun Mass Spectr 35(6):9013
Zhou Q, Tang D (2020) Recent advances in photoelectrochemical biosensors for analysis of mycotoxins in food. TrAC Trends Anal Chem 124:115814
Zhou H, Yang WT, Zhou X et al (2016) Accumulation of heavy metals in vegetable species planted in contaminated soils and the health risk assessment. Int J Environ Res Public Health 13(3):289
Zhou Y, Wu Y, Zeng W et al (2023) Determination of multi-pesticides residues in jasmine flower and its scented tea. Bull Environ Contam Toxicol 110(2):48

Chapter 12
Detection of Freshness and Other Quality Attributes of Horticultural Crops

Simple Kumar, Girish N. Mathad, and Rongsenyangla Ozukum

Abstract Horticulture crops include a wide range of fruits and vegetables that are important for supplying the world's food needs and for enhancing human health. The nutritional content, marketability, freshness, and overall sustainability of horticultural crops are directly affected by their quality and freshness. The term quality indicates the degree of excellence of a produce or its suitability for specific use. Quality-related characteristics are important for product utilization. The quality of produce includes sensory properties, nutritional content, chemical components, mechanical characteristics, functional attributes, and defects. Various instrument measurements are used for sensory evaluation in research and commercial studies, as they reduce errors among individuals and can provide accuracy to researchers, industry, and consumers. Innovative food packaging technology has been proven to improve the quality, safety, and sustainability of food. Recently, emerging food packaging technologies, such as antimicrobial packaging, nanotechnology, controlled-release packaging, biosensors, and radio frequency packaging identification (RFID), have attracted the attention of the food packaging industry. However, more research is necessary on nanomaterials and sustainable packaging materials used in the food sector to address health, safety, and environmental concerns.

Keywords Freshness · Horticulture crops · Sustainable · Quality

---

S. Kumar (✉)
Amity International Centre for Post-Harvest Technology and Cold Chain Management, Amity University, Noida, Uttar Pradesh, India

G. N. Mathad
Department of Tropical Agriculture and International Cooperation, National Pingtung University of Science and Technology, Pingtung, Taiwan

R. Ozukum
Adisan Rural Platform India Private Limited, Delhi, India

M. Thakur, T. Belwal (eds.), *Advances in Postharvest and Analytical Technology of Horticulture Crops*, https://doi.org/10.1007/978-981-97-7247-6_12

12.1 Introduction

Horticultural studies have included fruit and vegetable crops and medicinal, aromatic, and ornamental plants. These crops are comprised of various dietary nutritional components and medicines for humans. Fresh horticulture production requires lower temperatures (32–55 °F) and higher relative humidity (80–95%) to slow down respiration, metabolism, and transpiration rates. By reducing these processes, water activity is reduced, and the quality and nutritional components are maintained. Over the past 10 years, there has been a global increase in the freshness and quality of horticultural crops due to the social, economic, and environmental effects of food loss and waste (FAO 2019); large share of food loss and waste (approximately 44%) occurs from fruits and vegetables (Lipinski et al. 2013). In addition, it is estimated that 40–50% of food is lost and wasted in the food supply chain (Cassani and Gomez-Zavaglia 2022). After harvesting, fruits and vegetables perish very quickly owing to various factors, such as maturation, senescence, and climate; therefore, freshness is a key factor in consumer choice (Kumpulainen et al. 2016).

The impact of freshness on consumer satisfaction and market value: The quality of fruits and vegetables includes physicochemical characteristics pertaining to various horticultural commodities and consumer satisfaction. Generally, the quality of a commodity depends on the nature of the product, genotype, postharvest factors, and extrinsic factors, such as socioeconomic factors and marketing (Schreiner et al. 2013). Kumpulainen (2018) discussed about the meaning of "fresh" and "freshness" based on various literature, which includes the overall eating quality, physiological aging, raw, unprocessed, appearance, texture, method of packing, nutrition, and location of the crop.

Horticulture quality depends on various factors. These factors influence the deterioration and spoilage of the produce. If these factors are not handled well, postharvest losses can occur on a large scale. According to a study by Kader (2002), approximately one-third of fresh fruit is wasted before reaching consumers. Similarly, approximately 30–40% of fruit production is lost between harvest and final consumption. The major causes of postharvest loss include improper temperature control, harsh handling, inadequate packaging materials, and a lack of awareness regarding how to preserve food quality (Kitinoja et al. 2011).

Fruits and vegetables contain rich dietary sources, including macro- (e.g., carbohydrates, fats, and proteins) and micronutrients, as well as some phytochemical compounds (e.g., phenolic compounds, carotenoids, and glucosinolates) that help protect against certain diseases.

Currently, innovative food packaging technology is said to improve the quality, safety, and sustainability of food. Recently, emerging food packaging technologies, such as antimicrobial packaging, nanotechnology, controlled release packaging, biosensors, and radio frequency packaging identification (RFID), have attracted the attention of the food packaging industry. However, further research on the health, safety, and environmental concerns of nanomaterials used in food packaging systems is necessary. Fresh fruits and vegetables have a higher perishable rate and can easily deteriorate during postharvest handling along the supply chain until they reach the consumer. In developing countries, postharvest losses due to spoilage are approximately 50%, and even higher in developing countries. Almost all horticulture produces contain 65–95% water. In addition, metabolism continues after harvesting, resulting in vulnerability and deterioration.

Fresh vegetables can degrade and deteriorate due to biological, microbiological, physical, or environmental reasons. These issues are typically brought on by improper product handler training, poor storage arrangements, inappropriate handling techniques, inefficient quality control, and unfavorable environmental conditions.

To ensure that products fulfil their declared standards and satisfy the reasonable expectations of customers, food standards set the exact requirements. Standards are helpful for consumers, the food sector, and the governing agencies. Food standards may include requirements for a product's appearance, feel, nutritional value, safety, labeling, packaging, analytical methods, and sampling procedures.

Food standards are used to uphold consistency in product quality and safety, open up new markets and create existing ones, give all customers the same information about a product, avoid economic fraud, and build market presence.

12.2 An Overview of the Reasons for Postharvest Losses in Different Groups of Horticultural Produce

The prolonged shelf life of perishables, as well as the reduction of food losses, have been made possible by novel approaches to postharvest physiology and technology. Postharvest losses are still substantial, although reducing them would be the simplest, least expensive, and most successful course of action, as opposed to increasing food production (Pedreschi et al. 2013). Fruit development, ripening, and maturation are characterized by various physiological and biological changes that result in the coordinated development of complex characteristics. Table 12.1 shows the main causes of postharvest causes.

Table 12.1 Overview of main causes of postharvest losses

Horticulture produces	Causes for postharvest physiology	References
Fruits	Bruising, mechanical injuries; loss of water and decay; chilling injury	Pineiro and Diaz (2006)
Leafy vegetables	High respiration; loss of chlorophyll; mechanical injury	
Root vegetables	Sprouting; chilling injury; improper curing and mechanical injury	
Flower vegies	Discoloration; mechanical injury; water loss and decay	

Environmental factors such as light, CO_2, temperature, and water availability are the main limitations. These elements indirectly affect the bioactive molecules. The most important factor that controls the shelf life of fresh produce is temperature. As respiration occurs even after harvest, various physiological processes also occur. Respiration leads to deterioration, loss of nutrients, and changes in texture, flavor, and weight. These processes cannot be stopped; however, they can be controlled by changing the temperature and relative humidity during storage and transportation (Pedreschi et al. 2013). Temperature, humidity, and ventilation are the three most important factors that affect the quality of fresh produce. Recommended storage temperature (13–15 °C) and humidity (85–90%) for some horticulture produces are as follows: for mangoes, avocadoes, papayas, cabbage, lettuce, and carrots are 0–2 °C and 95–100%, respectively (Hofman et al. 2013).

12.3 Methods for Determining Quality of Fresh Commodities

The physical and chemical components of fruit and vegetables can be measured to assess their quality. There are destructive and nondestructive methods for estimating texture/firmness, taste, flavor, fragrance, and nutritional quality. Some nondestructive techniques, including those that use optical, vibrational, electrical, nuclear magnetic resonance, and gas analysis techniques, have the potential to be used commercially.

12.3.1 Sensory Evaluation Techniques

Sensory evaluation assesses the response to stimuli that arise from consuming food or food item (Meilgaard et al. 1999a, b). Sensory evaluation aims to address enquiries related to product quality, competitor products, and commonly raised questions by individuals involved in technical development, research, and other aspects. It serves quality assurance managers, marketing managers, and caters to the interests and concerns of various stakeholders (Mason and Nottingham 2002). It plays a crucial role in mitigating or managing the sources of undesirable errors. This is achieved by implementing measures, such as environmental control, employing appropriate experimental designs, and carefully selecting human subjects for product testing. These practices help ensure accurate and reliable results, while minimizing the impact of potential confounding factors (Valentin et al. 2012). Table 12.2 highlights the quality traits and measurement of fruits and vegetable freshness.

Table 12.2 Quality traits and measurement of fruits and vegetable freshness

Properties	Quality trait	Measurement of quality trait	References
External	Appearance, feel; damages	Visual appearance of color, shape, and size; Evaluation of texture and firmness of the product; visualization of defects and deterioration	Pattee (2012)
Internal	Smell/odor; taste; texture	Subjective assessment by smelling sweet, salt, bitter, and sour. Tenderness, firmness, chewiness, crispness, and overall mouthfeel	
Hidden	Nutrition; food safety	Protein, carbohydrates, fats, vitamins, and other components that are necessary for human well-being. Examination of food products such as pathogenic, microbial, chemical contamination, and other foreign materials	

12.3.2 Visual

The attractiveness of fresh produce is one of buyers' initial assessments of quality by consumers. When making a first purchase, the commodity's appearance is frequently the most important consideration (together with price), although later purchases may be influenced more by texture and flavor.

12.3.3 Color

Color in fruits and vegetables originates from natural pigments that can undergo changes during ripening. Pigments such as chlorophylls, carotenoids, water-soluble anthocyanins, flavonoids, and betalain are responsible for color changes. The external appearance of fruits is generally used as a gauge of ripeness (Shewfelt 2000). Inappropriate colors show an indication of loss of freshness, unripens, leads to turning away willing consumers, and buyers often prefer color for a particular item (Crisosto et al. 2003).

The tactile properties of fruits and vegetables, including their texture, can be perceived through a sense of touch when handling them by hand or when they are consumed and chewed. Unlike flavors, these characteristics can be easily quantified using instrumental methodology.

Plant materials typically consist of a substantial amount of water and other soluble substances enclosed within a semipermeable membrane and a cell wall. The texture of fruits and vegetables is affected by a variety of factors, including turgor pressure, the chemical structure of each plant cell wall, and the ability of the middle lamella to act as an adhesive to hold the cells together. The cell walls of fruits and vegetables consist of various components including cellulose, hemicellulose, pectic substances, proteins, and lignin. The exact structure of the three-dimensional network formed by these components is still not fully understood, but it is a subject of significant interest among scientists. This network largely determines the perception of consistency, smoothness, juiciness, and other sensory attributes of the fruit and vegetable tissues. Researchers are actively studying this topic to understand the relationship between the cell wall composition and sensory attributes of fruits and vegetables (Waldron et al. 2003).

12.3.4 Aroma and Flavor

According to a definition from Anthon (Kramer 1959), flavor encompasses a combined yet unified experience that involves sensations of taste, smell, and cutaneous perceptions such as warmth and color. It is commonly described as an aroma (odor) and taste.

12.3.5 Aroma Compounds

Aroma compounds, which are volatile in nature, are predominantly detected by the nose, whereas taste receptors located in the mouth are stimulated when food is chewed. Color and appearance are often the primary qualities that draw attention to fruit and vegetable products.

12.3.6 Flavor

Flavor is a sensory trait that combines the different sensations of taste, aroma, warmth, and mouthfeel. It has the most significant influence on its acceptability and on our desire to consume it again.

12.3.7 Taste

Taste is generally categorized into five primary tastes: sweet, sour, salty, bitter, and umami. Umami can be defined as a taste associated with the presence of various amino acids and nucleotides in the form of salts (Yamaguchi and Ninomiya 2000). During the evaluation of flavors, it is essential to consider not only desirable flavors but also “off-flavors.” Enzymatic reactions involving lipoxygenase or peroxidase, which produce reactive free radicals and hydroperoxides that are capable of catalyzing the oxidation of lipid molecules, may result in these off-flavors. Unwanted tastes that have been characterized as decaying, cardboard-like, oxidized, or tasting, such as a wet dog, may develop as a result of these interactions. Enzyme-catalyzed reactions occasionally produce palatable tastes. For instance, hydroperoxide lyase facilitates the production of characteristic tomato flavors (Anthon and Barrett 2006).

12.4 Sensory Methods of Quality Measurements

Comprehensive explanations of sensory techniques for assessing food quality, including fruit and vegetable quality, can be found in various sources, such as the Institute of Food Technologists (Meilgaard et al. 1999a, b; Prussia and Shewfelt 1993). As mentioned earlier, sensory evaluation is comprised of two main components: analytical and affective.

12.4.1 Analytical Measurements

It involves the creation of a vocabulary and the training of a panel (Meilgaard et al. 1999a, b). A lexicon refers to a compilation of terms utilized as descriptors, along with their precise definitions. In the context of sensory evaluation, if the testing material, such as tomatoes, has been thoroughly characterized, the lexicon will include specific terms relevant to describing its attributes (Krumbein et al. 2004). The descriptive panel assesses a wide range of varieties of a particular item to establish terms that accurately describe the diverse products to be examined. Lexicons can vary in size, ranging from extensive lexicons comprising up to 50 descriptors, to more minimalist lexicons consisting of as few as five descriptors. Once a lexicon is selected or established, panel members must undergo training and calibration to ensure the accuracy and precision of their evaluations. The training process can range from extensive sessions lasting up to 2 h per week over 6 months to less intensive training conducted within a 1- to 2-week period. Panelists, who deviate significantly from the consensus, may receive additional training or be excluded from the panel. After the training is complete, samples are evaluated using one of the many assessment techniques, such as the spectrum approach, in partitioned booths (Meilgaard et al. 1999a, b; Stone et al. 2020).

12.4.2 Affective Measurements

Affective testing encompasses various approaches; however, the most commonly employed techniques include focus groups, surveys, and consumer preference tests. Focus groups include interviews with a group of 8–10 participants who are current or potential consumers of the products under evaluation (Krueger 2014). Multiple iterations of focus groups are generally conducted on a specific topic until little or no additional qualitative information is obtained. Focus groups are highly effective in helping select appropriate survey questions. For surveys to be meaningful, a minimum of 50 participants, preferably over 200, were required. It is necessary to include basic demographic information and capture the regularity of the purchase or consumption of the items being evaluated in the survey. Survey results were segmented and divided into relevant categories of interest. Consumer preference tests (as shown in Table 12.3) serve as reliable benchmarks for any study because the opinions expressed by consumers in a survey questionnaire may not align with their actual preferences when presented with real samples. For instance, while many consumers may claim that certain items can be excessively sweet, the sample with the highest level of sweetness consistently received the highest preference ratings. This highlights the importance of conducting consumer preference tests to validate and verify consumer responses (Civille 1991).

The nine-point hedonic scale is the most commonly employed evaluation scale in sensory evaluation (Meilgaard et al. 1999a, b). Relying solely on a single sensory technique offers limited insight. Integrating one or more sensory methods can be a potent tool to evaluate the quality of fresh produce. By combining multiple approaches, a more comprehensive and robust understanding of the sensory properties and the overall quality of a product can be obtained. This integration enables a more accurate and nuanced evaluation, leading to better-informed decisions in fresh-cut produce quality assessment (Shewfelt 2000). Difference tests can be used to select the most favorable treatments to be included in a consumer preference test. By integrating these techniques, a comprehensive understanding of market segments, sensory attributes, and consumer preferences can be attained, facilitating informed decision-making in product development and marketing strategies.

12.4.3 Physicochemical Analysis

Firmness: Firmness is often measured using various instruments. For rapid assessment of significant changes in firmness, particularly for soft items, a subjective firmness measurement using fingertips might be helpful. The texture was used to

Table 12.3 Hedonic scales used in the evaluation of food quality

Grade	Score
Like extremely	9
Like very much	8
Like moderately	7
Like slightly	6
Neither like nor dislike	5
Dislike slightly	4
Dislike moderately	3
Dislike very much	2
Dislike extremely	1

gauge the firmness of the fruits. The heights of the probe and texture analyzer must be calibrated prior to assessing firmness.

Total Soluble Solids: Determination of the total soluble solid content (Brix) was performed using a destructive method. Random fruit samples (two fruits per session) were selected and juiced. The juice was filtered through Whatman filter paper. Two drops of filtrate were placed on the prism of a handheld refractometer. The measurements were performed in triplicate to ensure accuracy and consistency (Hoehn et al. 2003).

Titrable Acidity: The juice was extracted from the fruits and filtered using muslin cloth to obtain a clear beverage. Five milliliters of the filtrate were then mixed with distilled water until the total volume reached 50 mL. From this solution, a 5 mL aliquot was taken and titrated with 0.1 N sodium hydroxide, using phenolphthalein as an indicator, until a pink color appeared. Triplicate measurements were recorded, and titratable acidity (%) was calculated using the following formula (Xu et al. 2012):

$$\%\text{Acidity} = \frac{\text{V1NaOH} \times 0.64 \times \text{NNaOH}}{\text{Sample Weight}} \times 100$$

12.5 Advances in Postharvest Technologies for Maintaining Freshness

The production of vegetables and fruits is at a higher rate, and the major tasks are postharvest handling and maintenance of the freshness of produce Around 20–30% of fresh produce is wasted because of poor postharvest management. Recently, various developments have been made to maintain the freshness and improve the shelf life of horticulture.

12.5.1 Modified Atmosphere Packaging (MAP)

MAP is an effective packaging technique that extends the shelf life of fresh produce (Qu et al. 2022; Ghaani et al. 2016). Delays in maturation, withering, tissue damage, decrease in cell disintegration, and microbiological and enzymatic activity are all affected by changes in gas concentration within the MAP. The physiology of the product, material properties of the packaging, and MAP technique (active or passive) affect the gas concentration inside the package (Cortellino et al. 2015; Wilson et al. 2019). Active MAP relies on the injection of preset gas mixes into the package, which enables optimal gas equilibrium, whereas passive MAP primarily relies on natural product processes and film permeability (Ghidelli and Pérez-Gago 2018). In MAP, temperature management is crucial because it directly affects the rate of respiration, which in turn affects the shelf life of produce. The disadvantage of using

MAP is that if the oxygen concentration drops more than usual, it might cause tissues to ferment and release unfavorable off-flavors (Day, B.P.F 2003).

12.5.2 Controlled Atmosphere Storage (CAS)

This is a successful technology developed for storing horticulture. CAS uses lower levels of oxygen and higher levels of CO_2 in the storage atmosphere along with refrigeration (Bodbodak and Moshfeghifar 2016). CAS is the most appropriate method for extending shelf life and maintaining the freshness and quality of horticulture. It helps to reduce metabolic activity, control ripening and weight loss, delay softening, and control browning. CAS is effective on climacteric fruits, such as bananas, mangoes, papayas, apples, and apricots, and has greater potential for preserving vegetables with higher respiration rates (Valdez Fragoso and Mújica-Paz 2016).

12.6 Edible Coating and Films

The use of synthetic packaging materials has recently posed an environmental threat at the time of disposal. Hence, various coating techniques have been developed that control shelf life and are nonhazardous to nature. The materials used for coating must possess certain characteristics, such as acceptable sensorial properties, suitable barrier characteristics, high mechanical strength, stability of microbial, biochemical, and physicochemical properties, safety to use, lower coat, and ease of production (Diab et al. 2001).

Coating materials generally consist of starch, polysaccharide proteins, and cellulose, which are safe for human use. Carboxymethyl cellulose (CMC) is widely used in a wide range of applications. The materials used for the preparation were extracted from plants (wheat, soy protein, and corm zein) and animals (casein and whey protein). Pullulan, which is produced from *Aureobasidium pullulan*, is widely used as a coating material because of its highwater solubility, and is used as a hydrocolloid coating for strawberries and kiwifruit (Diab et al. 2001).

12.7 Nano-packaging of Food Products

Food packaging is an essential part of the food industry and helps retain its nutritional and organoleptic properties even after long-term storage, as well as preserving food and protecting it from physical and environmental conditions. Much research is being conducted on nanotechnology to enhance food packaging. Functional nanoparticles have been discovered to enhance the barrier quality, thermal stability, strength, and durability of packaging materials, thereby extending the shelf life of food goods (Shukla et al. 2019). To enhance the quality and shelf life of food products, nanoparticles can be added to traditional food packaging or fabricated into multilayer packaging systems such as nano-coatings by spraying and immersion (Tsagkaris et al. 2018). Nano-packaging helps nanotechnology to enhance packaging by improving its physical and mechanical characteristics. It includes active packaging with antibacterial, antioxidative, and UV absorption properties, as well as smart packaging that can control or monitor food conditions (Primožic et al. 2021).

12.8 Sustainability and Freshness Assessment of Horticultural Crops

Based on European Bioplastics, polymers derived from renewable sources are classified into various categories based on the raw materials and production methodology (Siracusa and Ingrao 2016). Materials obtained from agricultural and marine sources are characterized by high crystallinity and greater intermolecular interactions. Hence, an appropriate combination of mechanical shear, temperature, and additives is required.

Food packaging is one of the major contributors to environmental pollution; in fact, it makes up 50% of all sold packages and accounts for two-thirds of all packaging waste in terms of volume (Marsh and Bugusu 2007). The packaging industry is making efforts to reduce the weight and volume of packaging by removing unnecessary elements while maintaining the appropriate shelf life, safety, and protection functions. This was done in response to rising environmental concerns. Firms consider improving recyclability, reusability, composting, and energy recovery as additional strategies to reduce waste output. Utilizing novel materials such as biopolymers and extending the shelf life of products might significantly contribute to sustainable development because longer shelf lives result in less food and packaging waste (Restuccia et al. 2016). Horticulture crops include a wide range of fruits and vegetables that are important for supplying the world's food needs and enhancing human health. The nutritional content, marketability, freshness, and overall sustainability of horticultural crops are all directly affected by their quality and freshness. The significance of identifying freshness and other quality characteristics of horticultural crops, as well as the approaches used to make precise judgments, is discussed. To highlight the difficulties in preserving crop quality through packaging and storage, factors affecting freshness and quality, such as postharvest physiological changes and climatic conditions, were also addressed. Researchers and stakeholders in the horticulture and food sectors may exceed customer expectations by improving food safety, boosting economic viability, and contributing to a more sustainable and ecologically aware future by employing effective and environmentally friendly methods for freshness assessment.

12.9 Conclusion

A multidimensional process, the detection of freshness and other quality characteristics in horticultural crops, requires continuous research innovation and cooperation among researchers, industry stakeholders, and regulatory bodies. The horticulture and food sectors can not only fulfil consumer expectations but also contribute to a more resilient, resource-efficient, and environmentally aware global food system by prioritizing freshness and implementing sustainable practices across the supply chain. The desire for freshness goes beyond financial considerations and involves a common desire to safeguard food security and advance health, thus preserving the environment for future generations.

References

Anthon G, Barrett DM (2006) Standardization of a rapid spectrophotometric method for lycopene analysis. In: X International Symposium on the Processing Tomato 758. pp 111–128

Bodbodak S, Moshfeghifar M (2016) 2 - Advances in controlled atmosphere storage of fruits and vegetables. In: Siddiqui MW (ed) Eco-friendly technology for postharvest produce quality. Academic Press, Cambridge, MA, pp 39–76. https://doi.org/10.1016/B978-0-12-804313-4.00002-5

Cassani L, Gomez-Zavaglia A (2022) Sustainable food systems in fruits and vegetables food supply chains. Front Nutr 9:829061

Civille GV (1991) Food quality: consumer acceptance and sensory attributes. J Food Qual 14(1):1–8

Cortellino G, Gobbi S, Bianchi G, Rizzolo A (2015) Modified atmosphere packaging for shelf life extension of fresh-cut apples. Trends Food Sci Technol 46(2):320–330

Crisosto CH, Crisosto GM, Metheney P (2003) Consumer acceptance of 'Brooks' and 'Bing' cherries is mainly dependent on fruit SSC and visual skin color. Postharvest Biol Technol 28(1):159–167

Day, B. P. F. (2003). 10 - Novel MAP applications for fresh-prepared produce. In R. Ahvenainen (Ed.), *Novel Food Packaging Techniques* (pp. 189–207). Woodhead Publishing. https://doi.org/10.1533/9781855737020.2.189

Diab T, Biliaderis CG, Gerasopoulos D, Sfakiotakis E (2001) Physicochemical properties and application of pullulan edible films and coatings in fruit preservation. J Sci Food Agric 81(10):988–1000

FAO (2019) The state of food and agriculture 2019. Moving forward on food loss and waste reduction. FAO IJF, Rome, pp 2–13

Ghaani M, Cozzolino CA, Castelli G, Farris S (2016) An overview of the intelligent packaging technologies in the food sector. Trends Food Sci Technol 51:1–11

Ghidelli C, Pérez-Gago MB (2018) Recent advances in modified atmosphere packaging and edible coatings to maintain quality of fresh-cut fruits and vegetables. Crit Rev Food Sci Nutr 58(4):662–679

Hoehn E, Gasser F, Guggenbühl B, Künsch U (2003) Efficacy of instrumental measurements for determination of minimum requirements of firmness, soluble solids, and acidity of several apple varieties in comparison to consumer expectations. Postharvest Biol Technol 27(1):27–37

Hofman P, Bower J, Woolf A (2013) Harvesting, packing, postharvest technology, transport and processing. In: The avocado: Botany, production and uses. CABI, Wallingford, pp 489–540

Kader AA (2002) Pre- and postharvest factors affecting fresh produce quality, nutritional value, and implications for human health. Proceedings of the International Congress Food production and the quality of life. Sassari Italy, In, pp 109–119

Kitinoja L, Saran S, Roy SK, Kader AA (2011) Postharvest technology for developing countries: challenges and opportunities in research, outreach and advocacy. J Sci Food Agric 91(4):597–603

Kramer A (1959) Glossary of some terms used in the sensory (panel) evaluation of foods and beverages. Food Technol 13(12):733–736

Krueger RA (2014) Focus groups: a practical guide for applied research. Sage Publications, London

Krumbein A, Peters P, Brückner B (2004) Flavour compounds and a quantitative descriptive analysis of tomatoes (*Lycopersicon esculentum* Mill.) of different cultivars in short-term storage. Postharvest Biol Technol 32(1):15–28

Kumpulainen, T. (2018). The complexity of freshness and locality in a food consumption context (Doctoral dissertation, Doctoral Thesis in Food Sciences at the University of Turku, Finland).

Kumpulainen T, Sandell M, Junell P, Hopia A (2016) The effect of freshness in a foodservice context. J Culin Sci Technol 14(2):153–165

Lipinski B, Hanson C, Waite R, Searchinger T, Lomax J (2013) Reducing food loss and waste. World Res Inst. Accessed from: http://pdf.wri.org/reducing_food_loss_and_waste.pdf

Marsh K, Bugusu B (2007) Food packaging—roles, materials, and environmental issues. J Food Sci 72(3):R39–R55

Mason RL, Nottingham SM (2002) Sensory evaluation manual. JBUoQ. Accessed form: https://pdfcoffee.com/sensory-evaluation-manual%2D%2Dpdf-free.html

Meilgaard M, Civille GV, Carr BT (1999a) Sensory evaluation techniques, vol 3. CRC Press, Boca Raton, FL

Meilgaard MC, Carr BT, Civille GV (1999b) Sensory evaluation techniques. CRC Press, Boca Raton, FL

Pattee HE (2012) Evaluation of quality of fruits and vegetables. Springer, Cham

Pedreschi R, Lurie S, Hertog M, Nicolaï B, Mes J, Woltering EJP (2013) Post-harvest proteomics and food security. Proteomics 13(12–13):1772–1783

Pineiro M, Diaz LB (2006) Improving the safety and quality of fresh fruit and vegetables (FFV): a practical approach. In: I International Symposium on Fresh Food Quality Standards: Better Food by Quality and Assurance 741. pp 19–24

Primožic M, Knez Ž, Leitgeb M (2021) Nanotechnology in food science—food packaging. Nano 11:292

Prussia SE, Shewfelt RL (1993) Systems approach to postharvest handling. Academic Press, Cambridge, MA, pp 44–68

Qu P, Zhang M, Fan K, Guo Z (2022) Microporous modified atmosphere packaging to extend shelf life of fresh foods: A review. Crit Rev Food Sci Nutr 62(1):51–65

Restuccia D, Salomone R, Spizzirri UG, Saija G, Ioppolo G, Parisi OI, Picci N (2016) Industrial applications: Regulatory issues and life cycle assessment of food packaging. In: Antimicrobial food packaging. Elsevier, Amsterdam, pp 215–227

Schreiner M, Korn M, Stenger M, Holzgreve L, Altmann MJS (2013) Current understanding and use of quality characteristics of horticulture products. Sci Hortic 163:63–69

Seseña S, Sánchez I, Viñas MAG, Palop L (2002) Effect of freezing on the spontaneous fermentation and sensory attributes of Almagro eggplants. Int J Food Microbiol 77(1–2):155–159

Shewfelt RL (2000) Fruit and vegetable quality. In: Fruit and vegetable quality. CRC Press, Boca Raton, FL, pp 160–173

Shukla P, Chaurasia P, Younis K, Qadri OS, Faridi SA, Srivastava G (2019) Nanotechnology in sustainable agriculture: studies from seed priming to post-harvest management. Nanotechnol Environ Eng 4:1–15

Siracusa V, Ingrao C (2016) The use of polylactic acid in food packaging. In: Reference module in food science. Elsevier, Amsterdam, pp 1–5

Stone H, Bleibaum RN, Thomas HA (2020) Sensory evaluation practices. Academic Press, Cambridge, MA
Tsagkaris AS, Tzegkas SG, Danezis GP (2018) Nanomaterials in food packaging: state of the art and analysis. J Food Sci Technol 55:2862–2870
Valdez Fragoso A, Mújica-Paz H (2016) Controlled atmosphere storage: effect on fruit and vegetables. In: Caballero B, Finglas PM, Toldrá F (eds) Encyclopedia of food and health. Academic Press, Oxford, pp 308–311. https://doi.org/10.1016/B978-0-12-384947-2.00197-5
Valentin D, Pêcher C, Nguyen DH, Chambers D, Abdi H (2012). Integrating sensory evaluation into product development: An Asian perspective. In: 3. symposium international
Waldron KW, Parker M, Smith AC (2003) Plant cell walls and food quality. Compr Rev Food Sci Food Saf 2(4):128–146
Wilson MD, Stanley RA, Eyles A, Ross T (2019) Innovative processes and technologies for modified atmosphere packaging of fresh and fresh-cut fruits and vegetables. Crit Rev Food Sci Nutr 59(3):411–422
Xu K, Wang A, Brown S (2012) Genetic characterization of the Ma locus with pH and titratable acidity in apple. Mol Breed 30:899–912
Yamaguchi S, Ninomiya K (2000) Umami and food palatability. J Nutr 130(4):921S–926S

Chapter 13
Online Quality Control of Horticultural Crops During Processing

Rajesh Mor, M. L. Jat, Sonu Kumar, Hardeep, Rakesh Kumar, and Naleeni Ramawat

Abstract Size, color, form, texture, and defect are among the main qualities of horticultural products. Since a computer vision system produces precise, unbiased, and noninvasive results, it is utilized to assess processed food instead of a human inspector. Computer vision, HIS, MSI, infrared technology, ultrasound, tomographic imaging, MRI, CT X-ray (CT), Raman spectroscopy, NMR, and VOCs e-sensing have all been used by the horticultural and food processing industries to monitor quality online. The detailed examination of these most recent breakthroughs is provided in this chapter on machine vision systems. Because they can produce quick, effective, hygienic, reliable, and unbiased results, machine vision systems are widely employed in industry for inspection and evaluation. However, it can be inferred from the thorough survey carried out in this chapter that there is still a need for the development of robust online controlling methods with improved capabilities despite the fact that numerous approaches have been put forth by researchers for the evaluation of the quality of fruits and processed goods.

Keywords Horticultural crops · Raman spectroscopy · Horticultural crops · Online controlling methods

---

R. Mor · Hardeep
Department of Horticulture, Maharana Pratap Horticultural University, Karnal, Haryana, India

M. L. Jat · S. Kumar
Department of Horticulture, CCS Haryana Agricultural University, Hisar, Haryana, India

R. Kumar · N. Ramawat (✉)
Agricultural University Jodhpur, Jodhpur, Rajasthan, India

M. Thakur, T. Belwal (eds.), *Advances in Postharvest and Analytical Technology of Horticulture Crops*, https://doi.org/10.1007/978-981-97-7247-6_13

13.1 Introduction

The agriculture sector in developing countries today stands in the midst of a horticultural revolution. Agriculture is Indian economy's mainstay and it comprises 18.3% of the gross domestic products. Horticulture contributes 30.4% of the agriculture gross domestic product (GDP) using only 13.1% of gross cropped area (Anonymous 2022a). Over the years, total horticulture production in the country has surpassed the total food grains production, achieving a higher production record every year. India is currently producing about 320.48 million tons of horticulture produce, which has surpassed the food grain production, that too from much less area (25.66 MH for horticulture against 127.6 MH for food grains). Productivity of horticulture crops is much higher compared to productivity of food grains, i.e., 12.49 tons/ha against 2.23 tons/ha, respectively (Anonymous 2022b).

In such an era of digital revolution, it is the appropriate time to connect the agricultural landform with wireless technology worldwide to introduce and accommodate digital connectivity with farmer and introduce him to technologies that are sustainable. One such technology is "online quality control." With the advent of this modern, sophisticated technology, it will enable the users, both farmers and industry, to remotely monitor the crops and their processing. Smart farming helps agricultural processes, such as harvesting and crop yields, as the automation of sensors and machinery has made the farming workforce more efficient (O'Grady and O'Hare 2017). As per 2030 Agenda for Sustainable Development, the United Nations and international community have pledged a goal to end hunger by 2030. Recently, the World Health Organization reported that more than 800 million people will face food shortages worldwide (Fróna et al. 2019). In addition, the increasing global population is increasing the demand for quality food and thus the future of horticulture is expected to interconnect with AI and big data services. Online quality control of horticultural crops during processing is one such key technologies with methods focused on achieving sustainability for future generations.

Horticultural crops play a crucial role in promoting human nutrition and health, as they are recognized as significant source of beneficial phytonutrients, dietary fiber, and various micronutrients. The consumption of horticultural produce is primarily focused on fresh products, which poses a challenge due to their high perishability. Additionally, the availability of certain horticultural crops is limited to specific seasons, resulting in periods when they are not accessible. Unlike staple foods such as maize or rice, the prices of horticultural products experience substantial fluctuations influenced by the dynamics of supply and demand. (Arendse et al. 2021; Manning 2016; Cozzolino et al. 2015).

Fruits are characterized by their high water content, making them prone to rapid spoilage and the occurrence of postharvest losses across the value chain has been observed. To address this issue and reduce food waste while preserving the nutritional quality of horticultural produce, value-added products are developed. The assessment of the quality of agricultural commodities typically involves evaluating a combination of physical, chemical, and microbial properties (Magwaza and Opara

2015). The quality attributes that are encompassed comprise of external appearance that includes size, shape, color; the absence of defects or decay; texture that encompasses firmness, crispness, juiciness, and mealiness; taste that consists of sweetness, sourness, bitterness, and aroma; and nutritional value that includes dietary fiber, vitamins, minerals, and phytonutrients (Chen and Opara 2013; Opara and Pathare 2014). Additional measurable attributes such as lipid content, moisture, and protein content are also considered. The main aim of processing is to prolong the duration of perishable goods while retaining their nutritional, textural, and sensory attributes (Okere et al. 2021). During processing, it is essential to retain the color, flavor, texture, and nutrients of fruits and vegetables while prolonging their shelf life. However, processed products with complex value chains are susceptible to adulteration and contamination before reaching consumers (Sumonsiri and Barringer 2014; Arendse et al. 2021). The physiological quality of processed products undergoes changes during postharvest processing, storage, and transportation. Hence, it is of utmost significance to keep a close eye on the standard and safety of food products that undergo processing, all along the value chain. Food processing chains are exposed to risks such as pathogenic contamination, adulteration, and unintended introduction of chemical compounds as they undergo multiple stages (van Asselt et al. 2017; Okere et al. 2021).

In present times, the occurrence of issues pertaining to the quality and safety of food is a common phenomenon in everyday life. This has resulted in an upsurge in the demand for food items that are of high quality. At the outset, consumers tend to evaluate products based on a range of factors such as price, appearance, freshness, and nutritional value. To cater to the needs and expectations of consumers and to ensure that they are provided with delicious, visually pleasing, and dependable foods, the food industry must exercise greater caution and diligence at every stage of food production and processing (Müller and Steinhart 2007; Ma et al. 2016). This issue of safety related to the pre-exposure in the processing value chain presents a significant health hazard to individuals and consumers alike. Furthermore, it impacts the sensory perception of commodities, leading to a reduction in their nutritional value and functional properties once they have been processed (Okere et al. 2021).

In addition to technological advancements that improve the quality and safety of food products, such as cooling, freezing, drying, and edible coating, the recognition, identification, and classification of food play a crucial role in the processing and evaluation of food. These tasks are typically performed by skilled workers and are labor-intensive, time-consuming, and prone to human error, resulting in significant costs. Thus, there is an urgent need for a modern, rapid, accurate, and noninvasive technique to optimize food processing efficiency and reduce costs (Sun 2011; Ma et al. 2016).

Machine vision is an interdisciplinary engineering technology that combines various fields such as mechanics, optical instrumentation, electromagnetic sensing, digital video, and image processing. It is a comprehensive system that involves the capture, processing and analysis of two-dimensional images to construct meaningful descriptions of physical objects. Machine vision has found wide applications in examining, monitoring, and controlling diverse processes (Patel et al. 2012;

Razmjooy et al. 2012; ElMasry et al. 2012). Therefore, the development of an automated system for external quality inspection is urgent and necessary to replace manual inspection methods.

As the field of information science continues to expand at a rapid pace, it is worth noting that computer vision-based pattern recognition and image processing have become highly developed techniques for ensuring safety and quality in a variety of agricultural applications. Employing technology that emulates the human visual system, computer vision allows for the inspection of product quality through electronic image perception, interpretation, and character recognition, ultimately furnishing information for the operation of quality grading and sorting machinery. It is need of the hour to introduce computer vision technology (Bhargava and Bansal 2021; Ma et al. 2016; Narendra and Hareesha 2010), ultrasound in analysis, processing and quality control of food (Awad et al. 2012), volatile organic compounds e-sensing (Tiwari et al. 2020; Lin et al. 2022), machine vision system (Patel et al. 2012), vibrational spectroscopic techniques (Arendse et al. 2021), and noninvasive methods (Okere et al. 2021). Details of the above-mentioned methods for online quality control during processing were discussed in this chapter in the following sections.

13.2 Concept of Quality

The concept of "*quality*" originates from the Latin term "*qualitas*," which pertains to the attribute, property, or fundamental nature of an object (Camelo 2004). In the field of study, various authorities have presented concise definitions of quality, such as "fitness for use" (Ireson and Juran 1988), "conformance to requirements," "loss avoidance" (Okere et al. 2021), and "degree of excellence or superiority" (Kader 2000). These definitions aim to understand and optimize the entire system of value exchange (Straker 2001). However, the term "quality" has been used in different contexts and has generated numerous definitions depending on the specific area of study, encompassing a wide range of phenomena. In the agricultural industry, the assessment of quality for fresh and processed horticultural products involves considering a combination of values for multiple attributes or characteristics unique to each commodity. These quality characteristics of fresh and processed horticultural crops include sensory attributes (such as appearance, texture, flavor, and aroma), nutritional values, biochemical constituents, textural properties, and the presence or absence of defects. Consequently, defining a universal definition of quality for horticultural products is challenging due to the diverse perspectives on quality held by major stakeholders in the field of horticulture (Watada 1980; Abbott 1999).

13.3 Image Acquisition

Machine vision technology, which is a complex integrated system combining mechanical, optical, electronic, and software components, has been widely utilized in various fields for examination, monitoring, and control purposes (Patel et al. 2012). Image acquisition, a critical step in machine vision, involves capturing images using tools such as cameras, ultrasound, magnetic resonance imaging (MRI), electrical tomography, and computed tomography (CT). In the context of machine vision, cameras equipped with charged coupled device (CCD) or complementary metal oxide semiconductor (CMOS) image sensors are commonly used for image acquisition. A typical computer vision system comprises five essential components: illumination, an image capture board (such as a digitizer or frame grabber), a camera, and computer hardware. In fruit analysis, lighting systems are designed for front and back lighting. Front lighting is utilized for inspecting surface quality attributes such as color, texture, and skin defects. On the other hand, back lighting is employed to examine boundary quality attributes such as size and shape (Bhargava and Bansal 2021).

13.4 Computer Vision System

Conventional computer vision systems primarily focus on the RGB color model, which includes red, green, and blue colors. Images captured through RGB color cameras are typically based on the RGB wavelength. Computer vision systems allow for automatic grading and inspection of various features such as texture, shape, color, size, and defects. However, detecting certain defects that resemble human skin in terms of texture and color can be challenging. To overcome this challenge, Lorente et al. (2012) proposed a hyperspectral computer vision system that combines spectroscopic and imaging techniques. This system provides spectral information for each pixel of the spatial image (Bhargava and Bansal 2021).

Hyperspectral image acquisition can be achieved through three common methods: point scanning, line scanning, and area scanning approaches (Zhang et al. 2014). In contrast, multispectral imaging systems acquire two or more distinct waveband monochromatic images across the spectrum (Zhang et al. 2014).

The significance of appropriate and high-quality lighting is paramount in numerous vision applications. In order to achieve successful illumination in vision tasks, it is essential to comprehend the role of lighting in machine vision applications, the behavior of light on a given surface, and the available basic lighting techniques (Sun 2011). The process of digital image processing commences with the electronic capture of an image, which involves the camera and frame grabber. The camera converts photons into electrical signals, while the frame grabber digitizes these signals into a stream of data or bitmap image. There are various types of cameras, ranging from older pickup tubes such as vidicons to newer solid state imaging devices such as complementary metal oxide silicon (CMOS) cameras (Sun 2011).

Scanning technology in imaging encompasses area scan and line-scan sensors. These technologies differ in terms of resolution, exposure time, speed, and cost, and the major distinction lies in image formation. Area scan employs an array of photosensitive sensors to construct an image, whereas line-scan technology utilizes a single row of sensors that continuously scan moving objects, reconstructing the image line by line in software. Area scan technology is suitable for stationary or slowly moving objects, while line-scan technology is ideal for fast-moving conveyor line applications. However, capturing 3D-shaped objects such as fruits or processed foods using area scan may be ineffective if parts of the object fall outside the camera's field of view (Sun 2011).

13.5 Infrared

Infrared (IR) technology, specifically in the range of 700–1000 nm, offers unique advantages in vision technology. It allows for the detection of IR thermal radiation and the acquisition of spectral reflectance, providing valuable information about material structures. IR imaging has been widely utilized in the food industry due to its ability to correlate thermographic signatures with various physiochemical properties of food, including shelf life, fatty acids, and firmness (Ginesu et al. 2004; Gómez et al. 2006).

In IR technology, a 3D range image can be reconstructed by viewing light from different angles. This can be achieved through techniques such as triangulation, mirror interferometry, holographic interferometry, and Fresnel diffraction. Among these techniques, active triangulation is the most commonly used (Sun 2011). IR spectroscopy, a noninvasive method, is widely utilized for molecular analysis. Thermographic imaging is based on the principle that objects emit thermal radiation in relation to their temperature (Bhargava and Bansal 2021). The technique is dependent on the interaction between molecular bonds and electromagnetic radiation in the near-infrared (NIR) and mid-infrared (MIR) ranges. The NIR spectrum, which spans from 12,500/cm to 4000/cm or 800–2500 nm, displays broad overlapping bands that represent the vibrations of chemical bonds (C–H, O–H, and N–H) present in a molecule. This results in the emergence of broadband overtones. NIR spectroscopy offers several benefits, including its ability to penetrate food samples and provide spectral information on the physical and chemical properties of biological samples. In contrast, MIR spectroscopy provides more spectral data and higher spectral resolution due to its sensitivity to various chemical compositions within biological samples (Kelly et al. 2003; Nicolaï et al. 2007; Arendse et al. 2018).

13.6 Hyperspectral Imaging and Multispectral Imaging

Hyperspectral imaging (HSI) is a methodology that integrates spectroscopic and imaging components to capture images that contain both spatial and spectral data. This approach has found extensive applications in the domain of food techniques. The greatest strength of HSI lies in its capacity to acquire a multitude of monochromatic images with different wavebands simultaneously, thereby facilitating the extraction of a complete spectrum for each pixel in an image. The resultant data obtained from HSI systems consist of 3D structures, which include two spatial dimensions and one spectral dimension (Su and Sun 2018).

While HSI offers valuable information, it can be time-consuming and complex to acquire and analyze images under laboratory conditions. Caporaso et al. (2022) employed HSI to predict the aroma profile of roasted coffee using PLS regression within a spectral range of 1000–2500 nm. The study successfully projected primary aroma compounds, yielding prediction statistics such as R^2 values ranging between 0.21 and 0.71 and RPD values ranging between 0.84 and 1.87. The volatile compounds analyzed included aldehydes and pyrazines.

Multispectral imaging (MSI) is a technique that involves utilizing multiple spectral components of the electromagnetic wavelength in order to generate images of the same object region and scale. This approach is considered a modification of the hyperspectral imaging (HSI) methodology. MSI has proven to be particularly advantageous in real-time applications such as food processing plants and packing houses, where it provides both spatial and spectral information. The technique involves capturing two or more waveband monochromatic images using a limited number of discreet wavebands, typically less than ten, within the spectrum (Zhang et al. 2014).

Ion mobility spectroscopy (IMS) is a chemical analysis technique that utilizes the variance in ion migration velocity in the gas phase to characterize substances. IMS allows the migration of ions through a drift tube and monitors their drift time to detect and identify samples based on their drift times. IMS has been increasingly utilized in various fields, including food quality analysis, due to its high responsiveness and sensitivity. For example, IMS technology has been used to detect volatile compounds of fat in pork to determine feeding strategies, achieving an impressive prediction accuracy of 95.5% (Alonso et al. 2008).

13.7 Ultrasound

Ultrasound refers to sound waves with frequencies beyond the human hearing limit, typically exceeding 20 kHz. In the context of food processing, analysis, and quality control, ultrasound can be categorized into two groups based on their frequency range: high energy and low energy. Low-energy ultrasound, characterized by frequencies above 100 kHz and intensities below 1 W/cm^2, is utilized for noninvasive monitoring and analysis of food materials during processing and storage to ensure quality and safety. It has been employed in the nondestructive quality control of fruits at various stages, from preharvest to postharvest processing. High-energy ultrasound, on the other hand, involves intensities exceeding 1 W/cm^2 and frequencies ranging from 20 to 500 kHz. These high-intensity waves have disruptive effects on the physical, mechanical, or chemical/biochemical properties of foods, offering potential applications in food processing, preservation, and safety (Awad et al. 2012).

Ultrasound-based assessment of food quality relies on the analysis of data obtained from the reflection, transmission, refraction, absorption, and scattering of sound waves (Lefebvre et al. 1993). Recent studies have shown that ultrasonic velocity characteristics can be used to predict the percentage composition of components such as fats, water, and proteins in processed food products (Simal et al. 2003; Ma et al. 2016). Ultrasound has also been employed for moisture grading in food products and to assess the hydration and turgidity of orange peels (Bhargava and Bansal 2021).

13.8 Tomographic Imaging

Scientists often employ tomographic imaging when they need to observe the internal aspects of a sample. This imaging technique utilizes nuclear energy to create a two-dimensional distribution of an object by acquiring a sequence of one-dimensional projections. The necessary data is generated by scanning various sources of projection data. Multiple viewing angles are employed until the complete set of projection data is obtained. Tomographic imaging has found applications in real-time monitoring of food processing, storage, packaging, and distribution, among other areas (Bhargava and Bansal 2021).

Among the various modalities of tomography, nuclear tomography stands out due to its utilization of nuclear energy for imaging purposes. Nuclear tomography employs two categories of radiation sources. CT (computed tomography), which is categorized as "remote sensing," employs external radiation sources. Contemporary CT systems are typically equipped with a fan-beam arrangement that enables coverage of the entire object field of view, thereby reducing acquisition time. However, CT has limitations in imaging calcified objects due to its reliance on X-rays, while food objects are generally soft or semifluid. Nevertheless, with the progress in software and hardware, CT in conjunction with pertinent image analysis strategies has been effectively utilized to identify the magnitude and spatial distribution of salt gradients in dry-cured ham (Ma et al. 2016).

13.9 Magnetic Resonance Imaging

In contrast to CT, magnetic resonance imaging (MRI) provides the capability to visualize a wide range of food objects and is suitable for noninvasive and real-time monitoring during various food processing operations. MRI operates by detecting the sample within a magnetic field and emitting a characteristic nuclear magnetic resonance frequency. The hydrogen nuclei in the sample align their spin with the magnetic field, a phenomenon known as the Lamor effect. MRI can assess both the quantity and dynamic structural characteristics of water present in the sample. Since water is present in most food products, MRI has been widely used in studies involving food detection. For example, Borompichaichartkul et al. (2005) utilized MRI to capture images of corn kernels during the freezing process, enabling visualization of the location and state of water in the frozen corn. However, the high cost associated with MRI technology limits its extensive application in the food industry (Ma et al. 2016).

13.10 Microfocus X-ray Computed Tomography

Microfocus X-ray computed tomography (μCT) is an advanced imaging technique used for characterization and defect detection by reconstructing and rendering three-dimensional images (Kotwaliwale et al. 2014). The principle of X-ray attenuation is employed in μCT, which measures the density variation of the sample (Curry et al. 1990). By utilizing X-ray μCT, it is possible to observe, classify, and evaluate the physical and physiological compositions of biological matter with precision down to a few micrometers. Since fruits and vegetables contain a significant amount of moisture, X-ray absorption is influenced by water content, allowing for the visualization of structures and imperfections (Kotwaliwale et al. 2014). Arendse et al. (2018) conducted a study to determine the juice content of "Wonderful" pomegranate fruit cultivated in South Africa. According to the authors, the estimated juice volume was 142.7 ± 16.4 mL, accounting for 89.8% of the total aril volume (162.5 ± 16.2 mL).

13.11 Raman Spectroscopy

Raman spectroscopy is an analytical technique used in the analysis of food quality and authenticity, based on vibrational spectroscopy. It relies on the interaction between light and electron bonds within a molecule. When light interacts with a molecule, it excites the electrons to a virtual state, which then settle into an excited vibrational state (Lohumi et al. 2018). This phenomenon, known as Raman scattering, occurs when the incident photon collides elastically with the molecule of the sample (Yang and Ying 2011). The scattered light is collected, dispersed by a monochromator, and detected by a sensor (Ilaslan et al. 2015). As the molecule gains energy, there is a change in its rotational energy, resulting in a shift in wavelength of the radiation. This frequency difference is referred to as a Raman shift. The photons can shift toward longer wavelengths, known as Stokes lines, or shorter wavelengths, known as anti-Stokes lines (Mollazade et al. 2012). The spectra obtained from the relaxation and excitation of molecules by light radiation provide a fingerprint region that can be used for structural and qualitative analysis. In a study by Nekvapil et al. (2018), Raman spectroscopy was employed to evaluate the freshness of citrus juice using a spectral range of 1800–100/cm. The authors utilized two highly sensitive and portable systems. According to their findings, the coefficients of freshness (CFresh) values representing freshness varied between 2.8 and 3.5 for clementines (Day 1) and 1.5 for both mandarins and tangerines, respectively.

13.12 Nuclear Magnetic Resonance

In the late 1930s, Isidor Rabi made the initial discovery of nuclear magnetic resonance (NMR) by detecting the magnetic spin of atomic nuclei in gases. Since then, NMR has become a widely used noninvasive technique for analyzing the structure of organic materials (Nturambirwe 2012). NMR can be classified into three types based on their applications: magnetic resonance spectroscopy (MRS), magnetic resonance imaging (MRI), and low-field (LF) NMR (Erikson et al. 2012). The principle behind NMR is that atomic nuclei have a spin and electrical charge, which leads to their response when subjected to a magnetic field and electromagnetic pulse. Energy transfer occurs at specific wavelengths corresponding to radio frequencies when an external magnetic field is applied (Arendse et al. 2018). One of the key applications of NMR is the measurement of water content and distribution, particularly in relation to hydrogen nuclei, which exhibit a strong response to magnetic fields. Da Silva Grandizoli et al. (2014) used ^{1}H NMR at a frequency of 400.13 MHz to detect adulteration in grape juice, achieving a classification accuracy of 0.93% for the addition of other juices. Dielectric spectroscopy is an advanced and noninvasive technique that examines the interaction between electric fields and tested materials. It is based on the principles of dielectric mechanisms and polarization, which determine the dielectric permittivity of the material. Extensive research has been conducted on the dielectric properties of binary mixtures of solids and liquids (Lizhi et al. 2010). Dielectric spectroscopy utilizes electromagnetic fields to provide insights into the dielectric response of materials, particularly in evaluating moisture content in various foods (Okere et al. 2021). The analysis of food quality has seen the development of commercial instrumentation such as front-face fluorescence spectroscopy (FFFS) and synchronous fluorescence spectroscopy (SFS). Fluorescence spectroscopy involves the absorption of ultraviolet or visible light by a molecule, leading to the molecule becoming a fluorophore and emitting light at a higher wavelength as it releases energy. FFFS and SFS are utilized to detect adulteration in food products (Karoui et al. 2010).

13.13 Volatile Organic Compounds (VOCs) E-sensing

Plants emit volatile organic compounds (VOCs) during postharvest storage and horticultural commodities can undergo various natural processes such as senescence, sprouting, and ripening as well as be susceptible to microbial infection. These factors, along with different abiotic stress conditions such as temperature, humidity, and water, can lead to changes in the VOC profile of the stored crops, making VOCs useful as biomarkers in postharvest storage management (Tiwari et al. 2020). Different detection methods are employed to identify and analyze volatile compounds in processed food. These methods include gas chromatography (GC), electronic nose, near-infrared spectroscopy, colorimetric sensors, and other innovative technologies. These techniques offer advantages in terms of precision, ease of operation and, noninvasiveness (Majchrzak et al. 2018).

The concept of a colorimetric sensor array utilizes color changes as a detection technology. The sensor consists of a color-sensitive material attached to a substrate, which produces a color effect upon interaction with VOCs present in the sample (Rakow and Suslick 2000; Urmila et al. 2015; Lin et al. 2020a, b). Bulk acoustic wave (BAW) sensors are used for measuring mass, density, elasticity, and acoustic frequency changes and can function as biosensors or chemical sensors (Sherrit et al. 2003). Surface acoustic wave (SAW) sensors are often used in conjunction with BAW sensors. Electrochemical sensors, categorized as amperometric devices, measure biochemical reactions through the oxidation–reduction process. They can detect the concentration of various analytes, such as oxygen or glucose, based on current changes (Thvenot et al. 1999; Yu et al. 2006; Eggins 2008). Metal oxide sensors, commonly used in various industries and storage processes, are sensitive to oxidizing and reducing compounds. They detect changes in conductivity as gas molecules are absorbed (Yamazoe et al. 1983; Demarne et al. 1992). Metal-oxide field-effect transistors (MOSFETs) can adsorb and dissolve hydrogen, generating an electric field that can be modulated and detected (Baietto et al. 2010). Other sensors used include conducting polymer (CP) sensors, carbon black composite (CBC), metal oxide semiconductor (MOS), and quartz field asymmetric ion mobility spectrometry (FAIMS), which characterizes and monitors products based on the ionic strength and polarity of VOCs passing through an electric field (Kolakowski and Mester 2007). FAIMS has been applied in various fields, including food and agriculture, for flavor and microbial spoilage detection (Karpas 2013; Kafle et al. 2016). In specific studies, researchers have used solid phase microextraction and gas chromatography–mass spectrometry to analyze volatile components in fruits such as "Simmonds" avocados, pitanga fruit, and different apple varieties to determine their maturity and quality (Santos Silva et al. 2019; Guo et al. 2020; Lin et al. 2022).

13.14 Advantages

Computer vision systems have emerged as a convenient and efficient means to gather data that would otherwise be challenging to obtain manually (Lefebvre et al. 1993). The advancement of digital image analysis technology has further facilitated the widespread adoption of these systems, enabling the generation of precise descriptive data from visual information (Patel et al. 2012). With the growing need for automated and objective assessment of various products for quality control purposes and the increasing trend of automation in various industries, computer vision has emerged as a valuable tool for rapidly and objectively measuring visual features of products (Gerrard et al. 1996). Additionally, machine vision systems offer the advantage of nondestructive and unobtrusive information acquisition (Zayas et al. 1996), making them particularly suitable for tasks that rely heavily on visual inspection (Tao et al. 1995b). Another attractive aspect of machine vision systems is their capability to create permanent records of measurements, allowing for retrospective analysis at any given time (Tarbell and Reid 1991).

13.15 Disadvantages

The inspection and grading of produce by human graders is frequently a laborious, monotonous, repetitive, and subjective task (Park et al. 1996). This method is also variable, and decisions are not always consistent between inspectors or from day to day, in addition to being costly (Heinemann et al. 1994; Tao et al. 1995a).

13.16 Conclusion and Future Prospects

The paramount quality attributes of horticultural commodities consist of size, color, shape, texture, and defect. In order to supplant the manual examination of processed food, a computer vision system is employed that furnishes reliable, impartial, and noninvasive evaluations. This chapter concerning machine vision systems offers a thorough analysis of the recent advancements in computer vision, HIS, MSI, infrared technology, ultrasound, tomographic imaging, MRI, CT, X-ray μCT, Raman spectroscopy, NMR, and VOCs e-sensing and are applied to the horticultural and food processing industries to control quality online. The utilization of machine vision systems has become increasingly common in industry for the purposes of inspection and evaluation due to their ability to provide swift, cost-effective, hygienic, reliable, and unbiased assessments.

Despite this, it can be inferred from the comprehensive survey conducted in this chapter that despite numerous approaches being suggested by researchers for the evaluation of the quality of fruits and processed products, there remains a need for the development of strong online controlling methods with enhanced capabilities. There are still some challenges present in the adaptation of this technology, as evidenced by the relatively slow adoption of machine vision technology across all sectors. However, in order to achieve rapid and precise defect detection in-line, there exist various challenges that must be overcome. These challenges encompass tasks such as addressing the nonuniform distribution of lightness on the curvature surface, recognizing the stem–calyx, conducting comprehensive surface inspections, managing the time-consuming acquisition and processing of spectral images, efficiently selecting appropriate wavelengths for diverse applications, and distinguishing between different types of defects.

References

Abbott JA (1999) Quality measurement of fruits and vegetables. Postharvest Biol Technol 15:207–225 https://doi.org/10.1016/S0925-5214(98)00086-6

Alonso R, Rodríguez-Estévez V, Domínguez-Vidal A et al (2008) Ion mobility spectrometry of volatile compounds from Iberian pig fat for fast feeding regime authentication. Talanta 76:591–596. https://doi.org/10.1016/J.TALANTA.2008.03.052

Anonymous (2022a) Economy 2021–2022 Statistics and Growth Figures Ye. https://www.indiastat.com/data/economy/2021-22 (Accessed on 17 April 2023)

Anonymous (2022b) Home; Department of Agriculture & Farmers Welfare; MoA & FW; Government of India; India. http://agricoop.gov.in/en/Hirticulture (Accessed on 17 April 2023)

Arendse E, Fawole OA, Magwaza LS et al (2018) Fourier transform near infrared diffuse reflectance spectroscopy and two spectral acquisition modes for evaluation of external and internal quality of intact pomegranate fruit. Postharvest Biol Technol 138:91–98. https://doi.org/10.1016/J.POSTHARVBIO.2018.01.001

Arendse E, Nieuwoudt H, Magwaza LS et al (2021) Recent advancements on vibrational spectroscopic techniques for the detection of authenticity and adulteration in horticultural products with a specific focus on oils, juices and powders. Food Bioprocess Technol 14:1–22. https://doi.org/10.1007/s11947-020-02505-x

Awad TS, Moharram HA, Shaltout OE et al (2012) Applications of ultrasound in analysis, processing and quality control of food: a review. Food Res Int 48:410–427. https://doi.org/10.1016/j.foodres.2012.05.004

Baietto M, Wilson AD, Bassi D, Ferrini F (2010) Evaluation of three electronic noses for detecting incipient wood decay. Sensors 10:1062–1092. https://doi.org/10.3390/S100201062

Bhargava A, Bansal A (2021) Fruits and vegetables quality evaluation using computer vision: a review. J King Saud Univ Comput Inf 33:243–257. https://doi.org/10.1016/j.jksuci.2018.06.002

Borompichaichartkul C, Moran G, Srzednicki G, Price WS (2005) Nuclear magnetic resonance (NMR) and magnetic resonance imaging (MRI) studies of corn at subzero temperatures. J Food Eng 69:199–205. https://doi.org/10.1016/J.JFOODENG.2004.07.018

Camelo AFL (2004) The quality in fruits and vegetables. In: In manual for the preparation and sale of fruits and vegetables; agricultural services bulletin, 151st edn. FAO, Rome, pp 38–90

Caporaso N, Whitworth MB, Fisk ID (2022) Prediction of coffee aroma from single roasted coffee beans by hyperspectral imaging. Food Chem 371:131159. https://doi.org/10.1016/J.FOODCHEM.2021.131159

Chen L, Opara UL (2013) Approaches to analysis and modeling texture in fresh and processed foods – a review. J Food Eng:497–507. https://doi.org/10.1016/J.JFOODENG.2013.06.028

Cozzolino D, Segura-Carretero A, Arrá Ez-Romá D (2015) Infrared spectroscopy as a versatile analytical tool for the quantitative determination of antioxidants in agricultural products, foods and plants. Antioxidants 4:482–497. https://doi.org/10.3390/ANTIOX4030482
Curry TS, Dowdey JE, Murry RCJ (1990) Christensen's physics of diagnostic radiology, 4th edn. Williams Wilkins, Philadelphia, PA
Da Silva Grandizoli CWP, Campos FR, Simonelli F, Barison A (2014) Grape juice quality control by means of ^{1}H NMR spectroscopy and chemometric analyses. Quím Nova 37:1227–1232. https://doi.org/10.5935/0100-4042.20140208
Demarne V, Grisel A, Sanjinés R et al (1992) Electrical transport properties of thin polycrystalline SnO2 film sensors. Sensors Actuators B Chem 7:704–708. https://doi.org/10.1016/0925-4005(92)80389-F
Eggins BR (2008) Chemical sensors and biosensors, vol 28. Wiley, Hoboken, NJ
ElMasry G, Kamruzzaman M, Sun DW, Allen P (2012) Principles and applications of hyperspectral imaging in quality evaluation of agro-food products: a review. Crit Rev Food Sci Nutr 52:999–1023. https://doi.org/10.1080/10408398.2010.543495
Erikson U, Standal IB, Aursand IG et al (2012) Use of NMR in fish processing optimization: a review of recent progress. Magn Reson Chem 50:471–480. https://doi.org/10.1002/MRC.3825
Fróna D, Szenderák J, Rákos MH (2019) The challenge of feeding the world. Sustainability 11:5816
Gerrard DE, Gao X, Tan J (1996) Beef marbling and color score determination by image processing. J Food Sci 61:145–148. https://doi.org/10.1111/J.1365-2621.1996.TB14745.X
Ginesu G, Giusto DD, Märgner V, Meinlschmidt P (2004) Detection of foreign bodies in food by thermal image processing. IEEE Trans Ind Electron 51:480–490. https://doi.org/10.1109/TIE.2004.825286
Gómez AH, He Y, Pereira AG (2006) Non-destructive measurement of acidity, soluble solids and firmness of Satsuma mandarin using Vis/NIR-spectroscopy techniques. J Food Eng 77:313–319. https://doi.org/10.1016/J.JFOODENG.2005.06.036
Guo J, Yue T, Yuan Y et al (2020) Characterization of volatile and sensory profiles of apple juices to trace fruit origins and investigation of the relationship between the aroma properties and volatile constituents. LWT 124:109203. https://doi.org/10.1016/J.LWT.2020.109203
Heinemann PH, Hughes R, Morrow CT et al (1994) Grading of mushrooms using a machine vision system. Trans ASABE 37:1671–1677. https://doi.org/10.13031/2013.28255
Ilaslan K, Boyaci IH, Topcu A (2015) Rapid analysis of glucose, fructose and sucrose contents of commercial soft drinks using Raman spectroscopy. Food Control 48:56–61. https://doi.org/10.1016/J.FOODCONT.2014.01.001
Ireson WG, Juran JM (1988) Quality-control handbook. McGraw-Hill, New York
Kader AA (2000) Quality of horticultural products. Acta Hortic 517:17–18. https://doi.org/10.17660/ACTAHORTIC.2000.517.1
Kafle GK, Khot LR, Sankaran S et al (2016) State of ion mobility spectrometry and applications in agriculture: A review. Eng Agric Environ Food 9:346–357. https://doi.org/10.1016/j.eaef.2016.05.004
Karoui R, Downey G, Blecker C (2010) Mid-infrared spectroscopy coupled with chemometrics: a tool for the analysis of intact food systems and the exploration of their molecular structure-quality relationships-A review. Chem Rev 110:6144–6168. https://doi.org/10.1021/CR100090K/ASSET/CR100090K.FP.PNG_V03
Karpas Z (2013) Applications of ion mobility spectrometry (IMS) in the field of foodomics. Food Res Int 54:1146–1151. https://doi.org/10.1016/J.FOODRES.2012.11.029
Kelly JFD, Downey G, Fouratier V (2003) Initial study of honey adulteration by sugar solutions using midinfrared (MIR) spectroscopy and chemometrics. J Agric Food Chem 52:33–39. https://doi.org/10.1021/JF034985Q
Kolakowski BM, Mester Z (2007) Review of applications of high-field asymmetric waveform ion mobility spectrometry (FAIMS) and differential mobility spectrometry (DMS). Analyst 132:842–864. https://doi.org/10.1039/B706039D

Kotwaliwale N, Singh K, Kalne A et al (2014) X-ray imaging methods for internal quality evaluation of agricultural produce. J Food Sci Technol 51:1–15. https://doi.org/10.1007/s13197-011-0485-y

Lefebvre M, Gil S, Brunet D et al (1993) Computer vision and agricultural robotics for disease control: the potato operation. Comput Electron Agric 9:85–102. https://doi.org/10.1016/0168-1699(93)90031-U

Lin H, Cui KW, Juan JH et al (2020a) Discrimination of Chinese Baijiu grades based on colorimetric sensor arrays. Food Sci Biotechnol 29:1037–1043. https://doi.org/10.1007/S10068-020-00757-Z/TABLES/2

Lin H, Jiang H, Adade SYSS, et al (2022) Overview of advanced technologies for volatile organic compounds measurement in food quality and safety. Crit Rev Food Sci Nutr 0:1–23. doi: https://doi.org/10.1080/10408398.2022.2056573

Lin H, Jin LJ, Xiu MZ et al (2020b) Development of colorimetric detection of 2,4,5-trimethyloxazole in volatile organic compounds based on porphyrin complexes for vinegar storage time discrimination. Food Anal Methods 13:2192–2203. https://doi.org/10.1007/S12161-020-01819-Y/METRICS

Lizhi H, Toyoda K, Ihara I (2010) Discrimination of olive oil adulterated with vegetable oils using dielectric spectroscopy. J Food Eng 96:167–171. https://doi.org/10.1016/J.JFOODENG.2009.06.045

Lohumi S, Lee H, Kim MS et al (2018) Calibration and testing of a Raman hyperspectral imaging system to reveal powdered food adulteration. PLoS One 13:e0195253. https://doi.org/10.1371/JOURNAL.PONE.0195253

Lorente D, Aleixos N, Gómez-Sanchis J et al (2012) Recent advances and applications of hyperspectral imaging for fruit and vegetable quality assessment. Food Bioprocess Technol 5:1121–1142. https://doi.org/10.1007/S11947-011-0725-1/TABLES/3

Ma J, Sun DW, Qu JH et al (2016) Applications of computer vision for assessing quality of agri-food products: a review of recent research advances. Crit Rev Food Sci Nutr 56:113–127. https://doi.org/10.1080/10408398.2013.873885

Magwaza LS, Opara UL (2015) Analytical methods for determination of sugars and sweetness of horticultural products—A review. Sci Hortic 184:179–192. https://doi.org/10.1016/J.SCIENTA.2015.01.001

Majchrzak T, Wojnowski W, Dymerski T et al (2018) Electronic noses in classification and quality control of edible oils: a review. Food Chem 246:192–201. https://doi.org/10.1016/J.FOODCHEM.2017.11.013

Manning L (2016) Food fraud: policy and food chain. Curr Opin Food Sci 10:16–21. https://doi.org/10.1016/J.COFS.2016.07.001

Mollazade K, Omid M, Tab FA, Mohtasebi SS (2012) Principles and applications of light backscattering imaging in quality evaluation of agro-food products: a review. Food Bioprocess Technol 5:1465–1485. https://doi.org/10.1007/S11947-012-0821-X/FIGURES/10

Müller A, Steinhart H (2007) Recent developments in instrumental analysis for food quality. Food Chem 102:436–444. https://doi.org/10.1016/J.FOODCHEM.2006.10.015

Narendra GV, Hareesha KS (2010) Quality inspection and grading of agricultural and food products by computer vision-a review. Int J Comput Appl 2:43–65

Nekvapil F, Brezestean I, Barchewitz D et al (2018) Citrus fruits freshness assessment using Raman spectroscopy. Food Chem 242:560–567. https://doi.org/10.1016/J.FOODCHEM.2017.09.105

Nicolaï BM, Beullens K, Bobelyn E et al (2007) Nondestructive measurement of fruit and vegetable quality by means of NIR spectroscopy: a review. Postharvest Biol Technol 46:99–118. https://doi.org/10.1016/J.POSTHARVBIO.2007.06.024

Nturambirwe JFI (2012) Non-destructive measurement of internal fruit quality using SQUID-NMR techniques. Cape Town, Stellenbosch University, Stellenbosch, South Africa

O'Grady MJ, O'Hare GMP (2017) Modelling the smart farm. Inf Process Agric 4:179–187

Okere EE, Arendse E, Nieuwoudt H et al (2021) Non-invasive methods for predicting the quality of processed horticultural food products, with emphasis on dried powders, juices and oils: a review. Foods 10:1–31. https://doi.org/10.3390/foods10123061

Opara UL, Pathare PB (2014) Bruise damage measurement and analysis of fresh horticultural produce—A review. Postharvest Biol Technol 91:9–24. https://doi.org/10.1016/J.POSTHARVBIO.2013.12.009

Park B, Chen YR, Nguyen M, Hwang H (1996) Characterizing multispectral images of tumorous, bruised, skin-torn, and wholesome poultry carcasses. Trans ASABE 39:1933–1941. https://doi.org/10.13031/2013.27672

Patel KK, Kar A, Jha SN, Khan MA (2012) Machine vision system: a tool for quality inspection of food and agricultural products. J Food Sci Technol 49:123–141. https://doi.org/10.1007/s13197-011-0321-4

Rakow NA, Suslick KS (2000) A colorimetric sensor array for odour visualization. Nature 406(6797):710–713. https://doi.org/10.1038/35021028

Razmjooy N, Mousavi BS, Soleymani F (2012) A real-time mathematical computer method for potato inspection using machine vision. Comput Math Appl 63:268–279. https://doi.org/10.1016/J.CAMWA.2011.11.019

Santos Silva J, Damiani C, da Cunha MC et al (2019) Volatile profiling of pitanga fruit (Eugenia uniflora L.) at different ripening stages using solid-phase microextraction and mass spectrometry coupled with gas chromatography. Sci Hortic 50:366–370. https://doi.org/10.1016/J.SCIENTA.2019.02.076

Sherrit S, Bao X, Bar-Cohen Y, Chang Z (2003) BAW and SAW sensors for in situ analysis. Proc SPIE Int Soc Opt Eng 5050:81–91. https://doi.org/10.1117/12.484252

Simal S, Benedito J, Clemente G et al (2003) Ultrasonic determination of the composition of a meat-based product. J Food Eng 58:253–257. https://doi.org/10.1016/S0260-8774(02)00375-8

Straker D (2001) What is quality? Part 1&2. In: Quality world, may issue. J Chart. Qual Institute, London

Su WH, Sun DW (2018) Multispectral imaging for plant food quality analysis and visualization. Compr Rev Food Sci Food Saf 17:220–239. https://doi.org/10.1111/1541-4337.12317

Sumonsiri N, Barringer SA (2014) Fruits and vegetables – processing technologies and applications. In: Food processing: principles and applications, second edn. Wiley, Hoboken, NJ, pp 363–381. https://doi.org/10.1002/9781118846315.CH16. 9780470671146

Sun DW (2011) Computer vision technology for food quality evaluation. Academic Press, San Diego, CA

Tao Y, Heinemann PH, Varghese Z et al (1995a) Machine vision for color inspection of potatoes and apples. Trans ASABE 38:1555–1561. https://doi.org/10.13031/2013.27982

Tao Y, Morrow CT, Heinemann PH, Sommer HJ (1995b) Fourier-based separation technique for shape grading of potatoes using machine vision. Trans ASABE 38:949–957. https://doi.org/10.13031/2013.27912

Tarbell KA, Reid JF (1991) A computer vision system for characterizing corn growth and development. Trans ASABE 34:2245–2255. https://doi.org/10.13031/2013.31864

Thvenot DR, Toth K, Durst RA, Wilson GS (1999) Electrochemical biosensors: recommended definitions and classification (Technical Report). Pure Appl Chem 71:2333–2348. https://doi.org/10.1351/PAC199971122333/MACHINEREADABLECITATION/RIS

Tiwari S, Kate A, Mohapatra D et al (2020) Volatile organic compounds (VOCs): biomarkers for quality management of horticultural commodities during storage through e-sensing. Trends Food Sci Technol 106:417–433. https://doi.org/10.1016/j.tifs.2020.10.039

Urmila K, Li H, Chen Q et al (2015) Quantifying of total volatile basic nitrogen (TVB-N) content in chicken using a colorimetric sensor array and nonlinear regression tool. Anal Methods 7:5682–5688. https://doi.org/10.1039/C5AY00596E

van Asselt ED, van der Fels-Klerx HJ, Marvin HJP et al (2017) Overview of food safety hazards in the european dairy supply chain. Compr Rev Food Sci Food Saf 16:59–75. https://doi.org/10.1111/1541-4337.12245

Watada AE (1980) Quality evaluation of horticultural crops – the problem. HortScience 15:47–47. https://doi.org/10.21273/HORTSCI.15.1.47

Yamazoe N, Kurokawa Y, Seiyama T (1983) Effects of additives on semiconductor gas sensors. Sens Actuators A Phys 4:283–289. https://doi.org/10.1016/0250-6874(83)85034-3

Yang D, Ying Y (2011) Applications of Raman spectroscopy in agricultural products and food analysis: a review. Appl Spectrosc Rev 46:539–560. https://doi.org/10.1080/05704928.2011.593216

Yu D, Blankert B, Bodoki E et al (2006) Amperometric biosensor based on horseradish peroxidase-immobilised magnetic microparticles. Sens Actuators B Chem 113:749–754. https://doi.org/10.1016/J.SNB.2005.07.026

Zayas IY, Martin CR, Steele JL, Katsevich A (1996) Wheat classification using image analysis and crush-force parameters. Trans ASABE 39:2199–2204. https://doi.org/10.13031/2013.27725

Zhang B, Huang W, Li J et al (2014) Principles, developments and applications of computer vision for external quality inspection of fruits and vegetables: a review. Food Res Int 62:326–343. https://doi.org/10.1016/j.foodres.2014.03.012

Part IV
Challenges in Postharvest and Analytical Technology of Horticulture Crops

Chapter 14
Development of On-Farm Storage and Processing Technologies for Horticulture Produce

K. C. Dileep, Priyanka Rana, Kanchan Bhatt, Manisha Thakur, Monika Thakur, and Anshu Sharma

Abstract Horticulture, as an expanding sector of agriculture, plays a vital role in the cultivation of diverse crops with high productivity but limited shelf life, necessitating their consumption in fresh form. Effective postharvest management is crucial to minimize losses and enhance the livelihoods of small-scale farmers. In order to address the significant issue of postharvest losses (PHL) in horticultural crops, on-farm processing and storage have emerged as effective measures for mitigating these losses. This chapter provides a comprehensive overview of the protocols employed to prevent decay in horticultural crops, specifically focusing on implementing necessary measures at the farm level to reduce PHL. The primary objective of this chapter is to enhance understanding of the technologies and interventions available at the farm level, which can effectively minimize PHL. By delving into the significance of horticultural crops, identifying factors influencing PHL, reviewing control techniques, and exploring potential solutions, this chapter seeks to contribute to the improvement of efficiency and sustainability in horticultural production systems. Furthermore, the chapter sheds light on the various challenges encountered in on-farm storage and processing. These challenges encompass a range of factors such as infrastructure limitations, capital constraints, technical knowledge and skill gaps, quality control and food safety concerns, limited energy availability, seasonal variations, weak market linkages, and complex policy and regulatory environments.

Keywords Postharvest losses · On-farm processing · Evaporative cooling · Cold storage · Modified atmosphere storage

---

K. C. Dileep · P. Rana · K. Bhatt · M. Thakur · M. Thakur · A. Sharma (✉)
Department of Food Science and Technology, Dr. YS Parmar University of Horticulture and Forestry, Nauni, Solan, Himachal Pradesh, India

M. Thakur, T. Belwal (eds.), *Advances in Postharvest and Analytical Technology of Horticulture Crops*, https://doi.org/10.1007/978-981-97-7247-6_14

14.1 Introduction

Horticulture is a rapidly expanding branch of agriculture that deals with the propagation, cultivation, harvest, processing, and utilization of crops such as fruits, vegetables, flowering, or ornamental crops, and medicinal as well as aromatic plants. The horticulture crops are a source of variation in both agricultural output and dietary intake for people. These crops are the front-runners to improve the well-being of small and marginal farmers in the era of commercial and high-value agriculture (Khandetod 2019). When compared to field crops, the horticultural crops have substantially higher production per unit area (Saha et al. 2021). However, the horticultural crops have comparatively shorter shelf life and are majorly meant to be consume fresh. Three crucial factors that regulate the fresh product's quality are as follows: sensory quality (aroma, hardness, color), safety (pathogens), and nutritional value (content, bioavailability of bioactive substances). Further, unlike dry grains, they are challenging to manage during the postharvest period due to their short shelf life and internal characteristics. These crops have a high water content and are vulnerable to mechanical harm and desiccation (wilting, shriveling). They are also prone to bacterial and fungal infections, which can lead to pathological collapse (Kader 2013). In addition to the perishable nature of the produce contributing to postharvest losses, lack of knowledge and finance are other reasons why the horticulture industry is unable to embark on its needed expansion. (Hailu and Derbew 2015). Although the changes that happen to fresh fruit cannot be halted, they can be minimized by regulating the storage and transportation conditions such as low storage temperatures, relative humidity control, and appropriate packing during transportation (Ahmad et al. 2015). If these elements are not appropriately handled, significant postharvest losses occur.

Losses that occur during all or part of the stages after harvesting (handling, storage, processing, packaging, transportation, and distribution) as a result of quality degradation are known to be postharvest losses. Pests, microbial infections, natural ripening processes, and environmental factors including heat, drought, and incorrect postharvest management are the main causes of these losses, which result in a general decline of crop quality (Olayemi et al. 2010). These losses are quantitative as well as qualitative. Reduction in the physical weight of food accessible for human consumption and other uses is referred to as quantitative PHL (FAO 1980). The spillage, grain breakage, rodent and pest damage, deterioration due to temperature variations, chemical changes, and humidity content all contribute to quantitative losses (Aulakh et al. 2013). Whereas qualitative PHL refers to a loss in the food's intrinsic qualities such as nutritive value, edibility, caloric value, and acceptability. Sources of qualitative PHL include microbial contamination, rodent and pest attacks, humidity levels, chemical changes, cracked grain, contamination by dangerous fungus, and pesticide residues. It will be considered a quantitative loss if food quality declines to the point where it is unfit for human consumption and is eventually rejected (FAO 1980).

In place of solely relying on increasing production in a world with limited natural resources and in an era of high and volatile food prices, it is becoming increasingly understood that reduction in postharvest losses can provide an environmentally sustainable and cost-effective contribution to food security and income improvement (Aulakh et al. 2013). Reducing postharvest losses could significantly increase the world's food supply, preventing the need to increase production in the future. In order to increase the amount of food that can be produced from the current crop, postharvest losses and quality degradation must be reduced (Kasso and Bekele 2018). The inherent characteristics signifying the overall quality of produce can only be maintained and not improved for the anticipated period that is referred to as the shelf life of commodity. As a general approach for reduction of postharvest losses, practices such as harvesting the produces during the coolest time of day to maintain low product respiration, avoiding wounds, bruising, crushing, proper cleaning of harvesting equipment or containers followed by precooling, keeping the harvested produce in a cold storage, appropriate sorting of the fruits, and using clean packaging material can be utilized (Das 2020). Therefore, to reduce postharvest losses and maintain the qualitative attributes of fresh product, advances in postharvest technology such as good harvesting practices and packing techniques are essential (Elik et al. 2019). Farmers can reduce losses due to physical damage, spoilage, and inferior crop quality by putting effective farm level control measures in place, such as proper handling practices, prompt harvesting, sorting, and grading, adequate storage techniques, and pest and disease management. Furthermore, farm level control guarantees crop quality preservation, prolongs the shelf life of harvested commodities, increases marketability, and ultimately boosts the profitability and sustainability of horticulture crops. In addition to contributing to ensure food security and lowering waste in the agricultural supply chain, it is essential to maintain the value and market attractiveness of horticulture commodities.

This chapter is dedicated to overview the protocols used to forestall the decay of horticultural crops with major emphasis on the importance of implementing the necessary measures at farm level to mitigate the postharvest losses. The vital problem of postharvest losses in horticultural crops is addressed in this chapter, along with information on farm-level treatments that can successfully reduce these losses. It prominently offers a thorough understanding of technologies and farm-level interventions that help to lower post-harvest losses and provides needed insights into enhancing the effectiveness and sustainability of horticultural production systems by addressing the significance of horticultural crops, identifying factors influencing PHL, reviewing control techniques, and examining appropriate solutions.

14.2 Creation of Farm-Level Evaporative Cooling Systems

An evaporative cooling system utilizes the cooling effect resulting from the evaporation of water (Lal Basediya et al. 2013). As water is applied to the porous surface, it undergoes a change in phase from liquid to gas because it absorbs the heat from the surrounding environment, thus providing cooling effect. This method of cooling only requires water as a coolant that runs freely over the porous surface and, since the process is driven by heat from the surrounding environment, requires no additional energy (Elansari et al. 2019). The cooling process relies on the natural flow of air, which is to be speeded up using fan-assisted systems. Evaporative cooling systems are comparatively cheap since they require very little or no energy input, can be constructed using locally available materials (e.g., bricks, charcoal, and sand) and water as a coolant, and thus can be suitable for low-value produce (Lal Basediya et al. 2013). This cooling system provides limited control over the temperature due to its dependency on the local climatic conditions. Temperatures above 15 °C and having very high relative humidity are achieved in most cases. Hence, evaporative systems are mostly suitable for non-chilling fruit and vegetables, especially tropical/subtropical crops, such as tomatoes, mangoes, bananas, sapotas, plums, grapes, capsicums, cluster beans, peas, radishes, peaches, carrots, cucumbers, beats, ladies fingers, green peppers, cauliflowers, and leafy vegetables (Elansari et al. 2019). As the availability of water is a critical factor in the operation of evaporative cooling systems, a continuous supply of reasonably sanitized and soft (i.e., no or low levels of calcium or magnesium) water in sufficient quantity, along with suitable water pumping technology and water storage reservoirs, are required. The efficiency of any evaporative cooling system depends mainly on the humidity of the surrounding air. In case of air that is completely saturated with water, evaporation does not take place and hence imparts no cooling. The evaporatively cooled storage structures work on the principle of adiabatic cooling caused by evaporation of water, made to drip over the bricks or cooler pads (Camargo 2007). In farm-level cooling, the surrounding air is hot and dry with the wind blowing, hence the systems work differently, thereby preventing any postharvest losses by maintaining lower temperature naturally in the farm itself. The evaporative cooling systems can be classified into direct and indirect cooling systems. In a direct cooling system, the water evaporates into the air to be cooled, simultaneously humidifying, and the thermal process is adiabatic saturation. The main feature of this process is that it is mostly efficient with higher temperature. Direct evaporative cooling system is commonly used in association with domestic cooling systems that cool the air by evaporating the water, thereby increasing the moisture level of the surrounding air. The cooled moist air is moved directly to the commodities. In direct cooling operation, the wet bulb temperature remains unchanged, but there is reduction in the dry bulb temperature. This type of system provides an efficiency of 55–70%. Direct evaporative systems are mostly suitable in hot and dry climates, whereas the wet-bulb temperature is 68 °F. The drop in temperature will generally be only a small fraction of the total evaporative reduction that is possible (Odesola and Onyebuchi 2009). The water leaks through the material and evaporates from the surface of the material due to the warm air flowing through it. This method slowly cools down the water. In contrast, in case of indirect cooling heat exchangers, moist air produced by the system is cooled to a temperature lower than dry air and then used to cool the environment. The two-stage systems also referred to as indirect–direct evaporative (IDEC) cooling systems, where indirect cooling is paired with a second direct evaporative cooling stage that can meet the entire cooling load for many buildings in arid to semiarid

Table 14.1 Advantages and disadvantages of evaporative cooling systems (Source: Zhao et al. 2008)

Advantages	Limitations
Can be made from locally available materials	It requires constant water supply to the pads
No special skill required to operate	Water with high mineral content leave mineral deposits, leading to damage
It can be fitted to the house hold.	Requires large space
Environment-friendly storage system	No dehumidification required
Does not require energy input	High dew point (humidity) condition decreases the cooling capability
Retain nutritive value	
Prevent the drying out of the produce	
Less expensive to install and operate	

climates. Table 14.1 presents the advantages and disadvantages of evaporative cooling systems.

This system typically has an effectiveness of 100–115%. All methods of indirect evaporative cooling require power to run both water pump and fans. For this reason, indirect evaporative cooling has limited applications at the farm-level operations. It is primarily used to cool dwellings and rooms. On the other hand, indirect evaporative cooling cannot be used in all environments and the reduction in temperature that can be achieved with this system is not as great as the reduction that can be achieved with conventional mechanical cooling systems. Further, the freshly harvested horticulture commodities require high humidity to prevent desiccation; so, the direct evaporative cooling system is preferred over indirect evaporative cooling systems. The factors that must be considered in order to choose the right cooling system needs the relative humidity of the area, a good supply of water, speed of wind, availability of the materials, and skills needed to build the cooler. Several types of materials, namely metals, fibers, ceramics, zeolite, and carbon, has the potential to be used as heat and mass transfer medium in the indirect evaporative cooling systems (Zhao et al. 2008).

14.3 Evaporative Cooling Systems Developed Across the World

Evaporative cooling (EC) occurs when air that is not already saturated with water vapor is blown across any wet surface. Thus, evaporative coolers consist of a wet porous bed through which air is drawn and cooled and humidified by evaporation of the water. Different types of evaporative cooler have been discussed below in detail.

One adaptation on the basic pot design is the Janata cooler, developed by the Food and Nutrition Board of India (Anonymous 1985). They are simple evaporative coolers that can be used at the domestic level. It consists of a storage pot placed inside a bigger pot that helps holds the water. The inner pot stores food that is to be cooled. The pot is then covered with a damp cloth that is dipped into the reservoir of water. Water evaporates, keeping the storage pot cool. The bowl is placed on wet sand to isolate the pot from the ground.

The India Agricultural Research Institute develops a cooling system (zero-energy cool chamber) that can be built in any part of the country using locally available materials (Roy and Pal 1994). The chamber operates by evaporative cooling and is constructed using double brick with sand-filled cavity walls. The construction of the chamber is simple with the floor made from a single layer of bricks, and then a cavity wall made of bricks around the outer edge of the floor with a gap of 75 mm between the inner wall and the outer wall. This cavity is then filled with sand. The structure must be protected from sunlight with the help of roof to provide shade. The water must be sprinkled twice daily.

The various other evaporative cooling structure in places across the world are the solarium-wind aspirator ventilated evaporative cooling structure constructed at the Central Potato Research Station (CPRS), Jalandhar, India (Chouksey 1985): metallic EC chamber fabricated by Habibunnisa et al. (1988); two-stage evaporative cooler (TSEC) developed by Jain (2007); Dadhich et al. (2008) constructed an evaporative cool chamber that uses baked bricks and riverbed sand; Jha (2008) developed a 5-ton capacity evaporative cooled storage structure (ECSS) at pilot scale.

14.4 On-Farm Cold Storage and Modified Atmosphere Storage Facilities

Postharvest food loss is caused by a lack of understanding about postharvest management, a low level of technology, climate change, and a lack of investment in food value chains (Losses and Waste 2011; Yahia 2019). Reducing postharvest losses is a critical step toward achieving food and nutrition security by 2030 (Affognon et al. 2015). Fruits and vegetables have the highest magnitude of losses following harvest due to their short shelf life and the lack of suitable postproduction infrastructure for these crops (Yahia 2019). As a result, effective postharvest technologies must be used to maintain both the quantity and quality of fresh fruits and vegetables along the value chain. Cold chain solutions, such as cold storage facilities, are critical to reducing losses following the harvest of fresh fruits and vegetables.

A cold supply chain for perishable food items is described as the continuous handling of the product in a low-temperature environment during the value chain's postharvest processes (Kitinoja 2013). A food cold chain pathway after harvesting includes precooling, mass storage, transportation, retail cooling, and residential refrigeration before consumption. A cold chain is not required to involve all the preceding phases, but it must include at least one of them. The storage of (large) quantities of products after processing and initial postproduction handling is referred to as bulk cold storage. Bulk cooling can occur on farms, in manufacturing facilities, in collection/grading centers, or in processing facilities. Precooling of goods prior to substantial cooling is required to achieve target temperature decreases faster than direct incorporation into bulk storage.

The recommended storage temperature for fruits and vegetables generally falls within the range of 0–4 °C. In this particular range of temperatures, the process of ripening is decelerated, thereby reducing the likelihood of bacterial proliferation. Accordingly, the freshness and quality of the produce are conserved. Refrigeration plays a critical role in maintaining these safe storage temperatures, thereby ensuring the consumable safety of fruits and vegetables over an extended duration. It is noteworthy that the optimal storage temperature may vary depending on the specific type of produce. Table 14.2 furnishes the minimum safe temperatures for the transportation and storage of some fruits and vegetables.

14.4.1 Bulk Cooling Systems for Storage of Fruits and Vegetables

Vapor compression, sorption, and evaporative cooling systems are frequently used cooling techniques for bulk storage (Table 14.2). A variety of factors, such as temperature and relative humidity range, as well as energy and refrigerant use, distinguish the systems significantly. Many factors influence the installation of a cold storage system and the site selection (Elansari et al. 2019): temperature and relative humidity required; easily accessible location and close to produce collection points; continuous power supply and water the availability of capital and skilled labor; and the economic value produced by cold storage to justify investment and operating

Table 14.2 Lowest safe temperatures of fruits and vegetables for transport and storage

Lowest safe temperature (°C)	Fresh fruits/vegetables
3	Asparagus, cranberry, jujube
4	Cantaloupe, potato, apple (such as Mcintosh and Yellow Newton), avocado cultivars (Booth and Lula), lychee, tamarillo
5	Cactus pear, cowpeas, guava, longan, durian, feijoa, pepino, kumquat, orange, mandarin, lima bean
7	Chayote, okra, snap bean, pepper, olive, pomegranate, pineapple
10	Carambola, cucumber, plantain, squash (soft rind), passion fruit, papaya, lime, melons (casaba, crenshaw, honeydew, Persian), eggplant, grapefruit, mango (ripe), tomato (ripe), taro, watermelon
13	Banana, jackfruit, mature green mango and tomato, lemon, sapota, mangosteen, breadfruit, cherimoya, pumpkin, hard-rind squash, yam, sweet potato, ginger

costs. Furthermore, the chosen technology must be environmentally friendly and sustainable, with zero or minimal greenhouse gas emissions (Ravishankar et al. 2020). Off-grid electricity, zero or renewable energy, and natural refrigerants should be facilitated whenever possible. As a result, solar photovoltaic systems, wind power, and biomass energy are gaining momentum in the rapidly expanding renewable energy market (Bishoge et al. 2018). Hence, sustainable energy can be effectively harnessed to power off-grid cold storage facilities.

14.4.2 Vapor-Compression Cooling System

Vapor-compression cooling is an electric-driven system that relies on refrigerant fluid phase transitions (Fenton and Callahan 2019). During execution, a moving refrigerant fluid (halogenated or natural refrigerants) is subjected to varying pressures. When exposed to low pressure environments, the refrigerating fluid evaporates while absorbing and removing heat from its surroundings in order to give a cooling effect. The gaseous refrigerant is subsequently compressed and condensed, reverting to its liquid condition while expelling the heat already received by the environment. This refrigeration method has a wide temperature range and a relative humidity of 80–90%, making it ideal for chilling even highly valuable food. This type of cooling system has a reduced initial investment cost but may have high operating costs based on whether the cold space is prefabricated or owner-built, new or used, and high expenses for upkeep. The coefficient of performance (COP) of this system is relatively high, which is the ratio of refrigerating capacity to power consumed. The usage of halogenated refrigerants, which are readily available and relatively inexpensive but are not ecologically benign, has been a weakness of this technology. Organic possibilities, such as hydrocarbons, carbon dioxide, and anhydrous ammonia, are now commercially available, but remain relatively expensive in developing nations (Lange et al. 2016).

14.4.3 Sorption Cooling System

A sorption cooling system uses thermal energy and is based on the physicochemical attraction of a natural refrigerant and an adsorbent or absorbent. Ammonia-water or lithium bromide-water combinations are commonly employed in absorption systems, while silica gel-water or zeolite-water working pairs are commonly used in adsorption systems. To produce a cooling effect, the cooling fluid is placed in a low-pressure environment and allowed to evaporate at room temperature while accumulating heat from the air around it. After that, the absorbent or adsorbent substance absorbs or adsorbs the gaseous refrigerant. Consequently, the evaporator's pressure drops, allowing a greater amount of refrigerant to evaporate. Thermal energy is then employed to evaporate the refrigeration fluid from the adsorbent or absorbent while restoring the conditions to their original state. Sorption cooling, such as vapor-compression systems, can provide a broad temperature range and low/medium relative humidity. As a result, it is appropriate for cooling sensitive and even highly valuable products. Because of their flexible utilization of reduced-grade thermal power sources (such as solar and industrial waste heat) as well as minimal maintenance requirements, sorption cooling systems have greater capital prices but relatively cheap operational expenses. Furthermore, when compared to vapor-compression systems, sorption systems have a lower COP (Lange et al. 2016).

14.4.4 Controlled and Modified Atmosphere Storage

Controlled atmosphere (CA) storage is most likely the most effective technique introduced to the fruit and vegetable firm in the twentieth century. Low oxygen (O_2) and high carbon dioxide (CO_2) levels in the storage atmosphere are typically used alongside with refrigeration. Inside a food storage room in CA storage, the gas composition is constantly monitored and changed to maintain the ideal concentration within extremely narrow tolerances. Since CA refrigeration is heavily capital-intensive and difficult to operate, it is best suited for goods that can be stored for an extended period, such as apples, kiwifruit, and pears. The exclusive use of CA storage for apples was due to its superior benefits over cooling for this specific item. CA storage is good for treating certain apple storage problems and diseases, while other illnesses may be increased or created by CA storage, its benefits outweigh its drawbacks (Prange and DeLong 2006). Also, CA storage has specific applications such as insect control and disinfection. CA packaging is the enclosing of food in a gas impermeable package within which the gaseous environment in terms of CO_2, O_2, N_2 (nitrogen gas), water vapor, and other trace gases has been modified and is selectively regulated to extend shelf life. However, by combining O_2 and C_2H_4 absorbers in packaging with CO_2 release agents, the packaged product could be categorized as CA packaging at least in the early stages of storage. Hypobaric storage is a related technique that involves storing the food in an atmosphere where pressure, humidity, and air temperature are precisely controlled, as well as the pace at which air in the storage environment is changed and strictly regulated (Burg 2004). Other than air, no other gases are required for CA storage or modified atmosphere storage. The overall pressure within the hypobaric chamber is significant since the O_2 concentration is proportional to it. Despite extensive research into the use of hypobaric settings for chilled storage of meat foods and horticulture goods, it has not been widely used commercially for food storage or transportation.

Modified atmosphere (MA) storage technology and controlled atmosphere (CA) storage are pioneering techniques that are widely used for the preservation of agricultural products, particularly fruits and vegetables. These approaches are used to augment low temperature management in many fresh products to delay ripening, decrease physiological problems, and control rot (Smith et al. 1987; Kader et al. 1989). MAP means "the packaging of a perishable product in an atmosphere which has been modified so that its composition is other than that of air," whereas the gaseous composition of fresh MAP foods is continually changing owing to chemical reactions and microbial activity, therefore controlled atmosphere storage (CAS) includes carefully monitoring and adding gases to maintain a fixed concentration of gases surrounding the product. Permeation through the package material may also result in gas exchange between the pack headspace and the external environment (Coles et al. 2003). MAP is divided into passive and active condition. The most important parameters in passive MAP are crop respiration rate and packaging film permeability. Because of food material respiration, oxygen consumption correlates to carbon dioxide emission in the atmospheric packing. Over a set period, the gas composition in the fresh product package achieves a definite equilibrium between the rate of respiration and permeability of packaging film. The total amount of carbon dioxide produced and oxygen absorbed by respiration is the same as that penetrated via the membrane exchange in an equilibrium state. The major gases utilized in MAP are oxygen, carbon dioxide, and nitrogen. These gases are used to regulate product organoleptic qualities with safe shelf-life extension.

14.5 On-Farm Processing Systems

Processing procedures are conducted in order to enhance the value of agricultural and horticultural products subsequent to their harvest. The main objective of processing is to minimize the qualitative and quantitative deterioration of the produce after harvest, while also to clean the raw materials by removing extraneous matter and immature ones, thus preparing them for secondary processing. The Food and Agriculture Organization (FAO) categorizes processed foods into three groups: primary, secondary, and tertiary. Primary processing encompasses fundamental tasks such as cleansing, categorizing, and packaging, which are commonly applied to fruits and vegetables. Secondary processing involves altering the basic product to a stage preceding its final preparation, which may involve actions such as drying, extracting juice, pulping, cutting fruits and vegetables, subjecting them to brief boiling, and freezing. Tertiary processing culminates in the production of high-value consumable foods, such as bakery products, instant foods, and health beverages.

14.5.1 Primary Processing

Precooling, as the initial step in temperature management for postharvest fruit and vegetables, plays a vital role in preserving their quality by effectively slowing down biological processes. It serves as the first and most crucial line of defense in maintaining the freshness and overall quality of the produce (Duan et al. 2020). Precooling is currently widely acknowledged as the most effective therapy; however, it is rarely utilized on a regular basis. Forced air cooling, hydrocooling, liquid ice cooling, and vacuum cooling are commonly employed precooling techniques. Although these methods have technical distinctions, they all involve the rapid transfer of heat from the product to a cooling medium, which can be air or water. For farm-level precooling operations in the countries like India, the mentioned precooling methods are affordable and can be effectively established except vacuum cooling. As per climate, the suitability of packaging the fruit with crushed or flaked ice to give quick cooling combined with high RH or storing the produce under shade and throughout nights must also be assessed.

After field heat removal, washing, cleaning, and trimming of fruits and vegetables are crucial. Farm-level washing eliminates contaminants, pesticide residues, pathogenic and spoilage microorganisms, ensuring product safety and quality (Gil et al. 2009). Common sanitizers such as chlorine and its derivatives are used, but electrolyzed water (EW) is an advanced washing option. EW is produced by electrolyzing a chlorinated salt solution, generating free chlorine, reactive oxygen species (ROS), ozone, and radicals. EW effectively sanitizes washing water (Pinto et al. 2015). Sufficient water availability and proper waste water disposal facilities are necessary for farm-level washing. Chemical-free waste water can be repurposed for irrigation. Ultrasound, ozone, cold plasma, and zinc acetate combined show promise in reducing pathogens, enhancing sensory quality and extending shelf life. Ultrasonication reduces microbial load by using acoustic forces to remove contaminants from the produce surface. Cold plasma damages bacteria's macromolecules, leading to their inactivation. Combining ozone with ultrasonication improves ozone dissolution and directly destroys microorganisms' genetic material (Zhou et al. 2022). These methods do not require water for decontamination, eliminating waste water disposal issues. However, they cannot remove soil on the produce surface, and their effectiveness diminishes in the presence of soil. Additionally, affordability may be a constraint for farmers in certain regions like India.

Sorting and grading are the next steps after washing, traditionally done by hand, growers now rely on professional packers with grading experience. Grading adds value to the product and fetches better market prices. Various criteria such as physical, technological, organoleptic, and commercial factors are considered for grading. Graders and sorters are categorized based on shape, size, surface roughness, aerodynamic properties, magnetic properties, specific gravity, and optical properties. Advancements include screen graders for rotating fruits, electronic color grading, computerized weight graders, and nondestructive technologies such as NIR spectrometers, magnetic resonance imaging, X-ray imaging, ultrasounds, and volatile emission analysis (Raj et al. 2016). Sorting involves assessing ripeness, quality, decay, disease, and injury to categorize items. Artificial intelligence is emerging as a replacement for traditional methods, with on-farm sorting machines utilizing imaging technologies such as RGB cameras, hyperspectral cameras, CCD cameras, NIR sensors, spectroscopy, and thermal cameras (Zhou et al. 2023). However, AI systems can be costly and less efficient as they rely solely on superficial appearance of fruits.

After sorting and grading, the next step is waxing, which applies a safe and edible protective coating to fruits and vegetables. This coating prevents moisture loss and regulates respiration. Wax can be applied by dipping the produce in a water wax emulsion or by spraying followed by drying. As the coating dries, it forms a thin film that creates a modified internal environment, reducing water loss and regulating gas exchange (Maina et al. 2019). Waxing now serves additional functions such as incorporating anti-browning agents, colors, flavors, nutrients, spices, and antimicrobial compounds. These additives prolong shelf life, reduce pathogen growth, minimize waste, and are environmentally friendly. Advancements include using edible materials such as lipid, protein, and carbohydrates for coating preparation. Coatings are cost-effective and user-friendly, derived from renewable resources and waste products (Kore et al. 2017). However, further research is needed for commercialization of these biodegradable coatings.

Conventional on-farm bulk packaging such as bamboo baskets, wooden baskets, and gunny bags have been replaced by corrugated fiberboard boxes to reduce forest load. However, traditional packages are still used for many horticultural products. Individual shrink wrapping and Lino bags are also used nowadays. Advancements in packaging include active and intelligent systems. Sachets and pads containing emitters (antioxidants, antimicrobial agents, carbon dioxide, food additives, flavorings, etc.) and absorbers (oxygen, ethylene, carbon dioxide, flavors, etc.) are commercially used to maintain freshness and quality. Intelligent techniques such as RF-ID, biosensors, and time–temperature indicators (TTIs) are employed for package condition monitoring during transportation (Falagan and Terry 2018). More research is needed to understand the dynamic physiological reactions of fresh produce to controlled atmosphere packaging (CAP) and modified atmosphere packaging (MAP). This research aims to determine optimal conditions for different cultivars, considering factors that impact product quality and longevity.

14.5.2 Secondary Processing

Removing moisture from perishable foods through dehydration is an effective method to prolong their shelf life. The primary aim of dehydration in food preservation is to decrease moisture content, which inhibits spoilage caused by microorganisms and reduces deterioration reactions. The drying process is intricate, involving the simultaneous transfer of heat and mass, necessitating precise control. Various dried vegetables, such as potato, carrot, broccoli, cauliflower, green pea, and spinach, are available in the market in different forms such as whole, sliced, or ground. Lowering moisture content induces alterations in shape, density, and porosity. When rehydrated, dried vegetables should retain desirable sensory and nutritional qualities. During the process of drying and storage, foods can experience browning, lipid oxidation, loss of color, and changes in flavor (Ahmed 2018).

Sun drying, an ancient method for dehydrating foods, is cost-effective but slow, causing losses and quality reduction. Solar drying is a better alternative, using a glass cover for heat retention and protection. Indirect solar dryers collect heat through solar collectors for drying. Locally made dryers suit rural areas, while mixed-mode dryers are affordable and farmer-friendly (Mustayen et al. 2014). Advanced drying techniques have been devised to reduce drying durations, enhance product quality, and enable superior process management. These methods encompass tray drying, tunnel drying, conveyor belt drying, as well as innovative freeze, vacuum, osmotic, cabinet or tray, fluidized bed, spouted bed, ohmic, and microwave dryers. The selection of a specific drying approach depends on factors such as product type, dryer accessibility, dehydration expenses, and desired desiccated product quality. Energy consumption and the quality of the dried items are also vital considerations when choosing a drying method (Bazyma and Kutovoy 2005). The drying characteristics of food materials are influenced by their composition, moisture content, thickness, geometry, airflow rate, and relative humidity (Ahmed 2018).

The demand for ready-to-use fresh fruits and vegetables in the market has detected noteworthy growth in recent years. This growth can be attributed to the changing preferences of consumers, particularly in their inclination toward the consumption of fresh-cut produce for salad preparation. The quality of fresh-cut fruits and vegetables is influenced by different factors, which include their visual appeal, texture, taste, and nutritional value, as noted by Kader (2002). The importance of each of these quality parameters varies depending on the specific kind of produce or product, as well as the way in which it is consumed—whether fresh or cooked, and with or without flavor enhancements such as dressings and dips. When people buy fresh-cut fruits and vegetables, they usually judge the quality based on how the product looks and how fresh it is. The layer of acknowledgment is significantly influenced by economic and psychological factors, while the overall characteristic of the item is primarily determined by its inherent properties. Over time, factors such as color, firmness, and taste of fruits and vegetables undergo changes, which ultimately impact their quality (Rico et al. 2007). Cutting fruits and vegetables triggers physiological changes such as ethylene production, increased respiration, membrane deterioration, water loss, microbial spoilage, chlorophyll loss, pigment formation, acidity reduction, sweetness increase, flavor compound formation, tissue softening, enzymatic browning, lipolysis, and lipid oxidation. To maintain quality, fresh-cut produce requires treatments such as blanching, chemical (chlorine, calcium lactate, hydrogen peroxide, ozone, preservatives), and physical (HPP, irradiation, UV treatment, hurdle technology) methods. Proper packaging and storage systems such as MAP, edible coatings, and frozen storage are also crucial (Giannakourou and Tsironi 2021).

Freezing slows down degradation reactions and prevents microbial activity, making it an excellent method for long-term storage of fruits and vegetables. It helps maintain the original color, flavor, and nutritional value of the produce (Alabi et al. 2020). After harvest, fruits and vegetables are stored at low temperatures for buffering before processing like freezing. Slicing is done for convenience or faster thermal processes. Blanching is crucial before freezing to deactivate enzymes, especially lipid oxidase to prevent off-odors and peroxidase to reduce browning. Some vegetables are air-dried to remove surface water. Dehydration enhances frozen product quality by lowering freezing point and forming smaller ice crystals. Frozen foods are stored below −18 °C as per the regulations, but optimal storage is at the intersection of freezing and glass transition lines (−20 °C to −40 °C) to minimize damage (Van der Sman 2020). The choice of freezing equipment depends on factors such as raw material type, size, shape, desired product quality, production rate, space availability, investment, and cooling medium. Freezing equipment is classified into three categories: direct contact (plate freezer, scraped surface freezer), air cooling (still air freezers, air blast tunnels, belt freezers, spray freezers, fluidized bed freezers, impingement freezers), and liquid cooling (immersion-type freezers, cryogenic freezers) (Muthukumarappan et al. 2018).

14.6 On-Farm Processing and Storage as Sustainable Approach to Reduce the Postharvest Losses

On-farm processing and storage present a sustainable methodology to reduce postharvest losses (PHL) in agricultural and horticultural crops. Through the implementation of effective measures at the agricultural level, farmers can minimize losses and enhance the overall sustainability of their endeavors. This approach presents numerous advantages and contributes to the efficient utilization of resources while addressing the challenges associated with PHL. One notable advantage of on-farm processing and storage involves the reduction of physical losses that occur during transportation and handling. By undertaking the processing and storing of crops at the agricultural site, cultivators can minimize damage and uphold the quality of their produce. This, in turn, helps prevent losses due to bruising, rotting, or other types of physical damage that may arise during the transportation to distant markets.

On-farm processing allows farmers to add value to their crops by transforming them into processed products. This not only extends the shelf life of the crops but also opens new market opportunities. For example, fruits can be processed into juices, jams, or dried snacks, while vegetables can be pickled, canned, or turned into frozen products. By diversifying their product range, farmers can tap into different market segments and reduce their reliance on fresh produce, which is more susceptible to spoilage. Moreover, on-farm storage enables farmers to bridge the gap between production and demand. It allows them to store surplus produce during times of high production and release it gradually into the market during periods of low supply. This helps to stabilize prices, reduce wastage, and ensure a more consistent income for farmers. Additionally, on-farm processing and storage empower farmers by giving them greater control over their produce and its value chain. By directly processing and storing their crops, farmers can eliminate or reduce the dependency on intermediaries, thereby improving their profitability and market access. This can enhance the economic resilience of small-scale farmers and contribute to the overall sustainability of agricultural systems.

To guarantee the sustainability of on-farm processing and storage, it is imperative to take into account specific considerations. Farmers must be afforded the opportunity to utilize suitable technologies and machinery for processing and storage that are economical, energy-conserving, and ecologically sound. It is vital to provide farmers with extensive training and capacity-building programs to enhance their knowledge and proficiency in postharvest management practices, including processing methodologies, storage management, and quality control.

14.7 Challenges in the Development of On-Farm Processing and Storage Systems

The advancement of processing and storage systems in agricultural settings faces various challenges, including insufficient infrastructure and financial resources, limited expertise, concerns about quality management and food safety, inadequate energy availability, fluctuations in production levels across seasons, weak connections to the market, and complex policy and regulatory frameworks.

These challenges have been mentioned in Fig. 14.1 and are discussed below in detail:

- **Financial constraints:** Establishing processing and storage facilities on a farm necessitates a substantial capital outlay. Farmers might encounter difficulties in acquiring the essential funds for procuring equipment, erecting structures, and up-keeping the infrastructure. Primary financial options in rural areas typically involve loans obtained from regional rural banks or local landlords known as zamindars. However, limited credit accessibility or exorbitant interest rates can compound the already existing financial restraints, adding to the challenges faced by farmers in this regard.
- **Infrastructure and logistics:** Establishing on-farm processing units faces challenges in infrastructure, particularly in remote areas lacking essential resources such as water and electricity. The diverse nature of agriculture, with variations in geography and small-scale farming, makes universal solutions impractical. Small farmers, constrained by limited land holdings, find it challenging to allocate space for processing. This necessitates context-specific strategies to overcome the unique obstacles and establish on-farm processing and storage facilities effectively.
- **Technical knowledge and skills:** The operation of processing and storage facilities demands specialized technical knowledge and skills. In rural areas, farmers often face a significant challenge due to the lack of expertise and technical know-how in food processing and handling. It becomes essential for farmers to acquire proficiency in various aspects, including food processing techniques, equipment handling and maintenance, quality control, and storage management. However, the availability of training opportunities is limited, and there is a shortage of skilled personnel in rural areas. This shortage further compounds the difficulties in effectively managing these facilities. The absence of trained and skilled manpower in rural areas presents a notable obstacle. It is often challenging to find workers with the necessary high-level skills in these regions. Providing on-the-

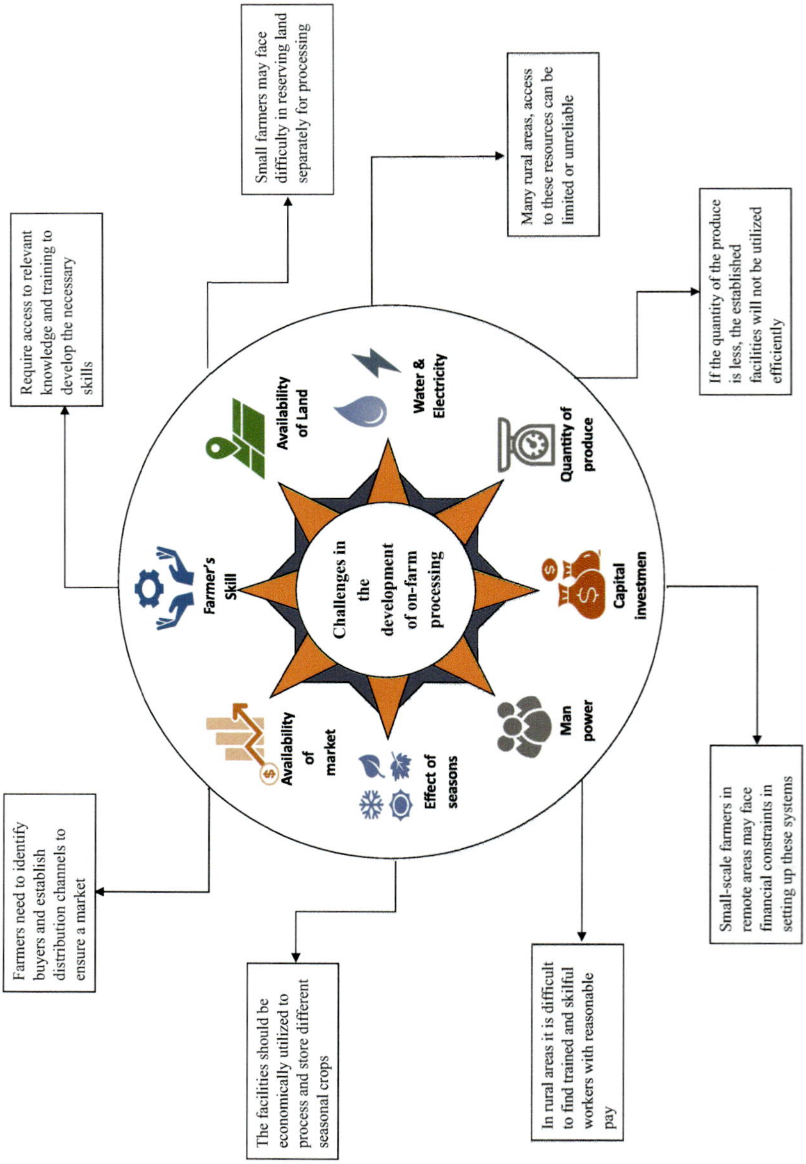

Fig. 14.1 Challenges faced during the development of on-farm processing and storage systems

job training becomes necessary, but it poses a serious problem for entrepreneurs as the workers may have limited education and must be taught in their local language for better comprehension. In fact, the scarcity of skilled, semiskilled, and unskilled workers has emerged as a critical factor in on-farm processing. Overcoming this challenge requires concerted efforts to enhance training opportunities, promote skill development programs, and address the language and education barriers that hinder the acquisition of necessary expertise.

- **Market access and demand:** The success of establishing on-farm processing facilities relies on a robust market for processed products. Farmers need to assess demand, secure market contracts, and establish distribution channels. Challenges arise, especially for those lacking industry networks. Building strong relationships with buyers, be it wholesalers, retailers, or direct consumers, is vital for a steady market. This involves negotiations, contractual agreements, and delivering products that meet buyer expectations. Effective marketing strategies, encompassing branding, packaging, advertising, and participation in local markets, are crucial for reaching the target consumers. Developing these strategies requires planning, market knowledge, and understanding consumer preferences. Overall, addressing challenges in market assessment, buyer relationships, and marketing is essential to ensure the viability of on-farm processing endeavors.
- **Seasonal variations and perishability:** Agriculture is inherently affected by seasonal variations, and many farm products have a limited shelf life. To optimize resource use and reduce product loss, processing and storage facilities must consider these factors. Effectively managing perishable commodities and aligning production schedules with these considerations requires meticulous coordination and logistics. To address product perishability, facilities should employ techniques such as quick processing, proper packaging, and efficient cooling or preservation methods. This ensures products maintain freshness and quality for an extended period, reducing the risk of spoilage. Coordinating production schedules is crucial, involving careful planning to align with raw material availability and processed product demand. Synchronizing processing with peak production periods allows farmers to use resources efficiently and minimize waste.
- **Energy and waste management:** Managing energy consumption in processing and storage facilities is crucial for cost control, emphasizing the adoption of energy-efficient practices such as optimizing equipment usage and incorporating advanced technologies. The exploration of renewable energy sources, such as solar panels or wind turbines, can further reduce reliance on traditional grids, lowering operational costs and promoting environmental sustainability. Concurrently, addressing waste management through proper disposal, adherence to environmental regulations, and implementing strategies such as recycling and utilizing organic waste for compost or bioenergy production contributes to a circular economy, minimizing environmental impact. Compliance with regulations, including monitoring emissions and pollution control measures, ensures that these facilities operate sustainably and responsibly, aligning with environmental standards.

- **Adaptation to technological advancements**: The agricultural industry is undergoing a transformative shift with the integration of new technologies, automation, and digital solutions. For farmers establishing processing and storage facilities, staying abreast of emerging trends and adopting relevant technologies is crucial for enhancing efficiency, productivity, and competitiveness. Yet, keeping pace with rapid technological advancements poses challenges. Farmers can overcome this by seeking guidance from agricultural extension services, industry experts, and technology providers. Collaboration with research institutions, participation in trade shows, and continuous learning through workshops or online courses offer valuable insights. Financial constraints may hinder technology adoption, but exploring partnerships, grants, subsidies, and funding opportunities can provide support. Farmers should assess technology relevance and feasibility based on specific needs, available resources, and market demands. Conducting cost–benefit analyses and piloting small-scale implementations can help mitigate risks and determine the potential impact before full-scale adoption.
- **Regulatory compliance**: Farm-level processing and storage facilities are subject to numerous regulations and standards pertaining to food safety, hygiene, environmental protection, and labor practices. Compliance with these requirements is essential but can pose significant challenges, particularly for small-scale farmers who may lack the necessary knowledge and resources to navigate the complex regulatory landscape. Ensuring food safety and hygiene is of utmost importance when processing and storing agricultural products. Farmers need to adhere to regulations regarding proper handling, processing, and storage techniques to prevent contamination and ensure the quality and safety of the final products. This includes implementing appropriate sanitation practices, maintaining proper temperature control, and following guidelines for packaging and labeling. Labor practices in farm operations are governed by regulations and standards aimed at ensuring fair treatment and safe working conditions for farm workers. Small-scale farmers, facing resource limitations, may encounter challenges in comprehending and implementing these labor standards. Adhering to regulatory and standards requirements in farm-level processing and storage facilities is essential for ensuring consumer safety, environmental sustainability, and fair labor practices. It is imperative to support small-scale farmers by facilitating their understanding and navigation of these regulations, offering necessary resources and guidance to enable effective compliance.
- **Tackling these challenges:** It necessitates collaborative efforts from farmers, government agencies, and research institutions. Supportive policies, capacity-building programs, access to finance, infrastructure development, and market linkages are essential in surmounting these obstacles and promoting the growth of on-farm processing and storage systems. In India, the government has rolled out various initiatives to uplift small-scale food industries and startups. These endeavors aim to foster entrepreneurship, offer financial support, spur innovation, and facilitate market access for small-scale food businesses. A compilation of notable government initiatives by different ministries and their respective objectives is provided in Table 14.3.

Table 14.3 Initiatives by different ministries of Indian government to support the establishment of small-scale food processing units

Sl. no.	Schemes	Year of establishment	Ministry	Detail	References
1	One District, One Product (ODOP)	2018	Ministry of Food Processing Industries (MoFPI)	The government provides financial support, infrastructure development, and marketing assistance to the identified products to enhance their production and market reach.	Ministry of Food Processing Industries (2023)
2	Pradhan Mantri Kisan Sampada Yojana (PMKSY)	2017	Ministry of Food Processing Industries (MoFPI)	PMKSY is a comprehensive scheme launched by the government to modernize and strengthen the food processing sector. It provides financial assistance for setting up new food processing units, upgrading existing ones, creating backward and forward linkages, and promoting food safety and quality.	Ministry of Food Processing Industries (2023)
3	Agro-Processing Cluster Scheme	2017	Ministry of Food Processing Industries (MoFPI)	The Agro-Processing Cluster Scheme promotes the development of agro-processing clusters in different regions of the country. It provides financial assistance for the creation of infrastructure facilities such as common processing centers, storage facilities, testing labs, and packaging units. The scheme encourages small-scale food processing units to operate in clusters, leading to economies of scale, enhanced efficiency, and shared resources.	Ministry of Food Processing Industries (2023)

(continued)

Table 14.3 (continued)

Sl. no.	Schemes	Year of establishment	Ministry	Detail	References
4	Startup India	2016	Ministry of Commerce and Industry	Encourages and supports startups across various sectors, including the food industry. Under this program, food startups can avail of benefits such as funding support, tax exemptions, simplified regulatory procedures, and access to a network of mentors and investors.	Startup India (2023)
5	Pradhan Mantri Mudra Yojana (PMMY)	2015	Ministry of Finance	PMMY is a scheme that provides financial support to small-scale entrepreneurs, including those in the food industry, through collateral-free loans. It aims to encourage entrepreneurship and self-employment.	MUDRA (2023)
6	Pradhan Mantri Kaushal Vikas Yojana (PMKVY)	2015	Ministry of Skill Development and Entrepreneurship	Skill development initiative aimed at providing industry-relevant training to the country's youth, enabling them to acquire employable skills and increase their productivity.	Pradhan Mantri Kaushal Vikas Yojana (2023)
7	Make in India	2014	Ministry of Commerce and Industry	The Make in India campaign promotes manufacturing and investment in various sectors, including food processing. It aims to simplify regulations, attract foreign investment, and foster a business-friendly environment to support small-scale food industries and startups.	Make in India (2023)

(continued)

Table 14.3 (continued)

Sl. no.	Schemes	Year of establishment	Ministry	Detail	References
8	Food Safety and Standards Authority of India (FSSAI)	2011	Ministry of Health and Family Welfare	FSSAI is the regulatory body responsible for ensuring food safety and quality standards in India. It provides guidelines, licenses, certifications, and conducts inspections to support compliance by small-scale food industries and startups.	FSSAI (2023)
9	Micro, Small, and Medium Enterprises Development (MSMED) Act	2006	Ministry of Micro, Small, and Medium Enterprises	The MSMED Act provides a legal framework for the promotion, development, and enhancement of competitiveness of micro, small, and medium enterprises (MSMEs), including those in the food sector. It offers various benefits and support mechanisms for MSMEs, such as credit facilities, infrastructure development, and technology upgradation.	Ministry of Micro, Small, and Medium Enterprises (2023)

14.7.1 Conclusion

In conclusion, effective postharvest management is crucial in horticulture to minimize losses and improve the livelihoods of small-scale farmers. This chapter provides a comprehensive overview of protocols aimed at reducing post-harvest losses (PHL) in horticultural crops, with a specific focus on implementing necessary measures at the farm level. By exploring the significance of horticultural crops, identifying factors influencing PHL, reviewing control techniques, and examining potential solutions, this chapter contributes to enhancing efficiency and sustainability in horticultural production systems. It emphasizes the importance of understanding and implementing technologies and interventions that can effectively minimize PHL. Moreover, the chapter sheds light on the various challenges faced in on-farm storage and processing. These challenges, ranging from infrastructure limitations to regulatory environments, underscore the need for addressing issues such as capital constraints, technical knowledge gaps, and quality control concerns. Overcoming these challenges is essential for improving postharvest practices in horticulture and ensuring the preservation of crop quality.

This chapter also provides various govt schemes to strengthen small scale food processing units. Additionally, the chapter provides information on various government schemes aimed at strengthening small-scale food processing units, which further support the improvement of postharvest practices and the overall development of the horticulture sector. Future efforts should be directed toward developing sustainable and cost-effective solutions. The integration of advanced technologies for improving storage conditions and reducing energy costs. Additionally, the development of innovative packaging materials and techniques that extend the shelf life of horticultural crops will be crucial. Furthermore, promoting knowledge transfer and providing training programs to farmers on proper on-farm storage and processing techniques will be essential for widespread adoption. Encouraging the establishment of farmer cooperatives or community-based storage and processing facilities can facilitate resource-sharing and enhance efficiency. Additionally, collaborations between government agencies, research institutions, and private stakeholders can facilitate knowledge exchange, research, and development, ultimately leading to the advancement of on-farm storage and processing practices.

References

Affognon H, Mutungi C et al (2015) Unpacking postharvest losses in sub-Saharan Africa: a meta-analysis. World Dev 66:49–68

Agriculture Organization of the United Nations (1980) Statistics division. In: Assessment and collection of data on post-harvest food grain losses. Food & Agriculture Org, Rome, p 13

Ahmad MS, Siddiqui MW et al (2015) Factors affecting postharvest quality of fresh fruits. In: Postharvest quality assurance of fruits: practical approaches for developing countries. Springer, Cham, pp 7–32

Ahmed J (2018) Drying of vegetables: principles and dryer design. In: Handbook of vegetables and vegetable processing. Wiley, Hoboken, NJ, pp 381–405

Alabi KP, Zhu Z et al (2020) Transport phenomena and their effect on microstructure of frozen fruits and vegetables. Trends Food Sci 101:63–72

Anonymous (1985) Zero energy cool chamber. Research bulletin No. 43, IARI, New Delhi

Aulakh J, Regmi A et al (2013) Estimating post-harvest food losses: developing a consistent global estimation framework

Bazyma LA, Kutovoy VA (2005) Vacuum drying and hybrid technologies. Stewart Postharvest Rev 4(7):6–13

Bishoge OK, Zhang L et al (2018) The potential renewable energy for sustainable development in Tanzania: a review. Clean Technol 1(1):70–88

Burg SP (2004) Postharvest physiology and hypobaric storage of fresh produce. CABI, Wallingford

Camargo JR (2007) Evaporative cooling: water for thermal comfort. An interdisciplinary. J Appl Sci 3:51–61

Chouksey RG (1985) Design of passive ventilated and evaporatively cooled storage structures for potato and other semi perishables. In: Proc. Silver jubilee convention of ISAE held at Bhopal, India, pp 45–51

Coles R, McDowell D, Kirwan MJ (eds) (2003) Food packaging technology, vol 5. CRC Press, Boca Raton, FL

Dadhich SM, Dadhich H, Verma RC (2008) Comparative study on storage of fruits and vegetables in evaporative cool chamber and in ambient. Int J Food Eng 4:1–11

Das K (2020) Role of organic and environment friendly post-harvest management of organically grown horticultural produces. Int J Chem Stud 8:1553–1556

Duan Y, Wang GB et al (2020) Postharvest precooling of fruit and vegetables: a review. Trends Food Sci 100:278–291

Elansari AM, Yahia EM, Siddiqui W (2019) Storage systems. In: Yahia E (ed) Postharvest technology of perishable horticultural commodities. Woodhead Publishing, Sawston, pp 401–437

Elik A, Yanik DK et al (2019) Strategies to reduce post-harvest losses for fruits and vegetables. Strategies 5:29–39

Falagan N, Terry LA (2018) Recent advances in controlled and modified atmosphere of fresh produce. Johns Matthey Technol Rev 62:107–117

Fenton DL, Callahan CW (2019) Refrigeration. In: Yahia EM (ed) Postharvest technology of perishable horticultural commodities. Woodhead Publishing, Sawston, pp 209–270

FSSAI (2023). Retrieved from https://www.fssai.gov.in/. Accessed 20 May 2023

Giannakourou MC, Tsironi TN (2021) Application of processing and packaging hurdles for fresh-cut fruits and vegetables preservation. Foods 10(4):830

Gil MI, Selma MV et al (2009) Fresh-cut product sanitation and wash water disinfection: problems and solutions. Int J Food Microbiol 134(1–2):37–45

Habibunnisa EA, Aror E, Narasimham P (1988) Extension of storage life of the fungicidal waxol dip treated apples and orange under evaporative cooling storage conditions. J Food Sci Technol 25:75–77

Hailu G, Derbew B (2015) Extent, causes and reduction strategies of postharvest losses of fresh fruits and vegetables–A review. J Biol Agric Healthc 5:49–64

Jain D (2007) Development and testing of two-stage evaporative cooler. Build Environ 42:2549–2554

Jha SN (2008) Development of pilot scale evaporative cooled storage structures for fruits and vegetables in hot and dry region. J Food Sci Technol 42:148–151

Kader AA (2002) Opportunities in using biotechnology to maintain postharvest quality and safety of fresh produce. HortScience 37:467–468

Kader AA (2013) Postharvest technology of horticultural crops-An overview from farm to fork. Ethiop J Appl Sci Technol 1:1–8

Kader AA, Zagory D, Kerbel EL et al (1989) Modified atmosphere packaging of fruits and vegetables. Crit Rev Food Sci Nutr 28:1–30

Kasso M, Bekele A (2018) Post-harvest loss and quality deterioration of horticultural crops in Dire Dawa Region, Ethiopia. J Saudi Soc Agric Sci 17:88–96

Khandetod YP (2019) Mechanization in horticulture crops: Present status and future scope. Advanced Agricultural Research and Technology Journal 3:92–103

Kitinoja L (2013) Use of cold chains for reducing food losses in developing countries. Population 6:5–60

Kore VT, Tawade SS, Kabir J (2017) Application of edible coatings on fruits and vegetables. Imp J Interdiscip Res 3:591–603

Lal Basediya A, Samuel DVK et al (2013) Evaporative cooling system for storage of fruits and vegetables-a review. J Food Sci Technol 50:429–442

Lange B, Priesemann C, Geiss M et al (2016) GIZ GmbH. Promoting Food Security and Safety via Cold Chains: Technology options, Cooling Needs and Energy Requirements

Losses FGF, Waste F (2011) Extent, causes and prevention. Food and Agriculture Organization of the United Nations, Rome

Maina B, Ambuko J, Hutchinson MJ et al (2019) The effect of waxing options on shelf life and postharvest quality of "ngowe" mango fruits under different storage conditions. Adv Agric 2019:1–9

Make in India (2023). Retrieved from https://www.makeinindia.com/. Accessed 20 May 2023

Ministry of Food Processing Industries (2023). Retrieved from https://www.mofpi.gov.in/. Accessed 19 May 2023

Ministry of Micro, Small & Medium Enterprises (2023) Know About MSME. Retrieved from https://msme.gov.in/know-about-msme. Accessed 20 May 2023
MUDRA (2023). Retrieved from https://www.mudra.org.in/. Accessed 19 May 2023
Mustayen AGMB, Mekhilef S, Saidur R (2014) Performance study of different solar dryers: a review. Renew Sustain Energy Rev 34:463–470
Muthukumarappan K, Tiwari B, Swamy GJ (2018) Refrigeration and freezing preservation of vegetables. In: Handbook of vegetables and vegetable processing. Wiley, Hoboken, NJ, pp 341–363
Odesola IF, Onyebuchi O (2009) A review of porous evaporative cooling for the preservation of fruits and vegetables. Pac J Sci Technol 10:935–941
Olayemi FF, Adegbola JA, Bamishaiye EI et al (2010) Assessment of post-harvest challenges of small-scale farm holders of tomatoes, bell and hot pepper in some local government areas of Kano State, Nigeria. Bayero J Pure Appl 3:39–42
Pinto L, Baruzzi F, Ippolito A (2015) Recent advances to control spoilage microorganisms in washing water of fruits and vegetables: The use of electrolyzed water. In: III International Symposium on Postharvest Pathology: Using Science to Increase Food Availability, pp 379–384
Pradhan Mantri Kaushal Vikas Yojana (2023) Home Page. Retrieved from https://www.pmkvyofficial.org/home-page. Accessed 20 May 2023
Prange RK, DeLong JM (2006) Controlled-atmosphere related disorders of fruits and vegetables. Stewart Postharvest Rev 5:1–10
Raj D, Senapati AK, Patel NL (2016) Recent advances in post harvest handling of horticultural crops
Ravishankar M, Bordat S, Aitken D (2020) Net zero cold chains for food. Carbon Trust
Rico D, Martin-Diana AB, Barat JM et al (2007) Extending and measuring the quality of fresh-cut fruit and vegetables: a review. Trends Food Sci Technol 18:373–386
Roy SK, Pal RK (1994) A low-cost cool chamber: an innovative technology for developing countries (tropical fruits storage). Poster paper on "Agricultural machinery and equipment; Handling, transport, storage and protection of plant products." Int Conference, Chiang Mai (Thailand), July19, 1993
Saha S, Kumar U, Saha TN et al (2021) Scope and importance of horticultural crops. In: Biointensive integrated pest management for horticultural crops. CRC Press, Boca Raton, FL, pp 1–11
Smith S, Stow J, Geeson J (1987) Production of modified atmospheres in deciduous fruits by the use of films and coatings. HortScience 22:772–776
Startup India (2023). Retrieved from https://www.startupindia.gov.in/. Accessed 19 May 2023
Van der Sman RGM (2020) Impact of processing factors on quality of frozen vegetables and fruits. Food Eng Rev 12:399–420
Yahia EM (ed) (2019) Postharvest technology of perishable horticultural commodities. Woodhead Publishing, Sawston
Zhao X, Shuli L, Riffat SB (2008) Comparative study of heat and mass exchanging materials for indirect evaporative cooling systems. Build Environ 43:1902–1911
Zhou W, Sarpong F, Zhou C (2022) Use of ultrasonic cleaning technology in the whole process of fruit and vegetable processing. Foods 11:2874
Zhou Z, Zahid U, Majeed Y et al (2023) Advancement in artificial intelligence for on-farm fruit sorting and transportation. Front Plant Sci 14:1082860

Chapter 15
Food Safety and Security Issues (HACCP and HAZOP) with the New Postharvesting Technologies and Their Uses

Niharika Shanker

Abstract The chapter explains how to identify and manage microbial dangers in novel postharvest technologies and their applications using a systematic, rule-based approach. The methodology, which is based on the Hazard and Operability Studies (HAZOP) concept that was first developed in the chemical industry, employs an artistic, collaborative approach to identify potential hazards and a diverse research group selected from inside the producer firm. It promotes a structured reflection on any deviations from procedures, which enables decision-makers to put preventive control measures in place. This chapter highlights the need of the Food Safety Management System (FSMS), Hazard Analysis Critical Control Points (HACCP), and Prerequisite Programs (PRPs), which form the foundation of HACCP, in order to prevent foodborne outbreaks. This chapter covers the use of the Food Safety Management System, Hazard Analysis Critical Control Points, and Prerequisite Programs, which form the basis of HACCP to manage foodborne disease outbreaks.

Keywords Food Safety Management System · Critical control points · Prerequisite programs

15.1 Introduction

In order to satisfy consumer demands for wholesome, sanitary food and environmentally friendly rules, the food processing industry of today is concentrated on integrating and deploying a variety of technologies on an industrial scale (Hotchkiss 1992). When opposed to more conventional advancements, these innovations are claimed to be more ecologically friendly and to have a minimally detrimental effect on the content of nutrients and non-nutrients. The implementation of Hazard

N. Shanker (✉)
Amity Institute of Food Technology, Amity University Uttar Pradesh,
Noida, Uttar Pradesh, India
e-mail: nshanker1@amity.edu

M. Thakur, T. Belwal (eds.), *Advances in Postharvest and Analytical Technology of Horticulture Crops*, https://doi.org/10.1007/978-981-97-7247-6_15

Analysis Critical Control Points (HACCP) ensures the safe manufacturing and production of food by analyzing all possible hazards associated with the introduction of new technology into existing conventional postharvest processing lines. Similar to this, Hazard and Operability Study (HAZOP) accredits operators utilizing these new technologies with safe working conditions: hazard analysis concepts HAZOP and HACCP in food processing planning. Presently, these ideas serve as thorough quality control instruments for the advancement of industrial procedures. By evaluating the health risks connected to food processing, distribution, and consumption and pinpointing crucial control points throughout the process, the HACCP program guarantees safe food handling and processing. Taking into consideration the impacts of contact and airborne contamination on health, HAZOP analysis is used to identify hazardous working conditions in particular industrial operation phases. In light of the advent of new postharvest technologies, this chapter addresses HACCP and its connections to HAZOP.

15.2 Early History—HACCP

Internationally recognized food safety management systems were first developed in the early 1960s as an effective way to support space exploration. The National Aeronautics and Space Administration (NASA), with support from Pillsbury, developed the initial idea for the program. The interim goal of this project was to develop a food product with the highest level of food safety for spaceflight (Fig. 15.1). The main motivating idea was that astronauts could not get sick in space, as this could seriously endanger their lives. Therefore, it is necessary to eliminate the risk of food poisoning due to hazards. Both institutions have developed groups that create crumb- and pathogen-free products that can withstand space travel and have a significantly longer shelf life by detecting all critical defect areas (Thakur and Belwal 2022).

15.3 Seven Principles of HACCP

The system consists of the following seven principles:

- ***Principle 1****: Conduct a hazard analysis.*
- This facility classifies food safety hazards and recognizes precautions and capabilities to control those hazards. Risks associated with post-harvest involves cultivation, harvesting, raw materials, raw materials, processing, manufacturing, sales, and marketing. The preparation and consumption of each food item is evaluated in detail. Potential microbiological, chemical, or physical contamination is identified. This includes both raw materials and finished goods received.
- ***Principle 2****: Identify critical control points (CCPs).*

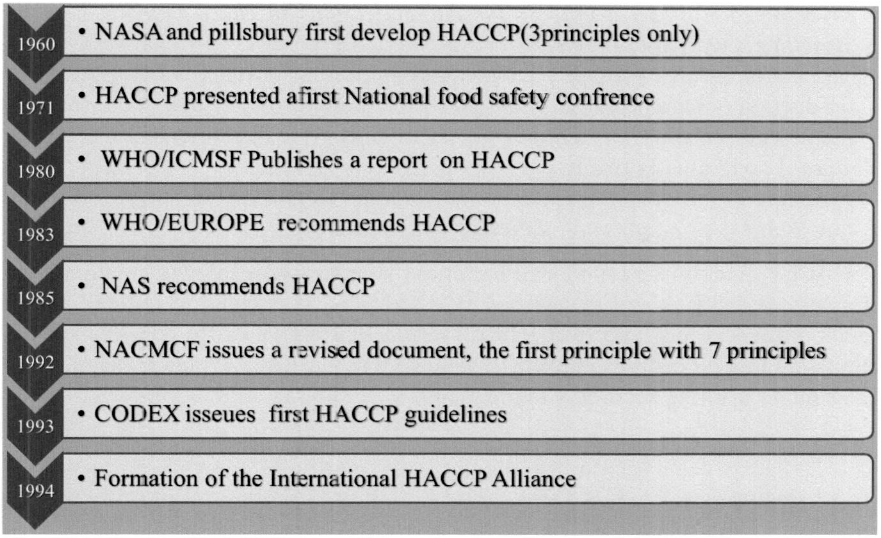

Fig. 15.1 Timeline of major events leading to the creation of HACCP. (1) National Aeronautics and Space Administration. (2) World Health Organization. (3) International Commission on Microbiological Specification for Foods. (4) National Academy of Science. (5) National Advisory Committee on Microbiological Criteria for Foods

- Critical controls identify points for controlling each of the hazards listed above. Critical control points are steps in a procedure where loss of control can result in unwanted health risks. Control of these points must be maintained to ensure product safety. Please carefully control the temperature after processing, where the product has critical control points, such as cooking, cooling, and reheating.
- ***Principle 3****: Initiate critical limits.*
- After identifying each critical control point, you should describe the boundaries of that point. This can be, for example, minimum or maximum temperature, or the addition of controlled amounts of pH, water activity, or available chlorine.
- ***Principle 4****: Organize a structure to monitor and control of the CCP.*
- Next, you need to establish specific procedures for monitoring critical control points. Specifying a maximum control point temperature is of little use unless you have a special way to collect data for critical control points. If a product relies on the addition of acid for safety reasons, pH or acidity limits must be established and monitored.
- ***Principle 5****: Initiate a helpful action while observing that a specific CCP is not under control.*
- Further step is to improve a strategy to take precautionary action while observing that critical control point limits have been exceeded. Curative measures are aimed at ensuring that products hazardous to health or products counterfeited as a result of deviations do not enter the market.

- ***Principle 6**: Initiate a process for authentications to check that the HACCP structure is working proficiently.*
- Validation ensures that the plan executes as designed. Success in ensuring the production of safe products. Industries should confirm their individual HACCP plans. This includes not only the plan itself but also the implementation of all key control point measurements.
- ***Principle 7**: Initiate the validation that contains all procedures and records suitable to these principles and their applications.*

HACCP guidelines need that all facilities maintain specific documentation, such as a hazard analysis and the written HACCP plan, documenting critical control points, critical limits, and oversight of verification activities. In conclusion, it is significant that procedures in place to validate that the HACCP plan is being monitored and is working according to the plan.

15.4 HAZOP

Hazard and Operability Analysis (HAZOP) is a systematic and structured study of operations/processes to evaluate and identify problems that may result in undesirable outcomes to systems, equipment, or personnel. This is a risk assessment tool that is carried out at the initial postharvest stage to ensure sufficient time to implement recommendations during harvest. HAZOP can also be run during the harvest phase to implement changes made during the process. Potential hazards and operational issues that could result in defective products are identified using HAZOP.

Deviations from processes are known to cause risk events. HAZOP is recognized by this directive. Quality risk management is one of the risk assessment tools in the postharvest field. HAZOP can assess risks in processes, systems, and equipment. Objectives for conducting HAZOP surveys include reviewing plantation harvest, milling and planning, determining what to build, interviewing suppliers, improving farmer safety, and reviewing operations. HAZOP execution has four phases as shown in Fig. 15.2.

HAZOP is important for agriculture when you want to predict situations that can prevent postharvest losses. This may allow you to take corrective action before an unexpected situation occurs. The HAZOP methodology allows you to better manage your procedures, processes, and services by investing in risk prevention

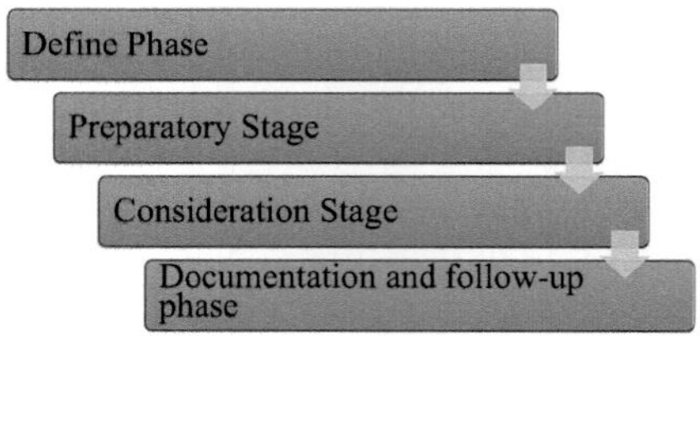

Fig. 15.2 Preliminary HAZOP execution phases

methods. Another point to note is that this method reduces internal costs. The sooner errors are detected, the sooner they can be prevented and corrected. Therefore, HAZOP is an excellent tool for improving internal safety and efficiency. This methodology allows us to stay one step ahead of possible undesirable situations. The acronym HAZOP stands for Hazard and Operability Study. This tool is a methodology for identifying risks associated with production processes. This is aimed at preventing accidents and ensuring occupational safety. However, HAZOP focuses on identifying potential problems early in project development or process execution and identifying what risks exist to the system and its execution.

15.5 Importance of Food Safety and Security

Food safety is a specific term that covers various aspects of food handling, processing, and storage to prevent postharvest damage and disease. The biochemical, microphysical, and microbiological parts of food security are brought together under this category. Critical to the chemical quality of food is the control of allergens, which can be fatal for some highly sensitive people. Other chemical properties of food, such as vitamins and minerals, are also major factors and greatly influence food quality, but are less important from a food safety perspective. The absence of foreign objects is important to avoid postharvest damage. Small physical elements such as glass and metal can be dangerous and cause serious injury to consumers. Pathogenic bacteria, viruses, and toxins produced by microorganisms can contaminate food and affect food safety (Hanning et al. 2012).

According to data from the National Institute of Allergy and Infectious Diseases (NIAID), 4% of adults and 5% of children in the United States suffer from food allergies. Up to 200 people are reported to die from food allergens each year (Neugut et al. 2001), but current death certificates indicate only about 20 such deaths per year. The Centers for Disease Control (CDC) suspects these deaths are underreported due to confusion over coding among people filling out death certificates. Comparative statistics on food poisoning caused by pathogenic microorganisms make microbial quality the most important aspect of food safety. Therefore, food safety mainly focuses on controlling the contamination of food by pathogens (Scallan et al. 2011).

15.6 Importance of PRP

The Prerequisite Program (PRP), a technique used in postharvest and food safety, takes into account typical circumstances for operating in nonhazardous situations throughout the manufacture of nutritious food. An established food safety program can be built upon this concept. In order to improve safe and appropriate sanitation conditions, prerequisite programs combine industry standards with general food

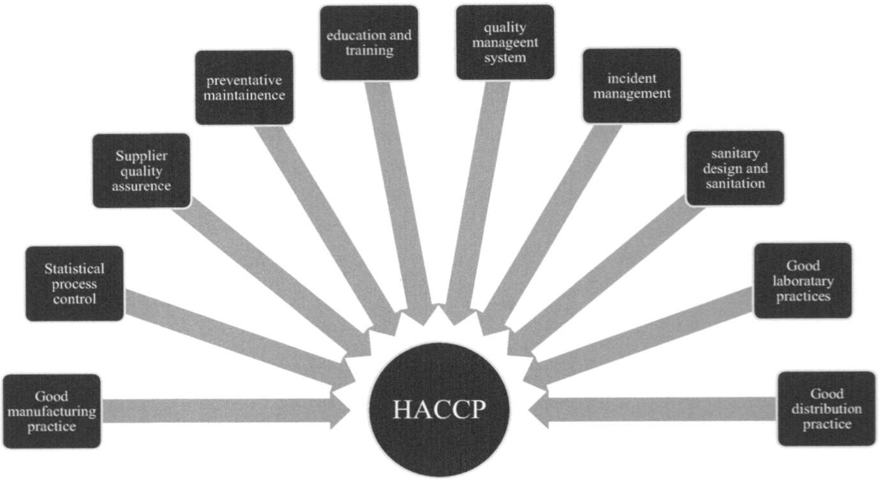

Fig. 15.3 Prerequisite programs (PRPs) adapted from functions of HACCP

management practices. These circumstances have brought up broad questions regarding food safety without the requirement for stringent testing procedures. The required curriculum is mostly designed in accordance with current Good Manufacturing Practices (cGMP). This application caused a delay in the release of new required programs because of the grocery store's development. This PRP includes food operations such as systematic cleaning and sanitation procedures, good personal hygiene, food room layout, and appropriate behavior during working hours.

The term "prerequisite program" is often misunderstood when compared to other programs that address food safety planning. Terms such as operational prerequisite program and critical control points are constantly misunderstood when it comes to PRP (Fig. 15.3) (Okpala and Korzeniowska 2021).

15.7 Examples of Common Prerequisite Programs

A HACCP system needs to be based on a strong foundation of essential programs in order to generate safe food. In order to preserve food while it is under control, each sector of the food industry must establish the required conditions. Traditionally, cGMP has been utilized to do this. Nowadays, creating and executing a successful HACCP plan is thought to require these circumstances and methods. The fundamental operating and environmental conditions needed to produce food that is both safe and healthy are established by prerequisite programs. Prerequisite programs that are frequently offered are as follows:

- ***Facility***: Facilities must be laid out, with harvesting and processing in accordance with hygienic design principles. This includes the supply of air, water, and energy.
- ***Supplier management***: Every facility must verify that their supplier has an effective GMP and food safety program in place. These may be subject to ongoing supplier assurance and supplier HACCP system reviews.
- ***Specification***: Written specifications should be available for all ingredients, products, and packaging materials.
- ***Production equipment***: All equipment must be constructed and installed in accordance with hygienic design principles. A preventive maintenance and calibration schedule should be established and documented.
- ***Cleaning and hygiene***: Cleaning and sanitation procedures for all equipment and facilities must be recorded and followed. A general hygiene plan should be developed.
- ***Personal hygiene***: All farmers/employees and others entering manufacturing facilities must follow personal hygiene requirements.
- ***Education***: All farmers/employees must receive written training in personal hygiene, GMP, cleaning and hygiene procedures, personal safety, and their role in the HACCP program.
- ***Chemical control***: To guarantee the appropriate handling and segregation of nonfood chemicals at the harvesting plant, documented protocols must to be established. This covers any insecticides, fumigants, cleaning agents, and baits used on or around the property.
- ***Receiving, storing, and shipping***: All raw materials and products must be stored under hygienic and suitable environmental conditions to ensure their safety and health.
- ***Traceability and recall***: All raw materials and products must be batch numbered and have a recall system in place to allow for quick and complete traceability and recall if required.
- ***Pest control***: An effective pest control program should be in place.

15.8 Symbiotic Importance of PRP and HACCP

A vital part of the world class food safety program is PRP. They serve as a solid foundation for other essential components and perform a significant supporting function. Although PRP was once thought of as a crucial component of HACCP, it is now known to (indirectly) actively contribute to food defense, the prevention of food fraud, and safe food processing technology and engineering (Lee et al. 2021). These complement HACCP as a preventive control system (PC) in a mutually beneficial way. PRP encompasses excellent mitigation practices in growing, harvesting, manufacturing, storing, distributing, retailing, food service, and agriculture and is applicable to all phases of the global food supply chain (Soon et al. 2020).

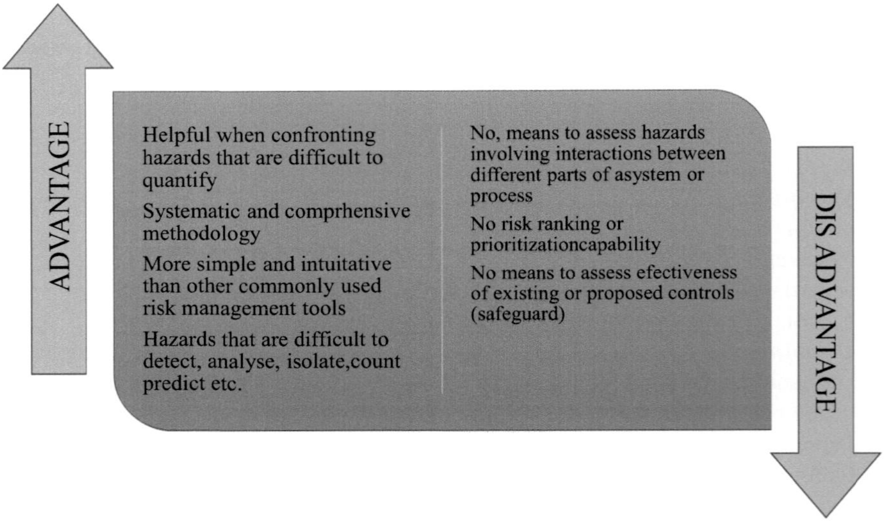

Fig. 15.4 Advantages and disadvantages of HACCP

The advantages and disadvantages of HACCP in relation to sound management techniques have been explained in detail in Fig. 15.4.

A HACCP plan's singular, focused goal is to guarantee food safety. The existence and efficacy of the necessary program must be assessed while creating and implementing a HACCP plan. Every necessary program needs to be recorded and examined on a regular basis. The necessary programs are not designed or managed using HACCP plans. He can, however, include some aspects of earlier initiatives into his HACCP plan. For example, many manufacturers have put preventative maintenance procedures for their processing equipment in place to avoid unplanned equipment failure and lost production. The HACCP team may determine that verification activities such as equipment calibration and routine maintenance should be included in the plan when creating a HACCP plan (Mortimore and Warren 2014). Successful implementation of a management's commitment to her HACCP approach is a requirement of the HACCP program. It is the duty of management to inform staff members about HACCP's benefits and drawbacks. Better resource management, quicker problem solving, and assurance of food safety are the advantages. An efficient HACCP plan needs to be established and put into practice before requirements and training can be created. State and federal laws and regulations apply to many of our terms and practices. Every necessary program needs to be created and maintained separately from the HACCP plan, reviewed on a regular basis, and documented. Certain programs could need verification. For instance, documentation that chemicals are safe for food and lower risk to a manageable level may be needed for cleaning and disinfection procedures.

15.9 Conclusion

A valuable framework for tracking the whole food system is offered by the HACCP system. HACCP is primarily used to lower the risk of foodborne illness from the point of harvest to the point of final consumption. HACCP is a preventive strategy intended to identify and lessen potential risks to food safety. Through constant observation and problem-solving throughout the process, you may spot dangers and act quickly to address them. By applying prevention and detection procedures to specific applications, HACCP allows businesses the flexibility to quickly adapt new techniques and technologies. The foundation of a good HACCP strategy is a strong commitment to: lower the risk of food poisoning; emphasize rigorous criteria for food safety; resolve security-related concerns efficiently and promptly. By maintaining accurate records and providing the necessary paperwork, you can offer legal protection. Increase commercial prospects by adhering to global regulations Trade standards contribute to the dependability of products. Under HACCP, it is now the manufacturer's or retailer's obligation to ensure food safety. By implementing this strategy, the likelihood of risk is decreased and consumers can feel secure knowing that the items they use are as safe as technology permits.

References

Hanning IB, O'Bryan CA, Crandall PG, Ricke SC (2012) Food safety and food security. Nat Educ Knowl 3(10):9

Hotchkiss JH (1992) Pesticide residue controls to ensure food safety. Critc Rev Food Sci Nutr 31(3):191–203

Lee JC, Daraba A, Voidarou C, Rozos G, Enshasy HAE, Varzakas T (2021) Implementation of food safety management systems along with other management tools (HAZOP, FMEA, Ishikawa, Pareto). The case study of Listeria monocytogenes and correlation with microbiological criteria. Foods 13(9):10

Mortimore SE, Warren BR (2014) Prerequisite programs: current perspectives in food manufacturing. Perspect Public Health 134:191

Neugut AI, Ghatak AT, Miller RL (2001) Anaphylaxis in the United States: an investigation into its epidemiology. Arch Inter Med 161:15–21

Okpala COR, Korzeniowska M (2021) Understanding the relevance of quality management in agro-food product industry: from ethical considerations to assuring food hygiene quality safety standards and its associated processes. Food Rev Int 39:879–1952

Scallan E, Hoekstra RM, Angulo FJ, Tauxe RV, Widdowson MA, Roy SL, Jones JL, Griffin PM (2011) Foodborne illness acquired in the United States — major pathogens. Emerg Infect Dis 17:7–15

Soon M, Brazier A, Wallace CA (2020) Determining common contributory factors in food safety incidents. A review of global outbreaks and recalls, 2008–2018. Trends Food Sci Technol 2020(97):76–87

Thakur M, Belwal T (eds) (2022) Bioactive components: a sustainable system for good health and well-being, vol I. Springer, Singapore, p 609. https://doi.org/10.1007/978-981-19-2366-1_5

Chapter 16
Effective Supply Chain Management: A Way Forward for Sustainability and Food Security

Vijay Yadav Tokala and Neeru Dubey

Abstract Food plays a significant role in human existence, and the survival of increasing world population depends on a reliable global food supply network. Having food accessible when you need it, in the right quantity, and of the right quality is essential, regardless of whether these food supply networks are national or international. In the last decades, the food industry has seen significant development. The movement of people for work and migration has led to a demand for food products from across the globe and retailers are keen to meet the rising demands. The design of retail spaces, technological developments in monitoring and tracking, shipping, operations management, and packaging have made it possible to import food products from anywhere in the world and make them available to customers all year long. The individual or organization has to have a thorough understanding of the many processes of the contemporary supply chain and the elements involved in the best performance at each step to achieve successful supply chain, while adhering to sustainable practices. The chapter discusses the effectiveness of food supply chains from beginning to finish as well as many elements that may impact it.

Keywords Food supply chain · Food safety · Food logistics · Food retailing

16.1 Food Supply Chain

The global food supply network is essential for maintaining prosperous human life. Whether these supply chains are national or multinational, having food available when you need it and in the proper quantity and quality is crucial. According to the most current estimate, more than 9.6 billion people will inhabit the planet by the

---

V. Y. Tokala (✉)
The Postharvest Education Foundation, La Pine, USA
e-mail: vijayyadav.t@postharvest.org

N. Dubey
Ernst and Young, New Delhi, India

M. Thakur, T. Belwal (eds.), *Advances in Postharvest and Analytical Technology of Horticulture Crops*, https://doi.org/10.1007/978-981-97-7247-6_16

year 2050 and feeding them would be one of the greatest concerns of humanity (John and Sabine 2012). Along with feeding the global supply chain has to also cater to the diversity developed into people due to increased exposure to different geographies. People all over the world now need a diverse variety of food that needs a robust and efficient supply chain. On estimate, about 30% of the entire amount of food produced for human consumption, or 1.3 billion metric tons, is lost or wasted globally. About 1.6 billion people could be fed by the food that is lost or wasted globally (FAO 2011). According to a recent FAO report, 13.8% of food is wasted after harvest on farms, during transit, storage, and processing, costing more than US$ 400 billion a year (FAO 2020). World leaders and industry professionals in the food industry have underlined the need for a paradigm shift in how we approach issues of food and nutrition security. There is a need for coordinated measures to make sure that food systems are effective, better managed, and sustainable in the endeavor to eradicate hunger and malnutrition. This will safeguard the food systems' long-term ability to generate food as well as causing less damage to the environment (HLPE 2014).

16.1.1 Actors of Food Supply Chain

The food supply chain refers to the network of activities, organizations, and parties that work together to transport food from its initial form as a raw material to our plates. It takes a complex network of interrelated entities working together to make food available, not just one chain of specific businesses. The producer (agriculture-associated organization) is where the food supply chain begins, and the food sourced at this point goes via different processing techniques. Numerous logistics and transportation firms support the migration from the point of production to the point of consumption. These businesses guarantee that the food will arrive to us promptly and in good condition.

The key actors of food supply chain are discussed in the following sections.

a. Producers

The food supply chain begins with the producer, who provides food in its unprocessed form, including grains, fruits, vegetables, fish, poultry, meat, and so on. Farmers who work in the agriculture sector are the producers. Small scale farmers to extremely large corporations operate in the farming industry. While some are new to the industry, others can be family farms that have been growing food for decades. Every nation needs a robust food production industry because it influences both the population's access to food and the industry's capacity to remain economically viable. There are parties in the supply chain that provide farmers with raw materials (seeds, agricultural equipment, pesticides, fertilizers, etc.). These "input providers" are often powerful global corporations with a great influence in the supply chain. Additionally, the producers contend with climate and weather patterns that are becoming more unpredictable, a lack of water and land degradation brought on by industrialization and urbanization. Farmers are now cultivating what they can sell for a good price to have economic sustainability as the margins for producers within the food supply chain are becoming narrower. It is a fair deal for farmers, but it does influence the essential diversity and also accessibility of core foods in the long run.

b. Processors
The businesses in the food supply chain known as processors are responsible for converting the raw materials given by food producers into finished goods to satisfy customer demands. This process is also called manufacturing. The fresh food from the producer will either be prepared at this point in the food supply chain for customers in a ready-to-eat format or used as a raw material to make other food items that consumers will purchase. Food processing is a crucial process that not only keeps the food sector's economy afloat by meeting consumer demands and needs but also contributes to waste reduction and increased food availability by extending the shelf life of raw food products. The food processors must collaborate closely with the businesses that make up the downstream supply chain and transport processed food to consumers.

c. Distributors and Retailers
Distributors are the companies that serve as a bridge between producers, processors, and markets. The processors provide the distributors with either fresh produce or processed food, which they subsequently distribute through a variety of routes to the ultimate customer. Distributors often purchase in bulk and use a network of distribution centers and warehouses to distribute the goods as and when they are needed further down the food supply chain.

The process of retailing involves showcasing the goods for the customer. This can take the shape of small neighborhood corner stores or huge supermarkets and hypermarkets that deal with millions of stockkeeping units. The customer can purchase a wide range of traditional and cutting-edge items from the food industry at the retailer level of the supply chain. Food processors compete with one another for shelf space in retail environments, while retailers compete with one another to draw more customers through their doors. Customers can choose from a variety of merchants, retail channels, and retail formats. Retailers are progressively developing creative business models that provide consumers a compelling value proposition based on pricing, quality, and service in an effort to set themselves apart from their rivals.

d. Consumers
The final link in the food supply chain is the consumer. The customers' purchase of the goods and provision of the required funds to move upstream through the supply chain is essential to the chain's economic sustainability. Every population on the earth depends on food, therefore competition within the food supply chain focuses on diversity and value addition rather than core products.

e. Hospitality Sector
The hotel industry is also a significant player in the food supply chain. Despite not being a retailer or distributor in the traditional sense, it serves as a vital conduit

between the production and the customer. Restaurants, motels, and takeout locations will purchase food as a raw ingredient from manufacturers or processors and then modify it to meet the needs of the ultimate customer. These organizations perform "made to order" service.

16.1.2 Factors Effecting Supply Chain

Farm products go downstream in one of two ways: directly to the customer as fresh produce (fresh fruit and vegetables, milk, grain, etc.) or in large quantities as a raw material in food processing factories. The availability of food, the makeup of the food product, and how it gets to our plates are all shaped by industrial, technological, economic, social, and political elements that have an impact on the food industry.

The general factors that affect the food supply chain are

- Production and geography of agriculture products
- Processing and maintenance of quality
- Choices of consumers and market demand
- Availability of infrastructure and logistics
- Local authorities and involvement of government bodies
- Small businesses

16.1.3 Major Barriers in the Agriculture Sector Development

a. Availability of the Funds

The production stage in agri-industry is usually with the least financial support/resources. The geographical dispersion of the producing sector, rising transaction costs, and a lack of adequate banking infrastructure in rural areas are all to be blamed for the lack of financing. In addition, the financing organizations lack the loan packages specifically designed for it.

b. Cooperation Among Supply Chain Participants

Coordination that is effective and efficient is lacking in the present agri-ecosystem. It is essential for building competitive advantage within the agri-supply chain. To satisfy the customer needs, coordination is required among producers, processors, distributors, and other supply chain participants. The quality and quantity of the produce and on-time delivery to consumers are the key performance indicators, which will be possible to achieve only via collaboration and strategic partnerships.

c. The Role of Public- and Private-Sector Organizations

To ensure the effectiveness of public spending, it is also necessary to encourage good governance and the inclusion of accountability measures in institutional and governance structures. Public private partnership investments should also be encouraged for competitive growth and specific developments.

16.2 Food Production

The food-producing or agricultural sector is a major global employer and source of revenue. This was the only thriving sector during the COVID-19 pandemic as it is necessary for every human being. The bulk of those employed in the agriculture industry work on small farms located in countries with substantial rural populations. The importance of the agriculture business sector or the food supply chain has increased dramatically as a result of the world's ongoing challenges such as population explosion, rising lifestyle health conditions, and climate changes. With the aim of "feeding the world," the emphasis has shifted to using the supply chain's food production stage as a foundation to develop value-added goods and innovations. The majority of nations are developing national policies and plans to encourage investments in agriculture value chain growth as the world debate on ensuring the food security and sustainability. This calls for a better comprehension of the factors affecting the industry (Chandrasekaran 2014).

16.2.1 Major Role Players in Agriculture Supply Chain

The agriculture supply chain/agribusiness refers to any business entity associated with agriculture, namely production, processing, input suppliers, trading, exporting, logistics, wholesalers, and retailers.

a. Input supplier:
The inputs to the manufacturing process must be of adequate quality and supplied on time for it to function successfully and efficiently. These inputs come in the shape of manpower, machinery, pesticides, fertilizers and seeds, organic fertilizers, biopesticides and seeds, among other things. Any break in the chain of input supplies might affect the availability of food. Most of the input providers are the large corporations that have made investments in food innovation (seeds, chemicals, etc.).

b. Owner of the land:
Land is a valuable resource in the majority of developing nations, and farming will only begin if a sufficient plot of land has been found. Individual farmers typically own relatively little land in low-income nations, making it challenging to manage productivity. On the other side, someone with a lot of land may sublease or rent it out for a little time.

c. Farmers and food production:
The production of food (including cattle, fish, and meat) is controlled at this stage of the agricultural supply chain, which is the most crucial since it takes into account

inputs and resource availability. There are around 450 million farmers worldwide, who may be divided based on the size of the farm, the crops farmed, and the degree of production process complexity.
The bulk of farmers in the supply chain are smallholders, and farm sizes range widely being 2–4 ha for small and marginal farmers, 4–8 ha for mid-sized farmers, and more than 8 ha for large-sized farmers. The most dangerous process in the agriculture value chain is farming. It is vulnerable to market volatility and unforeseen weather circumstances (KPMG, 2013).

d. Traders:
They are intermediate between the producers and processor/consumer. In Indian agriculture supply chain, the image of traders has been both positive (for maintaining continuity of the supply chain and sometimes investing in different links of the supply chain) and negative (for wrongfully utilizing their buying power to exploit farmers by paying very low prices). In Indian context, some states allow the distribution of certain commodities only through *mandis*, which have traders certified and licensed by the Agricultural Produce Market Committee (APMC) (NDTV 2015).

e. Direct markets:
They are also called as farmer markets and directly connect the farmers with consumer bypassing the need of the middlemen. This could be an open platform arranged by government or by farmer groups to reach out the possible consumers. Apart from these, logistic and transport companies also have a very important role in shifting the produce from one step to other.

16.3 Food Sourcing

Food sourcing is activity of securing the raw materials for either direct sale to customers or for processing. It entails locating the proper material source and assessing its conformity. However, once the source has been found, it is crucial to control the procurement and delivery of the content. "Procurement" is also the term used for the procedure that controls the sourcing and purchasing process. The terms sourcing and procurement are both used interchangeably in business and literature. In case of the SME (small- or medium-sized firm), a single person might be handling majority of steps of supply chain; however, for the large corporations, sourcing and purchasing, as well as other supply chain functions, would be filled by separate people. A variety of challenges within the food industry have made sourcing more difficult, which includes the influence of sustainable practices on public policy, the volatility of commodity prices, the availability of required quantities of the food, the effects of the erratic climate, traceability, and food safety.

Various sourcing tactics are used by organizations, depending on the industry and business strategy. Depending on the extent of their operations or services, businesses obtain their resources from several places. To obtain the lowest prices, some businesses may exclusively source locally, while others may source globally. Various product categories will require various sourcing approaches and reasons.

16.3.1 Multiple Sourcing or Single Sourcing

The majority of the organizations source their raw materials procures with the main aim to reduce operational costs and it can be from single source or from various locations with lower production costs, less labor requirement, greater transparency, and reliability and with less labor requirement. Depending on the commodity or service and its availability, the organization will decide in choosing suppliers. In the case of the single or limited supplier of certain commodity, the sourcing will be single. In some cases, the company opts for single sourcing with an objective to maintain close collaboration with the certain supplier. Single source is also preferred to ease the process of traceability and food safety of commodity.

The multiple sourcing mainly aims at keeping flexibility in case of failure of certain source to meet the demand or increased demand in the market. In majority of the cases, multiple sourcing allows buyer to have more negotiating power and ensure continuous supply chain.

16.3.2 Internal Sourcing and Outsourcing

Internal sourcing or insourcing includes production from subsidiary branch of the organizations or by internal production. While outsourcing involves purchase from external sources with the idea to reduce costs and utilize specific investments and expertise, the main challenge in outsourcing is lack of control over the specific operations, lower monitoring, and the inability to react to unprecedented fluctuations in the market. Sometimes the improper monitoring of some of the agriculture commodities can lead to serious quality and food safety issues.

16.3.3 Sustainable Sourcing

The companies strive to increase their profitability while delivering the message of their commitment toward sustainability without compromising the resource availability for future generations. They monitor the impact of their each operations on the environmental and societal factors. The important considerations while procuring include a code of labor rights, human rights, environmentally friendly practices, and anti-corruption.

16.4 Food Processing

This stage in the process is one of the important pillars of the supply chain as it changes the raw food generated in the agriculture field into a value-added form so that it may be served to customers with higher shelf life and ease of usage. This might be as simple as packing the raw food so that transportation can be facilitated, or it could be more complex like turning them into ready-to-eat meals for the retail setting. The main aim of the food processing is to extend the postproduction/post-harvest life of the food product, expand the dietary variety, and increase the value of the commodity.

The market for processed foods has grown in high-income nations and demand for frozen foods or foods that can be quickly heated in a microwave oven increased dramatically. This is due to changes in working habits, family lives, and increasing desire to explore new foods. However, when compared to consumers who opt for fresh foods, the customers favoring prepared meals or extensive processing foods is less. Minimal processed foods do have a great demand especially in urban areas.

As the processed food is prepared for ready consumption, the processors further must meet the consumer requirements by the way it looks, tastes, and uniqueness, as well as certain global standards, for example, chocolate coating done over the biscuits. Packaging is yet another essential factor to be considered during food processing. Over last couple of decades, there have been a range of novel developments in the packaging ensuring food safety, display of nutritional information, labeling, and brand development, leading to good marketing.

16.5 Food Packaging

The packaging of agriculture commodities is designed to ensure ease for handling, distribution, and marketing, while keeping the fragile commodities safe from damages from impact injuries, pests, and other external factors along the supply chain (Opara 2013). The main purposes of the packaging are to hold (containment), to prevent mechanical injuries to the produce along the supply chain (protection), to display instruction and product information (communication and traceability), to allow easy handling (convenience), and to be sturdy when piled up in the storage (strength) (Tokala 2022). Selecting the best packaging material for a certain product requires consideration of the physical, chemical, and morphological features of the produce as well as the functional characteristics of the packaging materials that are accessible (Han et al. 2018). A few aspects of the produce and the packaging to take into account are as follows (Paine & Paine, 1992):

- Product shelf life
- Storage system
- Transportation and handling techniques
- Marketing system

- Consumer preference
- Reusability/recyclability
- Price and effectiveness

16.6 Food Logistics

The flow of food through the supply chain until it arrives on the consumer's plate is known as food logistics. Logistics also includes the information flow from the customer back into the chain. Logistics operations, which include quantification, procurement, inventory management, transportation, and fleet management, as well as data collecting and reporting, comprise the operational part of supply chain management. Technological developments in food processing and transportation have made it feasible for enterprises to deliver larger volumes of food more quickly and over greater distances than ever before.

To guarantee effective transportation, it is crucial to map the infrastructure, including distribution centers, warehouses, and temperature-controlled storages, and to establish a logistics network. The single most crucial component in preserving the freshness of fresh produce is proper temperature regulation. The condition and integrity of the products will be damaged if there are interruptions in the cold chain, and the commodities will be more prone to quality decline. Perishables are kept at their best condition at destination markets by being kept at the ideal temperature to prevent exposure to temperature changes. Transport that is temperature-controlled can take advantage of a number of energy-saving initiatives. Typical energy-saving techniques include reducing the speed of the fans by 50% and cycling the refrigeration compressors on and off.

The way logistical services distribute goods to customers has been impacted by the rise of the internet and mobile technologies. Using linked networks makes it possible to track deliveries, utilize click and collect, change delivery timings, or reserve certain pickup times, among other things. The smart agri-food logistic system utilizes tracking and tracing technologies, with access to data being received through the sensors in the system. The key concepts in the smart agri-food logistic system include:

a. ***Tracking system***: It enables to trace the commodity and different steps in the supply chain. The sensors included along with tracking systems also provide dynamic information of the parameters such as temperature, humidity, ethylene, and light. The information obtained can be further processed to forecast shelf life, spoilage, and other quality parameters by the time the product reaches retail store or end-consumer.
b. ***Internet of things (IoT)***: This system includes several sensors along each step of supply chain to generate information, communicate each other, and act instantly to ensure the best quality possible till endpoint. Communication technologies

such as RFID, QR/bar codes, and near-field communications are used in this system, which can work only when strength of signals is good enough.

c. ***Business intelligence***: All the relevant data generated by the sensors and devices are processed to support in strategic and functional decisions to develop most efficient models for sourcing, processing, logistics, and distribution.

16.6.1 Logistics for the Perishables

The agricultural supply chain usually operates across four different ranges: at room temperatures (23–28 °C), low temperatures (7–18 °C), chilled temperatures (0–7 °C), and frozen temperatures (−18–0 °C). Commodities such as canned foods and dried grains are usually handled at ambient or room temperatures, while ice-cream and frozen fresh produce are kept at temperatures −15 °C or below. The majority of fresh fruit and vegetables are good to be kept at temperatures 7–18 °C. The quality of these commodities is seriously affected and even pose food security risk if their optimum temperatures are not maintained. Therefore, temperature monitoring becomes a legal requirement in the food sector (Bridgemohan et al. 2022). It is to be understood that handling different types of foods at same supply chain might cause risk of product interactions and adversely affect the quality of the commodities. Some of the key factors to be considered while designing cold chain for the perishables are as follows (Duggan 2014):

a. Maintaining product-specific continuous cold chain from farm to retail;
b. Mode of the transport should be decided based on the physiology of the produce, time and cost factor, and consumers' choice;
c. Traceability of the production, processing, and transport should be efficient to ensure food safety compliances;
d. Packaging best suited for the commodity is to be selected to protect them from any impact injuries and increase possible shelf life;
e. Appropriate technologies for monitoring temperature, relative humidity, spoilage, etc. will be helpful in collecting data for designing efficient supply chain;
f. Understanding the dynamic nature of the cold chain of certain commodity and analyzing effective supply design toward on-time delivery within the feasible costs.

16.7 Food Safety and Regulations

Food supply chain operates in a very dynamic environment and mismanagement at any step can seriously affect the food quality, making it unsafe for consumption. This further can lead to fatal outcomes and has already been evidenced in several instances around the world. Every aspect of the food supply chain needs to comply with certain regulations and standards to make sure the food reaching end-consumer is safe and retained the quality promised. Quality is a key factor influencing the market competition of the food product, and the organizations are supposed to be well-informed about the updated standards and regulations.

The main systems that ensure food safety of perishables along the supply chain are good agricultural practices (GAPs), food manufacturing practices (GMPs), sanitation procedures, and Hazard Analysis and Critical Control Points (HACCPs). Maintenance of appropriate sanitization throughout the supply chain is essential to reduce possibility of any contamination by harmful chemicals or microbes. The pathogenic microbes that can accumulate on the fresh horticulture produce pose greatest risk to human health from consumption, and the intensity of the microbial population on the produce is also an indicator for the poor handling, especially temperature abuse. The source of contamination along the supply chain could be from the infected field workers, contaminated soil and irrigation water, manure used, or even from customers at retail display. The only effective method to reduce actual food safety hazards and ensure food safety is to prevent contamination. Majority of the issues causing food contamination with microorganisms at the processing stage can be readily addressed through improved training for employees and the implementation of better hygiene protocols (US FDA 2004).

The Codex Alimentarius Commission (CAC) was formed by the Food and Agriculture Organization (FAO) and the World Health Organization (WHO) for developing the food standards, guidelines, and other related texts. The Hazard Analysis and Critical Control Points (HACCP) is popularly recognized as the main standard for safety and suitability of foodstuffs (US FDA 2022).

HACCP serves as the final phase of integrated food safety program, following good agricultural practices (GAPs), good manufacturing practices (GMPs), and Sanitation Standard Operating Procedures (SSOPs). HACCP programs should be simple with reasonable number of critical control points (CCPs) to serve the purpose effectively (US FDA 2022).

The seven major steps of HACCP are:

1. Perform a risk analysis
2. Select CCPs to address the identified hazards
3. Assign critical limits for each CCP
4. Create CCP monitoring specifications
5. Specify corrective measures to be used when a CCP exceeds critical limits
6. Establish record-keeping systems to document the HACCP program
7. Create processes to ensure the HACCP program is operating as intended

Successful adoption of food safety protocols requires a commitment from the grower and all staff to regular training, strict adherence to the practices, and systematic record keeping. Workers (family members, day laborers, and hired staff) must be properly and regularly trained in the use of tools, equipment, and chemicals.

16.8 Food Retailing

The retailer is usually the last but one step in the food supply chain before reaching the end-consumer. The retail environment not only connects the supply chain with end-consumer but also start the cash cycle in the food value chain. The successful supply chain process will depend on the retailer's capacity in selling the commodity to the consumer. The food retailer can create either physical infrastructure or online platform to display the products for customers to buy them. Following the manufacturer/processor, the below-mentioned parties make up the supply chain:

- Manufacturers sell their goods to distributors, who subsequently sell and distribute them to retailers, foodservice providers, and other distributors. Additionally, they will handle the food goods' physical logistics
- Distributors sell to wholesalers in bulk, and they then sell to retailers
- Traders enhance transactions between producers and retailers by acting as the representatives or agents of food manufacturers. They do not physically or legally hold the goods
- The networks and distribution facilities for large shops are their own. These locations get direct deliveries from manufacturers. After that, the merchant distributes the item to specific retail outlets

The retail business requires dynamic decisions to suit customers' changing behavior patterns. Retailers must collaborate closely with suppliers and depend on them to stay current on trends and promote innovation in order to maximize development inside the channel. Every day, consumers make decisions about their diets, including choices about what to eat, when to eat, where to eat, and with whom. Their social standing, culture, and influence on their physical and mental health are all reflected in the decisions they make. At one level, this activity is quite commonplace and predictable; at another, it is difficult, thrilling, and anything but predictable. Even the political, economic, social, and technical issues have an impact on consumer decision-making. For instance, the type of food we purchase is influenced by rising levels of wealth, the rise in one-person homes, variations in family size, and the proportion of women in the workforce. Culture, society, and individual qualities all influence consumer decisions.

16.9 International Food Supply Chain

With the advent of improved logistics, well-organized retail chains and changes in export/import regulations, the consumers are able to get fresh agricultural produce of their choice all year round. As the consumer demands are no more confined to local or seasonal produce, the supply chain systems are revamped to ship the fresh produce across the world in the best quality possible (Roekel et al. 2002). Advanced scientific technologies are allowing to keep the best quality of the produce, while innovative communication technologies are allowing traders to respond to the dynamic consumer demands and make sure the goods flow along the supply chain. The international food supply chains significantly contribute to the economic development of both exporting and importing nations.

Increase in the population size, changes in the patterns of economic capacity, health awareness, and dietary changes are few factors that are significantly affecting the international food trade. The national government interventions such as production subsidies, transport infrastructure development, and other market interventions along with international politics and trade agreements decide the pace of international food supply chain development. In some cases of international conflicts, the domestic interests are seriously affected to change with individual nations, making protectionist decisions toward the commodities of their priority. Such decisions will cause effect on the food availability, food inflation, and further social situations in the region.

The promotion of international food supply chain can be achieved by:

1. Capacity building of stakeholders in food sector on international trade protocols, food quality mandates, and standard operation procedures (SOPs)
2. Development of infrastructure needed for logistics
3. Providing incentives/financial support to farmer groups to motivate toward international trade
4. Encouraging R&D in public and private research institutes to maintain food quality, monitor logistics, and ensure traceability
5. Policy decisions to ease the global trade

16.10 Sustainability in Food Supply Chains

A couple of decades back, the topic of sustainability was more of a voluntary activity, but in the present generation, sustainable operations for businesses, especially those associated with agriculture, are a mandate. The announcement of Sustainable Development Goals by the United Nations including various aspects ranging from sustainable consumption and production to zero hunger, with a target to achieve them by 2030, shows the commitment of world organizations toward sustainability. The sustainability in the food supply chain should consider all the critical aspects of development, i.e., environmental (waste management, energy consumption, carbon emissions, environment reclamation, etc.), social (gender equality, fair pay, community development, labor welfare, etc.), and economic (return of investment, profitability, etc.).

There is growing awareness on the topic of sustainability among Indian as well as world population and popular brands and even retail outlets are under constant observation for their compliances. Any discrepancy, if found, would damage the image of the brand, followed by the negative effect on the business (Andrews et al. 2009). Few sustainable aspects of different steps of food supply chain are

Table 16.1 Sustainable aspects in different steps of food supply chain

Food supply chain stage	Sustainable practices
Sustainable production	• Assuring that ecology of the site is not disturbed • Water, soil, and other elements are conserved • Minimum or no use of synthetic chemicals as pesticides, fungicides, or weedicides • Ensuring fair working conditions, work safety, and well-being
Sustainable processing	• Reduce carbon footprint by following greenhouse gas protocol • Adapting energy reduction action plans • Reduce food wastes, develop recyclable packaging materials • Maintaining proper waste management practices • Efficient water management and proper treatment of effluents to make it least toxic
Sustainable retail	• Cutting energy costs by proper planning of lighting and air conditioning • Regular maintenance to make refrigeration systems more efficient • Work toward waste reduction and energy conservation

mentioned in Table 16.1. The standard protocols and third-party certification bodies are now available to prove the commitment of certain organization toward the sustainability. The performance benchmarking is done using parameters involving support to local growers, social development activities, responsible utilization of the natural resources, and so on. One of the popular tools to analyze the sustainability of the organization activities is "life-cycle analysis (LCA)." It considers every step of the life cycle of the product from production/collection of raw materials to pattern of consumption by end user (Sala et al. 2012).

16.11 Conclusions

Managers in the food and related sectors need a detailed examination given the importance of food supply chains and the hazards involved if they are not adequately handled. Global food security and the potential to feed more than 10 billion people by 2050 are hot topics of discussion. Numerous projects have been launched in an effort to address the problem. Sustainability in the industry and supply chain has received a lot of attention, along with food security, which focuses on safeguarding the future supply of food. The food system is under a lot of strain due to climate uncertainty, a lack of arable land and water, and other factors. The food industry is, however, experiencing a lot of positive changes. One of the characteristics of the food sector is its ability to innovate. There is a consistent process in the industry regarding innovations in food products, business models, packaging, technology, and so on, which adds value to the supply chain. To achieve an effective supply chain while observing the sustainable practices, the individual or entity needs to have a clear picture of different steps of the modern supply chain and factors involved in best performance at each step.

References

Andrews E et al (2009) Life cycle attribute assessment: case study of Quebec greenhouse tomatoes. J Ind Ecol 13(4):565–578

Bridgemohan P, Mohammed M, Tokala VY (2022) Cooling requirements of selected perishable crops during storage. In: Tokala VY, Mohammed M (eds) Cold chain management for the fresh produce industry in the developing world. Routledge, London, pp 19–32

Chandrasekaran N (2014) Agribusiness Supply Chain Management. CRC Press, Boca Raton, FL

Duggan B (2014) Advances in Cold Chain Tech Make Moving Perishables by Sea Highly Desirable, Supply Chain Brain, 2 May 2014 [Online] http://www.supplychainbrain.com/content/nc/technology-solutions/global-trade-management/single-article-page/article/advances-in-cold-chain-tech-make-moving-perishables-by-seahighly-desirable/

FAO (2011) Global food losses and food waste – extent, causes and prevention, by J. Gustavsson, C. Cederberg, U. Sonesson, R. van Otterdijk & A. Meybeck. Rome. http://www.fao.org/docrep/014/mb060e/mb060e00.pdf

FAO (2020) Tracking progress on food and agriculture – related SDG indicators 2020 – A report on the indicators under FAO custodianship. http://www.fao.org/sdg-progress-report/en/

Han JW, Ruiz-Garcia L, Qian JP, Yang XT (2018) Food packaging: A comprehensive review and future trends. Comprehensive Reviews in Food Science and Food Safety 17(4):860–877

HLPE (2014) Food losses and waste in the context of sustainable food systems. A report by the high level panel of experts on food security and nutrition of the committee on world food security, Rome. https://www.fao.org/3/i3901e/i3901e.pdf

John W, Sabine H (2012) World Population Prospects: The 2012 Revision. United Nations, Department of Economic and Social Affairs: Population Division, Population Estimates and Projections Section

KPMG (2013) The Agricultural and Food Value Chain: Entering a new era of cooperation [Online] http://www.kpmg.com/Global/en/IssuesAndInsights/ArticlesPublications/Documents/agricultural-and-food-value-chain-v2.pdf. Accessed 13 Feb 2015

NDTV (2015) Why Abolition of APMC is a Good Idea [Online] http://profit.ndtv.com/news/economy/article-whyabolition-of-apmc-is-a-good-idea-532601. Accessed 13 Feb 2015

Opara UL (2013) A review on the role of packaging in securing food system: Adding value to food products and reducing losses and waste. African Journal of Agricultural Research 8(22):2621–2630

Paine FA, Paine HY (1992) Handbook of food packaging. Blackie & Son, Glasgow

Roekel JV, Willems S, Boselie DM (2002) Agri-supply chain management: to stimulate cross-border trade in developing countries and emerging economies. World Bank, Washington, DC

Sala S, Farioli F, Zamagni A (2012) Life cycle sustainability assessment in the context of sustainability science progress (part 2). Int J Life Cycle Assess 18(9):1689–1697

Tokala VY (2022) Reducing losses of horticulture crops. Pach Mach World, Indian Institute of Packaging Machinery Manufacturers of India. 21(3):9–10

US FDA (US Food and Drug Administration) (2004) Good Manufacturing Practices (GMPs) for the 21st Century for Food Processing (2004 Study). https://www.fda.gov/food/current-good-manufacturing-practices-cgmps-food-and-dietary-supplements/good-manufacturing-practices-gmps-21st-century-food-processing-2004-study

US FDA (US Food and Drug Administration) (2022) HACCP Principles and Application Guidelines. https://www.fda.gov/food/hazard-analysis-critical-control-point-haccp/haccp-principles-application-guidelines

Chapter 17
Development of Institutional–Industrial Partnership in the Development of Postharvest and Analytical Technology for Horticulture Crops

D. S. Gaikwad, Guneshori Maisnam, Rajan Mahendra, and Sagar Wadkar

Abstract In this chapter, we explored how postharvest management technologies can help improve food security by reducing losses and waste globally. Specifically, we focused on the development of postharvest analytical technology for horticultural crops, which can greatly enhance progress in the agriculture sector. The use of this technology is essential for ensuring the safety and quality of these crops after they have been harvested. By collaborating with both academic institutions and industry partners, we can reap several benefits. Companies in the industry possess valuable knowledge, market insights, and expertise in crop management and processing, while academic institutions offer scientific expertise, research capabilities, and innovative solutions. Together, these partnerships achieved by any country's economy can help drive progress in the agriculture sector and improve food security for all.

Keywords Institutional–Industrial partnership · Analytical technologies · Horticulture crops

---

D. S. Gaikwad · G. Maisnam (✉)
Amity Institute of Organic Agriculture, Amity University Uttar Pradesh, Noida, Uttar Pradesh, India
e-mail: gmaisnam@amity.edu

R. Mahendra
Division of Food Science and Post Harvest Technology, ICAR-Indian Agriculture Research Institute, PUSA, New Delhi, India

S. Wadkar
National Cooperative Union of India (NCUI), New Delhi, India

M. Thakur, T. Belwal (eds.), *Advances in Postharvest and Analytical Technology of Horticulture Crops*, https://doi.org/10.1007/978-981-97-7247-6_17

17.1 Introduction

Postharvest technologies (PHT) establish an interdisciplinary science and techniques functional to agricultural commodities after harvest for the purpose of storage, conservation, marketing, utilization, preservation, packaging, distribution, processing, and quality control to encounter the food and nutritional necessities of customers in relation to their requirements. Further, due to high temperature in tropical countries, fruits and vegetables are more susceptible for instant happening of rotting, decay, and senescence. Both respiratory and transpiratory rates are proportional to temperature, increases and so that the produce quickly dries, wilts, and spoils unless properly preserved.

Fruits and vegetables, typically classified as horticultural produce, are particularly susceptible to spoilage because of their high moisture and nutritional content. In contrast to developed nations, where a robust infrastructure exists for managing the supply chain of these goods, developing countries face significant postharvest (PH) losses in this area (CGIAR 1997). These losses represent not just wasted produce but also a massive loss of valuable phytochemicals and nutrients. There are several potential strategies for reducing these losses, including implementing on-site storage and processing facilities, improving supply chain operations, and advancing processing techniques.

Developing a diverse array of processed goods from horticultural crops is one effective way to decrease PH losses. This strategy also leads to the production of large amounts of residue, which is typically discarded in developing countries, despite its high nutritional value. This is usually due to the lack of facilities to manage large quantities of this biomass, or the absence of established commercial applications for it. The many key factors are maintaining the quality of fresh fruits and vegetables (Lakshmi et al. 2017). The integrated effect shows the rate of deterioration and spoilage. Postharvest losses increase at their highest rate due to different factors. One-third of all fresh fruits are lost before it reaches the consumers (Kader 2002). The three main objectives of applying postharvest technology to harvested fruits and vegetables are (Lisa Kitnnoja et al. 2015):

- to maintain quality (appearance, texture, flavor, and nutritive value)
- to protect food safety, and
- to reduce losses between harvest and consumption.

The development of institutional–industrial partnerships in the development of postharvest analytical technology for horticultural crops can play a crucial role in advancing the agriculture sector. Postharvest analytical technology is essential for assessing the quality and safety of horticultural crops after they have been harvested. By collaborating with both academic institutions and industrial partners, several benefits can be achieved. These analytical technologies aid farmers, packers, and distributors in making informed decisions about storage, transportation, and marketing strategies, ultimately extending the shelf life, as well as maintaining the quality of horticultural produce for consumers (IIHR 2017).

17.2 Advantages of Collaborating with Both Academic Institutions and Industrial Partners

Postharvest analytical technology is essential for assessing the quality and safety of horticultural crops after they have been harvested. By collaborating with both academic institutions and industrial partners, several benefits can be achieved:

- ***Expertise sharing***: Academic institutions have researchers and scientists with specialized knowledge in various fields of agriculture, including postharvest technology. They can collaborate with industrial partners to share their expertise and conduct in-depth research to develop innovative technologies.
- ***Access to resources***: Industrial partners often have access to financial resources and advanced technology infrastructure. Collaborating with academic institutions allows them to leverage these resources for research and development, which might otherwise be limited for the institutions.
- ***Applied research:*** Institutional–industrial partnerships can facilitate applied research that addresses real-world challenges faced by the horticultural industry. Collaboration can lead to the development of practical and industry-specific solutions.
- ***Technology transfer:*** Through partnerships, academic institutions can transfer their research findings and technology to the industrial partners, which can then commercialize and implement the postharvest analytical technologies on a larger scale.
- ***Market-driven solutions:*** Collaborating with industrial partners ensures that the developed postharvest analytical technologies align with market demands and industry needs. This helps in creating solutions that are relevant, practical, and have a higher chance of adoption.
- ***Regulatory compliance***: Industrial partners have experience in navigating regulatory requirements, and their involvement can help in developing technologies that meet the necessary standards and regulations.
- ***Enzyme activity assays:*** Enzyme activity tests help evaluate the metabolic changes occurring in crops during postharvest storage and handling. For instance, peroxidase and polyphenol oxidase activity can indicate tissue damage and browning in fruits and vegetables.
- ***DNA-based techniques:*** Polymerase chain reaction (PCR) and other DNA-based methods can be utilized to identify pathogens, conduct traceability studies, and detect genetically modified organisms (GMOs).
- ***Electronic noses (e-noses):*** E-noses are devices that mimic the sense of smell and are used to detect and identify volatile compounds associated with spoilage or ripening processes.
- ***Imaging technologies:*** Technologies such as hyperspectral imaging and computer vision systems help identify defects, bruises, or surface damage in horticultural crops.

- ***Scale-up and implementation:*** Industrial partners have the capacity to scale up the developed technologies and implement them across their supply chains or distribution networks, thereby reaching a wider audience.
- ***Networking opportunities:*** Collaborative projects create networking opportunities between researchers, industry experts, and policymakers. This fosters a collaborative ecosystem where information and ideas can be exchanged freely.

These analytical technologies help farmers, packers, and distributors in making informed decisions about storage, transportation, and marketing strategies, ultimately extending the shelf life and maintaining the quality of horticultural produce for consumers. To foster successful institutional–industrial partnerships in the development of postharvest analytical technology for horticultural crops, it is essential to establish clear objectives, define roles and responsibilities, and have open communication between the parties involved. Additionally, seeking support from government agencies and funding bodies can also enhance the success and sustainability of such partnerships. Ultimately, the synergy between academic institutions and industrial partners can lead to transformative advancements in postharvest technology, benefiting both the agriculture industry and consumers.

17.3 The importance of Industrial–Institutional Partnerships in Postharvest Technology

Industrial partners bring practical knowledge, market insights, and experience in handling and processing horticultural crops. Academic institutions, on the other hand, contribute with their scientific expertise, research capabilities, and innovation. By combining these strengths, both parties can address complex postharvest challenges more effectively. Collaborative efforts enable the development of advanced postharvest technologies that may not be possible for either party individually. The expertise of academic researchers can help in designing and validating cutting-edge technologies, while industries can contribute to scaling up and commercializing these innovations (Kitinoja et al. 2011). Postharvest technologies play a crucial role in maintaining the quality and safety of horticultural.

Through partnerships, new techniques can be developed to reduce losses, minimize spoilage, and enhance food safety, thereby benefiting consumers and reducing food waste. Postharvest technologies that improve efficiency and reduce waste can contribute to more sustainable agricultural practices (Palumbo et al. 2022). By collaborating, institutions and industries can work together to address environmental concerns and promote sustainable agriculture. These partnerships create opportunities for networking and collaboration with other stakeholders in the industry, such as government agencies, NGOs, and international organizations. This can lead to a more holistic approach to addressing challenges in the agriculture sector. Today, developing a sustainable economy is a major problem for governments worldwide, including India. In order for a nation to prosper economically, industrialization is also crucial. No country can experience economic growth without appropriate, effective industries and industrialization that is driven by needs. In relation to this, a stable economy is built on a strong foundation when a nation uses its natural resources to complement its educated human resources. However, given that they are finite and the main raw material for the growth of any company, consideration should be given to the safer and more stable use of natural resources.

Achieving a sustainable economy is a significant challenge not only for India but also for other countries worldwide. Industrialization is a pivotal factor in a nation's economic growth. Without the presence of efficient and appropriately scaled industries, economic advancement is improbable. Part of this process involves effectively leveraging natural resources and educating human resources, both of which form the backbone of a stable economy. Nevertheless, it is crucial to manage the use of natural resources responsibly and sustainably, considering their limited availability and their pivotal role in industry development. Plant resources play an integral role in this equation. Among these, horticultural plants are noteworthy because they offer a high net return per unit area. This makes them a significant component for ensuring the sustainability of yield, consumption, and trade. Horticulture is a new area of agriculture that contributes 12.53% of India's total cropped land and 30.4% of the country's agricultural GDP. These plants produce almost 5.4 times as much as cereal grains (Tiwari et al. 2015).

Fruits, vegetables, flowers, plantation crops, medicinal and aromatic plants, and spices are just a few examples of the broad variety of horticultural crops. The Indian horticultural sector is heavily focused on exports. To tap into its potential, India should enhance the trade of both fresh produce and by-products from its horticultural industries, establishing strong connections with Europe, the Middle East, Japan, Singapore, Thailand, Malaysia, and Korea.

Starting from the mid-1980s, the Indian government directed considerable focus toward horticulture and related sectors such as processing, postharvest techniques, preservation, and packaging. The aim was to diversify agriculture and enhance its profitability. Despite these efforts, the horticultural industry in India is still at an early stage of development, and there is a notable lack of university–industry collaborations to foster horticultural advancements and attract foreign investments. To establish new horticultural-based industrial ventures, it is essential to modernize the economic infrastructure. Additionally, involving the private sector and restructuring existing businesses becomes crucial. Encouraging investment in export-oriented industrial segments and promoting corporate management are other necessary requirements to foster growth and development in the horticultural sector. Furthermore, emerging technologies such as electronics, biotechnologies, and nanotechnologies are originating from research conducted at universities, offering great potential for collaboration with industrial groups. This partnership between academia and industry can facilitate the transformation of academic research discoveries into highly valuable and sought-after commercial products.

To harness this potential, involving farmers in university training and preparing them to meet industrial demands is essential. Organizing them into entities such as Farmers Producers' Companies (FPOs) or Farmers Interest Groups (FIGs) further empowers them to play a significant role in the industrialization of horticulture, both at the local and national levels. Research focusing on postharvest processes began in 1976, with an emphasis on wine research in the fields of Plant Physiology, Biochemistry, Microbiology, and Processing. In 1978, the Postharvest Technology Division was established at this institute, originally known as the Division of Fruit and Vegetable Processing, Microbiology, and Postharvest Technology, and was later renamed to its current title. By 1991, the focus shifted toward exploring new frontier areas and basic research on postharvest handling of fruits and vegetables. The division's research efforts are currently carried out across four specialized laboratories: Handling and Storage, Processing, Microbiology, Cut Flowers.

The division boasts facilities for research on advanced storage systems, such as controlled atmosphere storage, and has a Product Development Laboratory for pilot scale processing. It was acknowledged as a Centre of Team of Excellence (TOE) under the National Agricultural Technology Project (NATP) for its work on Post-Harvest Management of Fruits and Vegetables. The division has also carried out externally funded projects on Post-Harvest Technology (PHT), including the PL-480 Scheme, Indo-USAID subproject on PHT of fruits and vegetables, All India Coordinated Research Project (AICRP) on PHT of fruits and vegetables, NATP-Mission Mode-subproject, Ad-hoc project on cut flower storage, and NATP (CGP) project on Post-Harvest Management (PHM) of cut flowers.

The research networks of international centers, along with their close collaboration with National Agricultural Research Systems (NARS), offer an excellent platform for widespread dissemination of postharvest technology. An illustration of this is seen through the role played by the International Rice Research Institute (IRRI) Crop and Resource Management Network (CREMNET) in the extensive evaluation of grain drying technology. This technology emerged from collaborative research involving the University of Agriculture and Forestry in Vietnam, the German Agency for Technical Aid (GTZ), and IRRI. In 1995–96, the Technical Advisory Committee (TAC) for the CGIAR (Consultative Group for International Agricultural Research) centers recognized a significant need and potential for increased engagement in the postharvest sector. Despite the recommendation, constraints in funding and existing program commitments limited the opportunities for implementation by CGIAR centers.

The Australian Centre for International Agricultural Research (ACIAR) can play a crucial role by encouraging centers to apply for ACIAR-IARC Restricted Project Grant or Special Purpose Grant funding to support postharvest research activities that mutually benefit the parties involved. There are opportunities to build upon ACIAR-funded research in areas such as grain drying and pest management, and to capitalize on new bilateral initiatives focusing on mycotoxin and pesticide residue research, as well as improvements in postharvest systems for tuber crops and bananas. India, as a developing country, is striving to enhance its economy, and industrialization plays a pivotal role in this process. Economic development is challenging without the establishment of efficient and relevant industries tailored to the country's needs. To achieve this, it is essential to focus on developing practical, science-based methods of postharvest and analytical technology for horticultural crops, which can be beneficially applied in other developing nations as well.

Horticultural crops encompass a diverse range, including fruits, vegetables, flowers, aromatic and medicinal plants, spices, and plantation crops. To identify effective postharvest technologies that can mitigate losses, several factors need examination. This includes characterizing indigenous crops based on their unique postharvest physiology, such as respiration rate, susceptibility to water loss, chilling sensitivity, and ethylene sensitivity. Moreover, understanding the distinctions between handling recommendations for well-known varieties and the specific requirements of local crop varieties is crucial. Additionally, evaluating the cost-effectiveness of scale-appropriate postharvest technologies for each crop in various local contexts is vital to ensure practical implementation and success. The horticultural industry in India is still in its early developmental stages, and there is a noticeable lack of effective university–industry partnerships for horticultural growth, particularly in attracting foreign investments. To establish new horticultural-based industrial entities, various steps are essential, including modernizing economic infrastructure, engaging the private sector, and reorganizing existing enterprises. Stimulating investments in export-oriented industrial sectors, promoting corporate management, and leveraging emerging technologies from university-based research, such as electronics, biotechnologies, and nanotechnologies, are also crucial for progress.

University–industry partnerships have the potential to transform university research findings into high-value commercial products, but it is equally important to involve farmers in this process. Educating farmers through university programs to meet industrial demands and gathering them in groups such as Farmers Producers' Companies (FPOs) or Farmers Interest Groups (FIGs) can significantly contribute to horticultural-based industrialization not only at local levels but also on a national scale. Developing appropriate postharvest technology relies on various disciplines that are essential for the overall success of horticulture, such as plant biology, engineering, agricultural economics, food processing, nutrition, food safety, and environmental conservation. Collaborating these disciplines ensures comprehensive advancements in the horticultural sector.

Fresh horticultural crops exhibit diverse morphological structures, compositions, and physiological characteristics, leading to varying postharvest requirements and recommendations for maximizing their shelf life. Due to their high-water content, these crops are prone to desiccation, mechanical injury, and susceptibility to bacterial and fungal attacks, resulting in pathological breakdown. Internal factors such as respiration rate, ethylene production and action, compositional changes in color, texture, flavor, and nutritive value, as well as external factors such as temperature, relative humidity, air velocity, atmospheric composition, and sanitation procedures, influence the rate of biological deterioration. Fresh fruits and vegetables play a crucial role in human nutrition, being rich sources of vitamins (Vitamin C, Vitamin A, Vitamin B6, thiamine, niacin), minerals, and dietary fiber. They also contain constituents such as flavonoids, carotenoids, polyphenols, and other phytonutrients, which may help reduce the risk of diseases, including cancer (Grusak and DellaPenna 1999).

However, postharvest losses, especially in Vitamin C content, can be significant and are exacerbated by factors such as physical damage, extended storage duration, high temperatures, low relative humidity, and chilling injury in chilling-sensitive crops (Jha et al. 2015).

India's location advantage is exemplified by the value of agricultural and processed horticultural product trade between India and the Gulf region. To diversify agriculture and enhance profitability, the Indian government has given considerable emphasis to horticulture, processing, postharvest practices, preservation, and packaging since the mid-1980s. However, the horticultural industry is still in its developmental phase in India, and there is a substantial gap that needs to be addressed to effectively leverage university–industry partnerships for horticultural development, particularly in attracting foreign investment. The establishment of new horticulturally based industrial organizations necessitates the modernization of economic infrastructure, the involvement of the private sector, restructuring of existing companies, stimulating investments in export-oriented industries, and promoting corporate management. Furthermore, the emergence of cutting-edge technologies such as electronics, biotechnologies, and nanotechnologies from university-based research presents a tremendous opportunity for collaboration with industrial groups. University–industry partnerships hold the potential to transform research findings into high-value, in-demand commercial products. However, to maximize the impact of such partnerships in horticulturally based industrialization, it is crucial to involve farmers actively. Educating farmers through university programs and preparing them for industrial demands, including online platforms, and bringing them together in groups such as Farmers Producers' Companies (FPOs) or Farmers Interest Groups (FIGs), can play a pivotal role not only at the local level but also in the national market.

The Australian Centre for International Agricultural Research (ACIAR) is a government statutory authority under the portfolio of Foreign Affairs and Trade in Australia. Established in June 1982, ACIAR aims to utilize Australia's agricultural expertise to benefit developing countries while also addressing Australia's agricultural challenges (Malik and Mazhar 2007). ACIAR is responsible for contributing to the International Agricultural Research Centres (IARCs) and funds research projects that help developing countries become self-sufficient and contribute to their well-being and economic growth (Bell 1998). The scope of ACIAR-funded research includes a broad range of activities relevant to animal and crop production, land and water management, fisheries, forestry, postharvest technology, and economic analysis of agricultural policies and technologies. ACIAR commissions research groups from Australian institutions, including universities, CSIRO, and state agriculture, forestry, and fishery departments, conduct research projects in collaboration with their counterparts in developing countries (ACIAR, 1997a). ACIAR has offices in various countries, allocates bilateral research resources to developing nations in Australia's area of interest on a regional basis, and aligns with the geographical focus of Australia's broader aid program (ACIAR, 1997b). An institutional–industrial partnership (IIP) in the development of postharvest and analytical technology for horticultural crops involves collaboration between research institutions and industry players to advance and commercialize technologies related to postharvest handling, storage, and analysis of horticultural produce.

17.4 Conclusion

Overall, an institutional–industrial partnership in the development of postharvest and analytical technology for horticultural crops is a collaborative approach that leverages the strengths of both research institutions and industry players. By combining expertise, resources, and market knowledge, these partnerships can drive advancements in technology, improve postharvest practices, and benefit the horticultural industry. Collaboration between research institutions and private companies is essential for the development of innovative technologies for the postharvest handling, storage, and processing of horticultural crops. The perishable nature of horticultural crops makes it crucial to preserve their quality and nutritional value through careful handling and preservation. Inadequate storage, transportation, and processing can lead to significant postharvest losses, which can impact farmers' profitability and consumers' access to nutritious food (Ranganna et al. 2003).

Institutional–industrial partnerships can facilitate the transfer of technology from research institutions to the private sector, leading to the commercialization of new products and services. These collaborations can also leverage the respective strengths and expertise of institutions and industries to address the challenges facing the horticultural sector and promote sustainable agriculture practices. Improved packaging materials, cold chain management systems, and analytical tools for quality control and safety testing are some of the innovative solutions that can be developed through institutional–industrial partnerships. These partnerships play a crucial role in advancing postharvest and analytical technology for horticultural crops, enhancing the quality and safety of horticultural products, as well as contributing to the growth and competitiveness of the agricultural sector.

References

ACIAR (1997a) ACIAR Co, Porale Plan 1997-2001. Australian Centre for International Agricultural Research, Canberra, p 12

ACIAR (1997b) ACIAR Pas/harvest Technology Strategic Plan. Australian Centre for International Agricultural Research, Canberra LLP

Bell M (1998) Report on IRR! Think Tank: How to increase the impact of engineering. ACIAR Pas/harvest Newslett 44:8

CGIAR (1997) Harvest and postharvest problems in agriculture, forestry and fisheries-The CGIAR contribution to research. Consultative Group for International Agricultural Research, Technical Advisory Committee (CGIAR TAC) Secretariat. FAO SDR/T AC: IAR/96/5 21 p

Grusak MA, DellaPenna D (1999) Improving the nutrient composition of plants to enhance human nutrition and health. Ann Rev Plant Biol 50:133–161

IIHR (2017) International Training Programme Post-Harvest Technology in Horticultural Crops for Executives of African and Asian 2017, ICAR-IIHR, Bengaluru. https://www.manage.gov.in/ftf-itt/prgReports/iihr.pdf

Jha SN, Vishwakarma AT, Rai A, Dixit AK (2015) Report on assessment of quantitative harvest and postharvest losses of major crops and commodities in India. ICAR-AICRP on Postharvest Technology, ICAR-CIPHET. Ministry of Food Processing Industries, Government of India

Kader AA (2002) Postharvest technology of horticultural crops. University of California, Division of Agriculture and Natural Resources

Kitinoja L, Kader AA (2015) Small-scale postharvest handling practices: a manual for horticultural crops. University of CA, Davis

Kitinoja L, Saran S, Roy SK, Kader AA (2011) Postharvest technology for developing countries: challenges and opportunities in research, outreach and advocacy. J Sci Food Agric 91(4):597–603

Lakshmi S, Pandey AK, Ravi N, Chauhan OP, Gopalan N, Sharma RK (2017) Non-destructive quality monitoring of fresh fruits and vegetables. Def Life Sci J 2(2):103–110

Malik UA, Mazhar MS (2007) Report on 'Evaluation of Postharvest losses in mango'. Australian Center for International Agricultural Research (ACIAR). Canberra, Government of Australia, Australia

Palumbo M, Attolico G, Capozzi V, Cozzolino R, Corvino A, de Chiara MLV, Cefola M (2022) Emerging postharvest technologies to enhance the shelf-life of fruit and vegetables: an overview. Foods 11(23):3925

Ranganna B, Ramakumar MV, Ilyas SM (2003) Agro-Processing Centres as an income generating activity-experiences in Karnataka. College of Agricultural engineering, University of Agricultural Sciences, Bangalore, India

Tiwari A, Singh AK, Singh S, Singh B, Pal AK (2015) University-Industry based partnership for horticultural development in India-needs and challenges. In: International Conference on Issues and Challenges in Doctoral Research, vol 25, p 21

Chapter 18
A Critical Look at Challenges and Future Scope of Phytonutrients in Horticultural Crops

Vinita Rajput, Sunil Kumar, Prity Pant, and Rishita

Abstract Horticultural crops include an abundance of naturally occurring compounds called phytochemicals, which are known to enhance health. These bioactive substances are vital to plants because they bolster their defenses against pests and diseases. Phytochemicals are a broad class of compounds that include lycopene, glucosinolates, flavonoids, ellagic acid, phenols, anthocyanidins, and carotenoids. Some phytochemicals serve as attractants, aiding in the process of pollination and greatly contributing to the vibrant hues, delicious flavors, and delightful aromas of plants. Fruits, vegetables, nuts, tea, and coffee are naturally high in these priceless phytonutrients. This chapter aims to provide a comprehensive overview of the various types of phytonutrients found in horticultural crops. It also delves into their bioactivity, elucidates the health benefits associated with them, and examines technologies designed to naturally increase the concentration of these compounds in plants.

Keywords Phytonutrients · Phytochemicals · Bioactivity · Defense mechanism

18.1 Significance of Phytonutrients

Nutritional security is next in importance after food security. It significantly altered peoples' eating patterns over the past few decades. Improper nutrition can lead to human diseases and health problems as one ages. Numerous previous studies have suggested that eating a balanced diet may reduce the risk of developing diseases like cancer. When it comes to chronic diseases like cancer, treatment is not always the

V. Rajput · S. Kumar
Krishi Vigyan Kendra Sirsa, CCS Haryana Agricultural University Hisar, Haryana, India

P. Pant (✉)
Food Technology, William Carey University Shilong, Shilong, Meghalaya, India

Rishita
CCS Haryana Agricultural University, Hisar, India

M. Thakur, T. Belwal (eds.), *Advances in Postharvest and Analytical Technology of Horticulture Crops*, https://doi.org/10.1007/978-981-97-7247-6_18

best course of action (Thakur et al. 2020; Thakur and Belwal 2023). What we eat on a daily basis has an impact on our health. An appropriate diet seems to lengthen life expectancy. The importance of including fruits and vegetables on a regular basis to our daily diet to support good health outcomes has been recognized by various food and health specialists all over the world (Handa et al. 2013). Due to the presence of several compounds known as phytochemicals that promote health, fruits and vegetables are (natural source of vitamins and minerals) are needed for the body's entire development which help us live healthy lives (Aune et al. 2017). Physiologically active substances originating from plants, or phytochemicals, have a wide range of functions in the human body. Because of their antioxidant function, which serves as a protective barrier against free radical damage, they are an essential part of the human body (Beecher 1999; Cabral and Klein 2017). Certain phytonutrients, such polyphenols and carotenoids, have antioxidant qualities and are better at scavenging free radicals, which lowers oxidative stress (Prior and Cao 2000; Zupo et al. 2022. Carotenoids, including lutein, lycopene, β-cryptoxanthin, zeaxanthin, and α and β carotene, revitalize body cells, enhance immunity, and lower the risk of diseases (Holden et al. 1999; Khan et al. 2010). In a similar vein, when included in a regular diet, the terpenes and phytosterols found in nuts and raw cold-pressed oils have been demonstrated to lower cholesterol. Plant-based foods including whole grains, fruits (such as grapes, citrus, berries, mango, papaya, and all other colored fruits), nuts, legumes, and vegetables (such as potatoes, garlic, kale, tomatoes, parsley, spinach, and broccoli) are rich in phytonutrients. Numerous phytochemicals with health-promoting bioactivities can be found in fruits and vegetables, such as flavonoids, glucosinolates, vitamins, carotenoids, and macro- and microelements (Handa et al. 2013). Horticultural crops include antioxidants that have been linked to the development of immunity against a range of pathogenic pathogens. Unexpectedly, phytonutrients found in fruits and vegetables also assist in mitigating the detrimental impacts of socioeconomic patterns, such as sedentary lifestyles and inactivity.

18.2 Consumption of Phytonutrients

Recognizing the significance of fruits and vegetables, the World Health Organization (WHO) recommends consuming five servings of fruits and vegetables daily, equivalent to 400 g (excluding potatoes, sweet potatoes, cassava, and other starchy roots). This daily intake can notably reduce the risk of developing chronic diseases. However, there remains uncertainty regarding the recommended daily allowance (RDA) for various nutrients found in fruits and vegetables. Consequently, it is often necessary to consume more quantities of fruits and vegetables to ensure an adequate intake of phytonutrients.

Globally, individuals often eat less fruit and vegetables than is advised, which may result in deficits in important vitamins and phytonutrients that are typically found in a diet that is rich in a variety of fruits and vegetables (Murphy et al. 2014). The amount of information about adult populations' consumption of phytonutrients is increasing, despite its small size. As many studies as possible have shown, eating a diet rich in fruits and vegetables can help prevent a number of diseases, such as cancer and other chronic conditions (Scalbert and Williamson 2000). When it comes to major deficits and diseases, those who consume smaller amounts of fruits and vegetables are more vulnerable than those who consume higher proportions of these food categories.

It is evident that dietary interventions involving the inclusion of phytonutrient-rich foods can serve as a crucial preventive strategy in mitigating diet-related diseases (Aune et al. 2017; Leitzmann 2016).

18.3 Types of Phytonutrients

To date, over 10,000 types of phytochemicals have been identified, each exerting a significant impact on human health based on their inherent nature, properties, and sources. These bioactive compounds are synthesized within plants through secondary metabolism, often in response to attacks by microorganisms or pests. They tend to accumulate in various plant parts in different qualities depending upon the intensity of infestation.

Numerous health-promoting attributes are demonstrated by the wide variety of phytonutrients, such as antioxidant activity, antibacterial activities, detoxification enzyme modulation, immune system activation, hormone metabolism regulation, and more (Mente et al. 2009). Past studies reveal that of the many phytonutrients found in fruits and vegetables, substances such as phenols, tannins, terpenoids, or isoprenoids have particularly significant health impacts (Zamora-Ros et al. 2010). The availability and classification of several phytonutrients in horticultural crops are shown in Table 18.1 and Fig. 18.1.

18.3.1 Phenolic Acids

Phenols, as secondary metabolites, play a crucial role in assisting plants to withstand stress during photosynthetic processes, counteracting reactive oxygen species, and defending against various threats, including pathogens and herbivores. Moreover, they contribute to attracting pollinators and serve other essential functions within plant systems.

According to surveys in the scientific literature, phenolic compounds are found in all plants and are thought to be the most common and structurally diversified phytoconstituents (Manach et al. 2004). There are about 8000 different types of polyphenols that have been found. The most notable types of polyphenols are flavonoids (which include flavonols, flavones, flavanones, and anthocyanidins), phenolic acids (vanillic acid and caffeic acid), and other important polyphenols (coumarins and tannins) (Darawsha et al. 2021; Scalbert and Williamson 2000).

Table 18.1 Availability of different phytonutrients in horticultural crops

Fruits/vegetables/spices	Major phytonutrients
Beets, cherries, cranberries, raspberries, strawberries, red peppers, watermelon, tomatoes, red beans, red onion, red radishes, pink/red grapefruit, red apples, red potatoes	Lycopene, quercetin, anthocyanidins, ellagic acid
Green apples, grapes, honeydew melon, artichokes, asparagus, broccoli, brussels sprout, cabbage, spinach, kale, peas, green peppers, green beans, kiwi fruit, pears, avocado, cucumbers, zucchini, basil, parsley, cilantro	Lutein, zeaxanthin, indoles, sulphoraphane, carotenoids, isoflavones
Carrots, orange and yellow peppers, yellow beets, squash, sweet potatoes, pumpkin, oranges, banana, apricot, cantaloupe, peaches, pineapple, ginger, turmeric	Carotenoids, lutein, zeaxanthin, lycopene, curcumin, gingerol
Blackberries, blueberries, plums, raisins, purple grapes, purple cabbage, purple carrots, eggplant, purple cauliflower, purple peppers, purple flesh potatoes, purple asparagus	Anthocyanidins, phenolics, flavonoids
Cauliflower, garlic, mushrooms, onions, parsnips, turnips, potatoes, rutabagas, white peaches, white nectarines	Glucocinolates

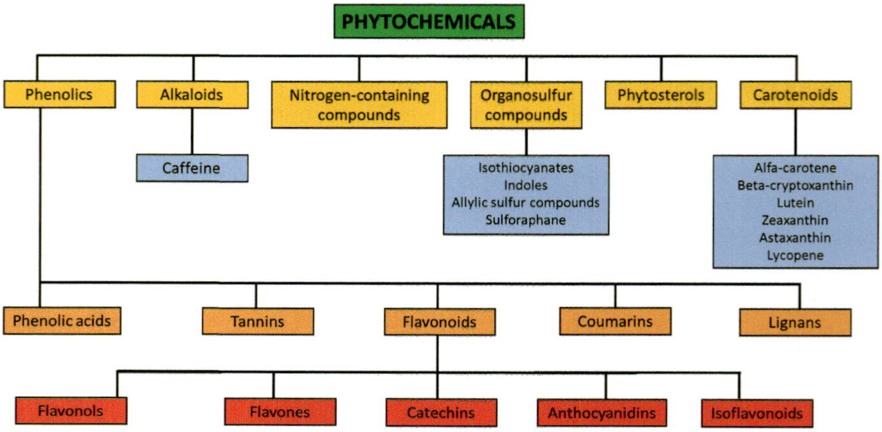

Fig. 18.1 Classification of phytonutrients (Source: Monjotin et al. 2022)

18.3.2 *Flavonoids*

Flavonoids constitute a diverse group of compounds categorized into various subclasses based on their chemical structures, including flavones, anthocyanins, flavanones, isoflavones, and flavonols, and are among the most abundant polyphenols found in plant-based foods. Specifically, reddish-hued pigments, prevalent in the skins of citrus fruits and red grapes, represent an extensively researched class of plant phenols.

Foods and drinks (red wine, black and green tea, fruit juices, tea, and coffee) that contain polyphenols include cereals (such as soy flour), vegetables (such as peppers, onions, and beans), oilseeds (such as rapeseed, canola, flaxseed, and olive seeds), and fruits (such as apples, apricots, black grapes, cherries, and berries). Notably, popular drinks from the Asian continent including tea, coffee, and cocoa frequently include flavonoids such as theine and caffeine. Among the many different types of flavonoids, reddish, blue, and purple-colored fruits and vegetables are rich in anthocyanidins. Endothelium-dependent microvascular reactivity is supported by quercetin, a flavonol that is widely distributed in plant-based diets, and hesperidin, a flavanone that is largely found in citrus fruits. Research has indicated that flavonoid consumption may be linked to longevity by reducing the risk of cardiovascular diseases (Mulvihill and Huff 2010).

18.3.3 Tannins and Ellagic Acid

Tannins, a type of water-soluble polyphenols found in plant-based foods, serve various purposes such as astringents, diuretics, anti-inflammatories, antiseptics, antioxidants, and hemostatics. These compounds contribute to blood clotting processes and play a role in maintaining blood pressure.

Various legumes, flaxseeds, peanuts, lentils, soy, and cruciferous vegetables (such as cabbage, cauliflower, kale, and radishes) are good sources of tannins. Plums, cocoa beans, carob beans, tea, wine, pomegranate bark, sorghum, and barley seeds are other foods high in tannins. Specifically, high molecular weight tannins found in persimmons have more antioxidative action than lower molecular weight tannins. It is noteworthy that the tannin content in fruits tends to decrease as the fruit ripens, while conversely, the carotenoid content associated with ripening is likely to increase.

18.3.4 Resveratrol

Resveratrol is a member of the stilbenoids subclass and is classified as a phytonutrient with notable antioxidant capabilities. It is interesting to know that grape skins and seeds contain the highest concentrations of this phytonutrient. Additionally, approximately 70 plant species, including peanuts, pistachio nuts, and strawberries, contain resveratrol, and albeit in lesser concentrations. This substance has a variety of uses, including antibacterial, antifungal, and antiviral effects.

18.3.5 Terpenoids

Terpenoids, sometimes called isoprenoids, are among the most abundant classes of secondary metabolites produced by plants. They comprise a number of subclasses, including sesquiterpenes (such as artemisinin), diterpenes (such as cafestol, cucurbitacin, limonin, nomilin, and lanosterol), tetraterpenes (such as lycopene), and hemiterpenoids (such as isovaleric acid).

These substances are essential to plant systems because they help with processes including controlling plant development, warding off illnesses and parasites, and drawing particular pollinating insects (Alves-Silva et al. 2016). For example, research has shown that terpenoids that are isolated from turmeric exhibit anticancer activities through the inhibition of metastasis or tissue invasion, induction of apoptosis, and impediment of cancer cell proliferation.

18.3.6 Carotenoids

There are around 600 different kinds of carotenoids found in plants, which give fruits, leaves, flowers, and other plant components their vivid orange to yellow colors. Numerous plants, algae, and photosynthetic bacteria produce these molecules, which are categorized as tetraterpenoids, which give rise to a broad variety of carotenoids. Due to their abundance in a variety of foods with a plant origin, the carotenoids α-carotene, β-carotene, β-cryptoxanthin, lutein, zeaxanthin, and lycopene have received the greatest scientific attention when it comes to health (Darawsha et al. 2021). Carotenoids are essential for photosynthesis and the maintenance of the equilibrium of photostasis.

Numerous studies have suggested the potential health benefits of carotenoids, particularly β-carotene, in reducing the risk of heart diseases. Lycopene, a pigment found in tomatoes, has been associated with antiprostate cancer properties. Lutein (a pigment found in lucerne) and zeaxanthin (a pigment in maize) may aid in reducing oxidative injury effects and prevent age-related macular degeneration. Moreover, β-carotene, a precursor of Vitamin A, supports eye health and helps counteract the effects of aging (Cheng et al. 2019). Carotenoids are naturally abundant in carrots, yellow/orange-colored fruits such as mangoes, papayas, and various leafy vegetables (Saini et al. 2022).

18.3.7 Alkaloids

Alkaloids are essential for the life and protection of plants. They efficiently protect against herbivores, bacteria, insects, and other dangers. Among the most well-known alkaloids include stimulants such as caffeine and nicotine, the analgesic morphine, and quinine, which is well known for its application as an antimalarial medication.

Tea leaves, guarana seeds, mate, and kola nuts all contain caffeine, which is one of the most often ingested psychoactive drugs in the world. Vinblastine and vincristine alkaloids, which are used to treat myeloblastic leukemia, malignant lymphoma, Hodgkin's disease, acute erythremia, and acute leukemia, are found in small amounts in *Catharanthus roseus* plants. Many plant-based alkaloids have been the focus of studies, highlighting their potential therapeutic uses in the areas of neurology, anticancer, and antimicrobial research.

18.3.8 Glucosinolates

Glucosinolates, which are sulfur-containing glucosides, are primarily present in vegetables that are part of the Cruciferae family, which includes mustard, radishes, Brussels sprouts, cauliflower, cabbage, and broccoli. Glucosinolates are well known for their ability to reduce inflammation and increase metabolism. It is also thought that they may have some anticancer effects. It is important to remember that glucosinolate concentrations tend to drop with prolonged storage. Thus, it is recommended to eat glucosinolate-containing vegetables fresh for the optimum intake of this phytonutrient (Monjotin et al. 2022).

18.3.9 Ellagic Acid

Ellagic acid is a cyclic phenolic molecule that is mostly present in walnuts, guavas, pomegranates, and berries. This substance is well renowned for its defense against infections, inflammations, and damage from free radicals. Furthermore, it helps prevent obesity, which aids in weight loss (Clifford and Scalbert 2000).

18.3.10 Curcumin

Curcumin is the pigment that gives the yellow color to turmeric and is responsible for its antimicrobial, anticancer, and antithrombotic properties. Curcumin prevents organ damage by suppressing inflammation-causing cells in the human body.

18.3.11 Capsaicin

Chili peppers and capsicum are common sources of capsaicin, a water-soluble phytonutrient that is a member of the alkaloid family. Studies show that endorphins, sometimes known as the "happiness hormone," are released when capsaicin is consumed. Moreover, psoriasis, gastric ulcers, and arthritis are just a few of the inflammatory diseases that capsaicin is known to reduce. Nonetheless, data points to the possibility that some people have allergies to capsaicin.

18.4 Enhancement in the Quality and Availability of Phytonutrients

The value of phytonutrients in our diets is being acknowledged more and more. However, because of our shifting eating patterns and the small amounts of fresh plant-based meals we often eat, we frequently do not get enough of these vital nutrients in our diets. The concentration of these nutrients in different horticultural crops is still quite low, depending on species, variety, and agro-ecological circumstances. Strict genetic and developmental restrictions over their metabolic pathways during plant growth and development are thought to be the cause of this lower level of health-promoting phytonutrients in plants.

The approaches of increasing the amounts of nutrients in plant sources could be investigated in order to close the gap between our bodies' nutritional requirements and the phytonutrient content found in foods. Previous research (Shukla and Mattoo 2009; Mattoo et al. 2010) has proposed a number of methods to increase the concentration of advantageous phytochemicals in fruits and vegetables, including changing cultural behavior (cultivation practices), manipulating genes, and natural selection. Additionally, using inedible plant tissues that include beneficial or bioactive phytochemicals may be a successful strategy for cultivators.

Two well-liked strategies are frequently mentioned when it comes to improving the phytonutrient content of horticultural crops: conventional breeding and molecular breeding. These methods are the mainstay of initiatives to increase the nutritional content of crops.

18.4.1 Traditional Breeding Approach

While yield and biotic resistance have always been prioritized in traditional breeding research, the emphasis on nutritional quality traits has typically trailed, significantly. Nonetheless, a commercial niche and demand for specialist, extremely nutritious crops have been created by growing public knowledge of the health advantages linked to phytochemicals. Despite this, there have been difficulties in considerably raising the concentrations of specific phytonutrients using traditional breeding techniques.

The availability of germplasm with greater levels of phytonutrients appropriate for inclusion in breeding programs is one of the main drawbacks of traditional breeding techniques. The identification, purification, and quantification of phytonutrients have been made possible by sophisticated analytical techniques, which have also aided in the identification of several possible germplasms (Mattoo et al. 2010). It is crucial to remember, though, that traditional breeding is a somewhat difficult and drawn-out procedure. Selected phytonutrients in several segregating progenies resulting from repeated crossings and backcrosses must be quantitatively assessed.

18.4.2 Molecular Breeding Approach

Molecular breeding has been useful in overcoming a number of challenges and producing novel cultivars with elevated concentrations of targeted phytochemicals. Thanks in great part to the discovery of quantitative trait loci (QTLs) and the availability of molecular markers associated with these QTLs, molecular breeding has expedited crop breeding procedures, considerably (Zhang et al. 2009).

It is challenging to obtain germplasms with higher phytonutrient accumulation, though, when there is a lack of thorough understanding of the metabolic routes for phytonutrients. Some of these problems have been quickly resolved by modern metabolomics technology (Myles et al. 2009). In addition, the field of molecular genetics has been essential in defining and pinpointing the genes that control the synthesis of phytonutrients in plants.

Notably, by focusing on the rate-limiting mechanisms involved in the synthesis of certain chemicals, advances in genetic engineering have made it possible to increase the quantities of phytonutrients in plant tissues or organs (Mattoo et al. 2010; Naqvi et al. 2009).

18.5 Conclusion

The importance of phytochemical-rich diets is shown by the indisputable effects they have on human health. These essential phytonutrients should be included in our diet in the right amounts. They are primarily found in natural foods, especially fruits and vegetables. It is necessary for the body to provide the antioxidants, vitamins, and other antiaging substances it needs for regular functioning and to maintain immunity.

Numerous techniques, including molecular approaches, gene transfer, gene editing, and traditional breeding methods, are being investigated to address the relatively low amounts or concentrations of phytonutrients in horticultural crops. The field of research provides ample opportunity to accurately measure the required intake levels of phytonutrients from fresh fruits and vegetables, improve phytonutrient synthesis in plants, and pinpoint pathways that aid in the development of resistance mechanisms in plants and humans. It is imperative to recognize that additional research is necessary to properly understand the long-term effects, potential adverse effects, and toxicity associated with overconsumption of these phytonutrients.

References

Alves-Silva JM, Zuzarte M, Marques C, Salgueiro L, Girao H (2016) Protective effects of terpenes on the cardiovascular system: current advances and future perspectives. Curr Med Chem 23:4559–4600

Aune D, Giovannucci E, Boffetta P, Fadnes L, Keum N, Norat T, Greenwood DC, Riboli E, Vatten LJ, Tonstad S (2017) Fruit and vegetable intake and the risk of cardiovascular disease, total cancer and all-cause mortality—A systematic review and dose-response meta-analysis of prospective studies. Int J Epidemiol 46:1029–1056

Beecher GR (1999) Phytonutrients' role in metabolism: Effects on resistance to degenerative processes. Nutr Rev 57(9.2):S3–S6

Cabral CE, Klein M (2017) Phytosterols in the treatment of hypercholesterolemia and prevention of cardiovascular diseases. Arq Bras Cardiol 109:475–482

Cheng HM, Koutsidis G, Lodge JK, Ashor AW, Siervo M, Lara J (2019) Lycopene and tomato and risk of cardiovascular diseases: a systematic review and meta-analysis of epidemiological evidence. Crit Rev Food Sci Nutr 59:141–158

Clifford M, Scalbert AV (2000) Ellagitannins—Nature, occurrence and dietary burden. J Sci Food Agric 80(7):1118–1125

Darawsha A, Trachtenberg A, Levy J, Sharoni Y (2021) The protective effect of carotenoids, polyphenols, and estradiol on dermal fibroblasts under oxidative stress. Antiox 10:2023

Handa AK, Anwar R, Mattoo AK (2013) Proceedings of FVHH 2013 International Symposium on Quality Management of Fruits and Vegetable for Human Health August 5–8, 2013, Golden Tulip Sovereign Hotel, Bangkok, Thailand, pp 29–35

Holden JM, Eldridge AL, Beecher GR, Buzzard M, Bhagwat S, Davis CS, Douglass LW, Gebhardt S, Haytowitz S, Schakel S (1999) Carotenoid content of US foods: an update of the database. J Food Compos Anal 12(3):169–196

Khan N, Adhami VM, Mukhtar H (2010) Apoptosis by dietary agents for prevention and treatment of prostate cancer. Endocr Relat Cancer 17(1):R39–R52

Leitzmann C (2016) Characteristics and health benefits of phytochemicals. Complement Med Res 23:69–74

Manach C, Scalbert A, Morand C, Remesy C, Jimenez L (2004) Polyphenols: food sources and bioavailability. Am J Clin Nutr 79(5):727–747

Mattoo AK, Shukla V, Fatima T, Handa AK, Yachha SK (2010) Genetic engineering to enhance crop-based phytonutrients (nutraceuticals) to alleviate diet-related diseases. Adv Exp Med Biol 698:122–143

Mente A, de Koning L, Shannon HS, Anand SS (2009) A systematic review of the evidence supporting a causal link between dietary factors and coronary heart disease. Arch Intern Med 169(7):659–669

Monjotin N, Amiot MJ, Fleurentin J, Morel JM, Raynal S (2022) Clinical evidence of the benefits of phytonutrients in human healthcare. Nutrients 14(9):1712

Mulvihill EE, Huff MW (2010) Antiatherogenic properties of flavonoids: implications for cardiovascular health. Can J Cardiol 26(suppl A):17A–21A

Murphy MM, Barraj LM, Spungen JH, Herman DR, Randolph RK (2014) Global assessment of select phytonutrient intakes by level of fruit and vegetable consumption. Br J Nutr 112:1004–1018

Myles S, Peiffer J, Brown PJ, Ersoz ES, Zhang Z, Costich DE, Buckler ES (2009) Association mapping: critical considerations shift from genotyping to experimental design. Plant Cell 21(8):2194–2202

Naqvi S, Zhu C, Farre G, Ramessar K, Bassie L, Breitenbach J, Perez CD, Ros G, Sandmann G, Capell T, Christou P (2009) Transgenic multivitamin corn through biofortification of endosperm with three vitamins representing three distinct metabolic pathways. Proc Natl Acad Sci U S A 106:7762–7767

Prior RL, Cao G (2000) Antioxidant phytochemicals in fruits and vegetables: diet and health implications. Hort Sci 35:588–592

Saini RK, Prasad P, Lokesh V, Shang X, Shin J, Keum YS, Lee JH (2022) Carotenoids: dietary sources, extraction, encapsulation, bioavailability, and health benefits—A review of recent advancements. Antioxidants (Basel) 11:795

Scalbert A, Williamson G (2000) Dietary intake and bioavailability of polyphenols. J Nutr 130(8S):2073S–2085S

Shukla V, Mattoo AK (2009) Potential for engineering horticultural crops with high antioxidant capacity, CAB reviews: perspectives in agriculture, veterinary science. Nutr Nat Res 4:066

Thakur M, Belwal T (2023) Bioactive components. Springer, Singapore. https://doi.org/10.1007/978-981-19-2366-1_5

Thakur M, Singh K, Khedkar R (2020) Phytochemicals: extraction process, safety assessment, toxicological evaluations, and regulatory issues. In: Prakash B (ed) Functional and preservative properties of phytochemicals. Elsevier, Amsterdam, pp 334–356

Zamora-Ros R, Andres-Lacueva C, Lamuela-Raventós RM, Berenguer T, Jakszyn P, Barricarte A, Ardanaz E, Amiano P, Dorronsoro M, Larrañaga N, Martínez C, Sánchez MJ, Navarro C, Chirlaque MD, Tormo MJ, Quirós JR, González CA (2010) Estimation of dietary sources and flavonoid intake in a Spanish adult population (EPIC-Spain). J Am Diet Assoc 110(3):390–398. https://doi.org/10.1016/j.jada.2009.11.024

Zhang W, Lorence A, Gruszewski HA, Chevone BI, Nessler C (2009) AMR1, an Arabidopsis gene that coordinately and negatively regulates the mannose/L- galactose ascorbic acid biosynthetic pathway. Plant Physiol 150(2):942–950

Zupo R, Castellana F, De Nucci S, Sila A, Aresta S, Buscemi C, Randazzo C, Buscemi S, Triggiani V, De Pergola G (2022) Role of dietary carotenoids in frailty syndrome: a systematic review. Biomedicines 10:632

Chapter 19
Sustainable Management of Soilborne Diseases of Horticultural Crops

Shalini Singh Visen, Deepshikha Thakur, and Saiteja Atluri

Abstract Soilborne diseases pose significant challenges in maximizing the yield potential of improved cultivars across various horticultural crops. Managing these diseases is particularly challenging due to the pathogens' ability to persist in the absence of a host for extended periods, a broad host range encompassing weeds, limited efficacy or impracticality of chemical control, difficulties in developing resistant plant varieties, the heterogeneous incidence of these diseases, and a lack of awareness regarding the epidemiological aspects of soilborne pathogens. These pathogens can endure for several years by forming persistent structures such as microsclerotia, sclerotia, chlamydospores, or oospores in the absence of a host plant. Accurately diagnosing soilborne diseases proves challenging, compounded by the microscopic size of the organisms and the nonspecific symptoms of infection.

The difficulty in in vitro culturing further complicates accurate identification. Given their inconspicuous nature and the absence of visible symptoms, soilborne organisms often escape the attention of growers and plant protection workers. Although cultural practices have been employed to reduce losses from soilborne and above-ground diseases based on observations and generational experience, claiming absolute "control" over these diseases in agriculture remains rare. As agriculture diversifies its crop portfolio, there is an increasing need to extend methods that effectively mitigate soilborne illnesses. Comprehensive research into the persistence and dissemination of soilborne diseases, the impact of environmental factors, the role of cultural practices, and understanding host resistance and susceptibility becomes paramount for effective disease control. This chapter aims to consolidate the latest insights on diverse practices for managing soilborne diseases, encompassing strategies such as crop rotation, soil solarization, soil amendments, hydroponics, soilless growing systems, grafting, biological control, utilization of natural compounds, and chemical control.

S. S. Visen
Amity Food and Agriculture Foundation, Amity University, Lucknow, Uttar Pradesh, India

D. Thakur (✉) · S. Atluri
Amity Institute of Organic Agriculture, Amity University, Noida, Uttar Pradesh, India

M. Thakur, T. Belwal (eds.), *Advances in Postharvest and Analytical Technology of Horticulture Crops*, https://doi.org/10.1007/978-981-97-7247-6_19

Keywords Soilborne pathogens · Sustainable management · Biological control · Cultural practices

19.1 Introduction

One to ten million microorganisms per gram of soil are commonly found in a typical soil sample, with fungus and bacteria accounting for the majority of the species. Certain plant-pathogenic bacteria that are present in soil can be harmful to economically important plant species. These soilborne plant diseases, according to Bruehl (1987), can either finish their life cycles in the soil or spend a portion of them on the plant's aboveground portions. Large amounts of microbial activity are found in the rhizosphere, the area where soil and plant roots interact. Plant roots discharge a variety of organic compounds into the surrounding soil in addition to absorbing water and minerals for growth.

Soilborne diseases need a susceptible host plant in order to initiate their parasitic phase. These microbes can also survive in the soil as saprophytes on leftover organic matter or as resistant, inactive forms for weeks or even years, depending on their biology. Both the saprophytic and parasitic phases of these bacteria are influenced by the biological and physicochemical characteristics of the soil.

The base of the stem or the plant's root system are often the targets of soilborne diseases. Withering illnesses can result from their ascent up the plant, by a variety of techniques, including transfer, aerial inoculation, or vascular bundle cell obstruction. The quality of crop products produced below ground (root or tuber rot, gall, proliferation, etc.) or the ability of the plant to absorb water and nutrients (root necrosis) and transfer them to the upper parts of the plant (vascular disease) are compromised by these diseases, which cause major harm to plants. These illnesses have a significant financial impact on agricultural productivity all around the world. Restoring beneficial organisms that compete with soilborne plant diseases entails either directly introducing these organisms or improving the soil's habitat by adding organic matter such as compost.

A sustainable and environmentally friendly management strategy is essential. Crop rotation, soil additives, soil solarization, hydroponic and soilless growth systems, grafting, biological control, the use of natural chemicals, and chemical control are a few of these.

19.2 Predominant Soilborne Pathogens

19.2.1 Fungi

Soilborne plant-pathogenic fungi cause a variety of diseases, such as root rot, stem rot, crown rot, damping-off, and vascular wilts, resulting in significant economic losses in the yield and quality of horticultural crops worldwide. By infecting the roots, the diseases prevent the plant from absorbing and using nutrients and water.

Wilting, leaf loss, limb and branch death, and, in extreme situations, plant mortality are some of the possible symptoms (Rahman et al. 2021). Few of the predominant fungal diseases have been listed in the following subsections.

19.2.1.1 *Rhizoctonia* Root Rot Disease

Rhizoctonia Root Rot Disease, also known as damping-off, head rot, wire stem, and crown rot. The fungus only affects older seedlings' outer cortical tissues, causing elongated tan to reddish-brown lesions. Bodah (2017) state that the area grows longer and wider in several horticultural crops, such as tobacco, apple, bean, blueberry, tomato, and peas, girdling the stem and killing the plant.

19.2.1.2 Stem, Collar and Head/Crown Rots

These kinds of illnesses are caused by a wide range of pathogens, such as those found in sunflower, tomato, bean, and apple (insert references for each in one bracket), as well as species of *Fusarium*, *Phytophthora*, *Sclerotium*, *Rhizoctonia*, and *Sclerotinia*. The most common indication of the disease is the degradation of the stem at ground level. Rot is often the cause of wilting indications, leaf death, and ultimately plant mortality. Under these circumstances, *Phytophthora* spp. may cause diseases such as pineapple heart rot, citrus root rot, potato and tomato blight, tomato crown and root rot, and specific fruit rots (Nsiah et al. 2021). Similar to this, in warm climates, *Rhizoctonia* spp. can induce cabbage head rot (Rani 2008; Garibaldi and Gullino 2009).

19.2.1.3 Wilts

Fusarium oxysporum and *Verticillium* spp. are the primary fungal taxa responsible for these infections (Lamichhane et al. 2020). Wilting of the leaves and internal necrosis of the vascular tissue in the plant's stem are two of their most recognizable signs. Similar symptoms can also be caused by some bacteria in brinjal, ginger, guava, banana, tomato, and banana (Garibaldi and Gullino 2009; Bora and Bora 2020).

19.2.1.4 Seedling Blights and Damping-off Diseases

The most common terms used for diseases of seedlings are seedling blight and damping-off. The fungi that commonly cause seedling death include *Pythium*, *Phytophthora*, *Rhizoctonia*, *Sclerotium rolfsii*, and less commonly *Fusarium* spp. These fungi can infect the seedling during the germination, preemergence, or postemergence phases of seedling establishment. In northern Vietnam, *Pythium*, *Rhizoctonia*, and *Sclerotium rolfsii* are commonly associated with seedling death of cucumber soybean and pea (Lamichhane et al. 2017; Khoury and Makkouk 2010).

19.2.1.4.1 *Pythium* Damping-off Disease

The species that causes damping-off symptoms are *Pythium debaryanum*, *P. ultimum*, *P. aphanidermatum*, and *P. graminicola* (Arora et al. 2021). This disease roughly occurs in a circular pattern. This is due to the tendency of fungi to grow radically from the point of origin, which is one of the field markers to distinguish the diseases between other agents that cause more or less the same symptoms in chili, tomato, cucumber, soybean, coriander, and fennel (Ashwathi et al. 2017; Arora et al. 2021).

19.2.1.4.2 *Phytophthora* Damping-off Disease

The most common species causing primarily low stem rots damping-off of vegetables, forest trees, and ornamentals are *P. cactorum*, *P. fragaria*, *P. palmivora*, and *P. syringae*. Unlike *Pythium*, *Phytophthora* is aggressive in warmer soil temperatures (15–23 °C), but still cool condition. Flooding, along with warm temperature favors disease development. Initially affected tissue develops a soft, watery brown rot. Drying of the affected plant parts are reported in citrus, chili, and cucumber (Savita and Nagpal 2012; Hashemi et al. 2019).

19.2.2 Bacteria

Bacteria are less responsible for soilborne diseases because of their nonspore-forming property. They cannot stay in the soil for very long, and they also need a wound or other natural opening to enter the plant before they can start an infection (Lebeda et al. 2014). The most common examples are *Ralstonia solanacearum*, causing bacterial wilt of tomato and *Agrobacterium tumefaciens* causal agent of crown gall (Spadaro and Gullino 2019). Filamentous bacteria such as *Streptomyces* spp. (Viswanath et al. 2018) are well adapted to such condition and can survive more easily in the soil and can also infect plants. Some examples are *Erwinia* (soft rot of potato), *Rhizomonas* (corky root of lettuce), and *Streptomyces* (potato scab, soft rot of sweet potato, bacterial wilt, and black leg are considered the most important diseases) (Yang 2020; Charkowski et al. 2020).

19.2.3 Viruses

Viruses typically need living plant tissue to thrive and are rarely caused by or able to persist in the soil when their host is absent for an extended period of time. However, they can also travel on fungi or nematodes and enter through water (Velasco et al. 2020). A virus controls a plant cell to make new virus cells after it enters the cell (Tiwari et al. 2020). When romaine lettuce plants are infected with the lettuce necrotic stunt virus, the lower leaves become stunted, yellowed, and occasionally spotted, but the newer leaves stay thick and green (Reddy 2010; Verma and Sharma 1999).

19.2.4 Nematodes

Commonly known as roundworms, nematodes are unsegment worms with rounded body and pointed at both the ends. Some nematodes are parasitic, e.g., nematodes sold to feed on beetle larvae in the lawn and some will thrive on or in the roots of agricultural crops. These parasites are very problematic, especially for root crops, such as carrots (Abawi and Widmer 2000). Root rot nematodes, the most commonly known one, causes distortion and swelling of roots due to which the plant's vigor is affected. Needle nematodes feed on roots tips, which in turn leads to extensive branching and swelling of the branches (Ayala-Doñas et al. 2020). Additionally, stubby root nematodes cause short and stubby roots (Lopez-Nicora et al. 2022).

19.3 Biology of Soilborne Pathogens

19.3.1 Survival

Soilborne pathogens can be either residents (where they can stay for relatively longer periods of time) or invaders (where they can persist for comparatively shorter periods of time), according to Nassarawa et al. (2022). Additionally, these infections may exist as saprophytes, which are organisms that consume organic waste as it breaks down. When an active, susceptible host is present and the right hospitable conditions are met, these saprobes will typically change into a pathogenic form (Liu et al. 2015).

19.3.2 Distribution of Pathogens in the Soil

The distribution of soilborne diseases can be horizontal or vertical depending on production techniques, historical cropping patterns, and a variety of other factors. The top 10 inches of the soil profile, which is where host roots, tissues, and other organic substrates are situated, are the inoculum for the majority of root infections. Inoculum distribution often accumulates on the horizontal plane in fields where a sensitive crop has been planted (Delgado-Baquerizo et al. 2020).

19.3.3 Factors Influencing Distribution

Numerous factors affect the activity of soilborne pathogens and the emergence of disease. The kind of soil, soil texture, pH, moisture, temperature, and nutrient levels are some of the key factors influencing the onset of the disease and the success of infection. However, poorly draining soil tends to be more conducive to the survival and spread of pathogens such as *Pythium*, *Phytophthora*, and *Aphanomyces*. This is similar to how *Verticillium* and *Fusarium* wilts can get worse in wetter soils than they do in dry soils. Very few illnesses are favored by dried soil, such as the common potato scab brought on by *Streptomyces scabies* (Calegar et al. 2021).

19.4 Sustainable Management of Soilborne Plant Diseases

For the sustainable management of soilborne plant diseases, a thorough understanding of the relationships among host plants, environmental variables, and pathogen biology is necessary (Calegar et al. 2021). Sachdev and Singh (2017) state that the presence of a viable inoculum, exposure of susceptible plants to the pathogen, and the existence of favorable environmental circumstances that encourage infection and pathogen proliferation are all necessary for the development of a disease. Effective disease management strategies require economic feasibility, and the choice of control measures requires careful consideration of factors such as illness incidence, severity, and potential crop loss (Sachdev and Singh 2017). Control measures must be implemented as soon as possible, particularly in cases of harmful diseases like *Phytophthora* root rot (Singh et al. 2017).

19.4.1 Cultural Control

Cultural control practices involve modifying crop agronomic practices to manage diseases, especially for low-unit-value, larger-area crops such as forests and temperate grains (Katan 2000). These methods are often economically practical for certain crops, considering the high labor costs associated with some agricultural practices. Cultural control strategies, being preventive and indirect, are akin to biological control and rely on a deep understanding of pathogen biology and host responses (Lecomte et al. 2016). Assessing the effectiveness of these practices can be challenging as they are often implemented in anticipation of pathogen attacks. In certain cases, disease spread can be prevented or reduced through quarantine, sanitation, and disinfection of soil, potting media, and irrigation water to eliminate inoculum (Lecomte et al. 2016).

19.4.1.1 Cultural Practices to Reduce Initial Levels of Inoculum

Introduction of pathogens into previously clean sites typically leads to disease epidemics (Katan 2000). To prevent or reduce disease spread to new areas, quarantine, sanitation, and soil disinfection practices are crucial (Lecomte et al. 2016).

19.4.1.1.1 Selection and Preparation of Planting Materials

The choice of cultivar significantly impacts disease control, as certain cultivars exhibit natural resistance to specific infections (Sarolia et al. 2018). Developing resistant varieties is a potent method for controlling diseases in crops. Ongoing plant breeding efforts are necessary to combat evolving pathogen races responsible for the breakdown of host resistance. Crop diversity and the use of mixed cultivars can decrease pathogen spread, while efficient seed-cleaning procedures contribute to disease control (Singh et al. 2017; Gupta and Chaturvedi 2022; Tigrel et al. 2022).

19.4.1.1.2 Destruction of Crop Residues

Agricultural residues serve as ideal substrates for many diseases, and cultural management techniques include removing, burning, or burying crop residues during intercropping seasons (Dung 2020). Deep burial of crop residues can control specific pathogens, such as *Scierotium rolfsi* causing groundnut blight. However, the success of burying residues depends on how pathogens are redistributed in the soil. Composting and burning stubble are also common practices, but the drawbacks, such as nutrient loss, environmental pollution, and increased soil erosion, must be carefully considered against disease prevention benefits (Chandra et al. 2012; Baggs et al. 2000; Koul et al. 2022; Mani 2022; Litterick et al. 2004).

19.4.1.2 Elimination of Alternate Hosts

Wild plants or weeds can serve as alternative hosts for plant parasites, allowing diseases to thrive during dormant periods and act as a source of inoculum for future growing seasons. For example, grasses such as *Hordeum* and *Leportnum* host the fungus *Gaeumannomaces graminis*, causing harm to wheat (Vikram et al. 2018). Some leguminous green manure crops harbor *Pseudomonas syringe* pv. *phaseolicoal*, the bacterium causing halo blight, while dicotyledonous weeds may contain *Meladogyne* spp. and other nematodes. Removing these alternative hosts can significantly reduce or eliminate inoculum for the following season (Zhao et al. 2013). However, the relationship between weeds and crop diseases is complex, and some weeds may carry viruses without showing symptoms. Eradicating certain weeds may not be practical, especially if they can grow into commercially valuable crops. Growers may need to collaborate to implement strategies such as changing crop rotation patterns or coordinated efforts to eliminate virus vectors (Stoffella and Kahn 2001; Wisler and Norris 2005; Mani 2022).

19.4.1.3 Fertilizer Application

Fertilizer application plays a crucial role in disease management. Ammonium bicarbonate has been found to inhibit the growth of *Saccharomyces cerevisiae* and *Hansenula wingei* (Curran and Montville 1989). The exact mechanism of how ammonium bicarbonate prevents fungal growth is unclear, but it may involve the compound's unionized ammonia. Additionally, phosphatic fertilizers contribute to increased host resistance by promoting plant production. Fertilizer application, combined with irrigation, enhances overall plant health, reducing disease severity. Phosphoric acids are effective in preventing *Pythium* outbreaks, and the mode of action involves interrupting the internal phosphorus utilization cycle essential for the pathogen's survival (Punja and Grogan 1982; Smyth and Cassel 1995). Adequate potassium levels inhibit the growth of various parasites, including fungi, bacteria, viruses, and nematodes. Potassium also indirectly influences disease development, promoting wound healing, reducing frost injury, and delaying crop maturity (Reddy 2014; Achary et al. 2017).

19.4.1.4 Providing Good Soil Drainage and Air Circulation Among Plants

Maintaining good soil drainage conditions reduces the activity of specific soilborne pathogens and nematodes, such as *Pythium* (Welbaum 2015). Practices such as managing irrigation intervals to prevent water dispersal of soilborne pathogens and promptly removing diseased or dead plants can minimize disease incidence (Krishnan et al. 2014; Ristvey et al. 2019). Flooding fields over an extended period or dry fallowing can decrease the prevalence of pathogens such as *Fusarium*, *Sclerotini sclerotiorum*, and nematodes. Irrigation has been shown to reduce charcoal rot related to *Macrophomina phaseolina* (Ray 2015; Jiru and Wegari 2022).

19.4.2 Crop Rotation

Many soilborne pathogens can survive in the soil or plant debris for several years. Crop rotation involves growing crops in a definite succession on the same piece of land, helping control soilborne inoculums by reducing the presence of the host. Crop rotation becomes less effective if the soil resident can generate resistant spores or survive as a saprophyte for more than 5 or 6 years. However, crop rotation can still be beneficial in lowering disease populations, especially for pathogens with specific host requirements (Kata 2002). Long-term rotations may be necessary to control pathogens such as *Streptomyces scabies*, *Phgmatotrichopsis omnivora*, and

Rhizoctonia solani. In some cases, fallow periods or growing nonhost crops can help control certain pathogens, as observed with sugar beet crops and the nematode *Heterodera schachtii* (Gong et al. 2021; Jiru and Wegari 2022).

19.4.3 Tillage Practices

Tillage practices play a crucial role in disease management by reducing the potential spread of diseases among crops. Deep soil layers help bury plant pathogens, rendering them less capable of causing diseases (Lamichhane et al. 2020). Seed bed preparation techniques alter physical soil qualities, including bulk density, moisture content, aeration, and temperature profiles, impacting disease frequency. Modifying the soil by creating hills, ridges, or raised beds enhances irrigation and drainage (Failla et al. 2021). Overall, tillage effects, seen as increased plant vigor, influence nutrient release mechanisms (Hummel et al. 2002). Healthy plants are observed to exhibit increased resistance to some pathogens, while the microclimate within crops may influence the spread of other pathogens (Litterick et al. 2004). Successive tillage operations can reduce inoculum levels of certain pathogens by exposing them to desiccation in the sunlight (Wezel et al. 2014). Subsoiling before planting has been found to enhance green pea yields in soil infested with root rot pathogens (Cook 2000).

19.4.4 Soil Amendments

Soil amendments, such as organic materials (e.g., sawdust, straw, and oil cake), have proven effective in controlling infections caused by various pathogens, including *Pythium*, *Phytophthora*, *Verticillium*, *Macrophomina*, *Phymatotrichum*, and *Aphanomyces* (Chindo et al. 2012). These amendments gradually suppress pathogenic microbes while promoting the growth of beneficial microorganisms in the soil. Lime application, raising soil pH to 8.5 (2500 kg/ha), has been successful in reducing cabbage club root. while sulfur application (900 kg/ha) lowers soil pH to 5.2, diminishing the prevalence of potato common scab caused by *Streptomyces scabies* (Kulkarni and Lingraju 2015). Utilizing decaffeinated waste and water hyacinth has demonstrated high efficacy in treating root knot nematode (Davis et al. 1974; Fig. 19.1). Regular additions of organic materials, including crop and cover crop residues, compost, and manure, contribute to increased soil organic matter (SOM), supporting greater microbial biomass and activity (Agarwal et al. 2021). The concentration of carbon in SOM varies with different cropping systems and cultural practices, with intensive tillage accelerating microbial respiration or erosion that removes carbon from the soil (Ahmad et al. 2021). Almeida et al. (2003) investigated the impact of tillage and soybean debris on the soilborne fungus *Macrophomina phaseolina*, causing charcoal rot in soybeans. Traditional tillage

Fig. 19.1 Root knot nematode (*Melidogyne incognita*) affected brinjal roots

methods exhibited higher final disease severity and area under the disease progress curve (AUDPC) values, particularly during drought conditions (De Corato 2020).

19.4.5 Soil Solarization

Solarization, a method involving covering damp soil with translucent polyethylene during bright summer days, can elevate soil temperatures to approximately 52 °C, compared to a maximum of 37 °C in un-mulched soil (Khan et al. 2012). This technique renders many soilborne fungi, nematodes, and bacteria near the soil surface inactive or causes their demise. Solarization, occurring over several days or weeks of consistent sunlight, reduces inoculum levels and the risk of disease development (Butler et al. 2014). Successful control of tomato bacterial canker, caused by *Verticillium* wilt and *Fusarium* wilt, has been achieved through soil solarization using transparent or black polyethylene sheets (Panth et al. 2020).

19.4.6 Biological Control

Biological control, where diseases are regulated by natural predators, can significantly impact pathogen populations. Antagonistic microorganisms present in suppressive soils prevent pathogen spread through food competition, direct parasitization, or the production of antibiotics and lytic enzymes (Tariq et al. 2020). Additionally, taller plants may act as trap plants or release substances into the soil to reduce inoculum. Scientists are exploring these organic antagonistic relationships to develop effective biological control strategies against various plant diseases (Pal and McSpadden 2006).

19.4.6.1 Suppressive Soils

Suppressive soils hinder the development of several soilborne pathogens, such as Pythium species, *Fusarium oxysporum*, *Gaeumannomyces graminis*, *Phytophthora cinnamomi*, and *Heterodera avenae*. The suppression mechanism is often linked to the proliferation of antagonistic microorganisms, including fungi such as *Trichoderma*, *Penicillium*, and *Sporidesmium*, as well as bacteria from *Pseudomonas*, *Bacillus*, and *Streptomyces* genera. In some instances, soil amendments with antagonistic microorganisms have successfully produced crops largely free from specific diseases, such as *Streptomyces* species controlling potato scab (Sultana and Ghaffar 2010). Other examples include *Trichoderma* managing Papaya root rot (Verma et al. 2007) and the biocontrol potential of nonpathogenic *F. oxysporum* strains against *Fusarium* wilt in various crops (Larkin and Fravel 1999). Bacterial bioagents, including *Pseudomonas*, *Bacillus*, *Burkholderia*, *Lysobacter*, *Pantoea*, and *Streptomyces*, have also demonstrated effectiveness in suppressing diseases and promoting plant growth (Souza 2015).

However, the success of these bioagents may depend on factors such as proper substrate availability and environmental conditions. For instance, *B. cepacia* has shown potential as an effective biocontrol agent against Fusarium dry rot of potatoes in storage under suitable conditions (Recep et al. 2009). Overall, understanding the dynamics of suppressive soils and harnessing the potential of biological control agents offer promising avenues for sustainable disease management in agriculture.

19.5 Conclusions

To effectively and economically control soilborne diseases, one must be fully informed about the crop, the disease affecting it, its history in previous years, the host's resistance levels, and the current environmental conditions. The synergistic or cumulative benefits of combining multiple illness management techniques can be observed. Since the epidemiologies of soilborne diseases differ greatly, this type of approach is very crucial. Sustainable agriculture systems are designed to maximize energy flows, soil organisms, and current water and nutrient cycles in order to produce as much food as feasible. Moreover, the aim of these systems is to produce food that is wholesome and devoid of impurities that may pose a risk to human health.

References

Abawi GS, Widmer TL (2000) Impact of soil health management practices on soilborne pathogens, nematodes and root diseases of vegetable crops. Appl Soil Ecol 15(1):37–47

Achary VMM, Ram B, Manna M, Datta D, Bhatt A, Reddy MK et al (2017) Phosphite a novel P fertilizer for weed management and pathogen control. Plant Biotechnol J 15(12):1493–1508

Agarwal P, Saha S, Hariprasad P (2021) Agro-industrial-residues as potting media, Physicochemical and biological characters and their influence on plant growth. Biomass Convers Biorefin:1–24. https://doi.org/10.1007/s13399-021-01998-6
Ahmad G, Nishat Y, Ansari MS, Khan A, Haris M, Khan AA (2021) Eco-friendly approaches for the alleviation of root-knot nematodes. In: Plant growth-promoting microbes for sustainable biotic and abiotic stress management. Springer, Cham, pp 557–575
Almeida AMR, Amorim L, Filho B, Torres E, Farias JRB, Benato LC, Pinto MC, Valentim N (2003) Progress of soybean charcoal rot under tillage and no-tillage systems in Brazil. Fitopatol Bras 28:131–135
Arora H, Sharma A, Sharma S, Haron FF, Gafur A, Sayyed RZ et al (2021) Pythium damping-off and root rot of Capsicum annuum L impacts, diagnosis, and management. Microorganisms 9(4):823
Ashwathi S, Ushamalini C, Parthasarathy S, Nakkeeran S (2017) Morphological, pathogenic and molecular characterisation of Pythium aphanidermatum: a causal pathogen of coriander damping-off in India. Pharma Innov 6(11):44
Ayala-Doñas A, Cara-García MD, Talavera-Rubia M, Verdejo-Lucas S (2020) Management of soil-borne fungi and root-knot nematodes in cucurbits through breeding for resistance and grafting. Agronomy 10(11):1641
Baggs EM, Rees RM, Smith KA, Vinten AJA (2000) Nitrous oxide emission from soils after incorporating crop residues. Soil Use Manag 16(2):82–87
Bodah ET (2017) Root rot diseases in plants: a review of common causal agents and management strategies. Agric Res Technol Open Access J 5(3):555661
Bora P, Bora LC (2020) Disease management in horticulture crops through microbial interventions: an overview. Indian J Agric Sci 90(8):1389–1396
Bruehl GW (1987) Soilborne plant pathogens. New Biological Books, Macmillan, NY. 368pp
Butler DM, Kokalis-Burelle N, Albano JP, McCollum TG, Muramoto J, Shennan C et al (2014) Anaerobic soil disinfestation (ASD) combined with soil solarization as a methyl bromide alternative: vegetable crop performance and soil nutrient dynamics. Plant Soil 378:365–381
Calegar DA, Bacelar PAA, Evangelista BBC, Monteiro KJL, Santos JPD, Almeida MM et al (2021) Socioenvironmental factors influencing distribution and intensity of soil-transmitted helminthiasis in the Brazilian Amazon: Challenges for the 2030 agenda. J Trop Med 2021:6610181
Chandra R, Takeuchi H, Hasegawa T (2012) Methane production from lignocellulosic agricultural crop wastes: a review in context to second generation of biofuel production. Renew Sustain Energy Rev 16(3):1462–1476
Charkowski A, Sharma K, Parker ML, Secor GA, Elphinstone J (2020) Bacterial diseases of potato. In: The potato crop. Springer, Cham, pp 351–388
Chindo PS, Bello LY, Kumar N (2012) Utilization of organic wastes for the management of phyto-parasitic nematodes in developing economies. In: Kumar S (ed) Management of organic waste. BoD, Norderstedt, pp 133–148
Cook RJ (2000) Advances in plant health management in the twentieth century. Annu Rev Phytopathol 38(1):95–116
Curran DM, Montville TJ (1989) Bicarbonate inhibition of Saccharomyces cerevisiae and Hansenula wingei growth in apple juice. Int J of Food Microbiol 8(1):1–9
Davis JR, Garner JG, Callihan RH (1974) Effects of gypsum, sulfur, terraclor and terraclor super-x for potato scab control. Am Potato J 51(2):35–43
De Corato U (2020) Agricultural waste recycling in horticultural intensive farming systems by on-farm composting and compost-based tea application improves soil quality and plant health: a review under the perspective of a circular economy. Sci Total Environ 738:139840
Delgado-Baquerizo M, Guerra CA, Cano-Díaz C, Egidi E, Wang JT, Eisenhauer N et al (2020) The proportion of soil-borne pathogens increases with warming at the global scale. Nat Clim Change 10(6):550–554
Dung JK (2020) *Verticillium* wilt of mint in the United States of America. Plant 9(11):1602

Failla S, Pirchio M, Sportelli M, Frasconi C, Fontanelli M, Raffaelli M, Peruzzi A (2021) Evolution of smart strategies and machines used for conservative management of herbaceous and horticultural crops in the mediterranean basin: a review. Agronomy 11(1):106

Garibaldi A, Gullino ML (2009, September) Emerging soilborne diseases of horticultural crops and new trends in their management. In VII International Symposium on Chemical and Non-Chemical Soil and Substrate Disinfestation, vol. 883, pp 37–47

Gong X, Shi J, Zhou X, Yuan T, Gao D, Wu F (2021) Crop rotation with cress increases cucumber yields by regulating the composition of the rhizosphere soil microbial community. Front Microbiol 12:631882. https://doi.org/10.3389/fmicb.2021.631882

Gupta S, Chaturvedi P (2022) Commercial scale tissue culture for horticulture and plantation crops. Springer, Singapore, pp 1–21

Hashemi L, Golparvar AR, Nasr Esfahani M, Golabadi M (2019) Correlation between cucumber genotype and resistance to damping-off disease caused by Phytophthora melonis. Biotechnol Equip 33(1):1494–1504

Hummel RL, Walgenbach JF, Barbercheck ME, Kennedy GG, Hoyt GD et al (2002) Effects of production practices on soil-borne entomopathogens in western North Carolina vegetable systems. Environ Entomol 31(1):84–91

Jiru EB, Wegari HT (2022) Soil and water conservation practice effects on soil physicochemical properties and crop yield in Ethiopia: review and synthesis. Ecol Process 11(1):21

Kata J (2002, August) Role of cultural practices for the management of soilborne pathogens in intensive horticultural systems. In XXVI International Horticultural Congress: Managing Soil-Borne Pathogens: A Sound Rhizosphere to Improve Productivity in 635, pp 11–18

Katan J (2000) Physical and cultural methods for the management of soil-borne pathogens. Crop Prot 19(8–10):725–731

Khan MA, Marwat KB, Amin A, Nawaz A, Khan R, Khan H et al (2012) Soil solarization: an organic weed-management approach in cauliflower. Commun Soil Sci Plant Anal 43(13):1847–1860

Khoury WE, Makkouk K (2010) Integrated plant disease management in developing countries. J Plant Pathol 92:S35–S42

Koul B, Yakoob M, Shah MP (2022) Agricultural waste management strategies for environmental sustainability. Environ Res 206:112285

Krishnan PR, Kalia RK, Tewari JC, Roy MM (2014) Plant nursery management: principles and practices. Jodhpur, CAZRI

Kulkarni S, Lingraju S (2015) Management of plant diseases under organic cultivation. In: Recent advances in the diagnosis and management of plant diseases. Springer, Cham, pp 11–16

Lamichhane JR, Dürr C, Schwanck AA, Robin MH, Sarthou JP, Cellier V et al (2017) Integrated management of damping-off diseases. A review. Agron Sustain Dev 37:1–25

Lamichhane JR, You MP, Laudinot V, Barbetti MJ, Aubertot JN (2020) Revisiting sustainability of fungicide seed treatments for field crops. Plant Dis 104(3):610–623

Larkin RP, Fravel DR (2002) Effects of varying environmental conditions on biological control of Fusarium wilt of tomato by nonpathogenic Fusarium spp. Phytopathology 92(11):1160–1166

Lebeda A, Křístková E, Kitner M, Mieslerová B, Jemelková M, Pink DA (2014) Wild Lactuca species, their genetic diversity, resistance to diseases and pests, and exploitation in lettuce breeding. Eur J Plant Pathol 138:597–640

Lecomte C, Alabouvette C, Edel-Hermann V, Robert F, Steinberg C (2016) Biological control of ornamental plant diseases caused by *Fusarium oxysporum*: a review. Biol Control 101:17–30

Litterick AM, Harrier L, Wallace P, Watson CA, Wood M (2004) The role of uncomposted materials, composts, manures, and compost extracts in reducing pest and disease incidence and severity in sustainable temperate agricultural and horticultural crop production—a review. Crit Rev Plant Sci 23(6):453–479

Liu Y, Fang S, Chesson P, He F (2015) The effect of soil-borne pathogens depends on the abundance of host tree species. Nat Commun 6(1):10017

Lopez-Nicora HD, Enciso-Maldonado GA, Caballero-Mairesse GG, Sanabria-Velázquez AD, Armandans-Rojas AJ et al (2022) Distribution and abundance of nematodes in horticultural production in Paraguay. Plant Health Prog 23(4):466–475

Mani M (2022) Pest management in horticultural crops under protected cultivation. In: Trends in horticultural entomology. Springer, Cham, pp 387–417

Nassarawa SS, Luo Z, Lu Y (2022) Conventional and emerging techniques for detection of foodborne pathogens in horticulture crops: a leap to food safety. Food Bioprocess Technol 15(6):1248–1267

Nsiah FB, Oteng AS, Oppong A, Nunoo I, Brobbey L (2021) Phytophthora diseases prevalence, its effects and controls in Ghana. IntechOpen. https://doi.org/10.5772/intechopen.99130

Pal KK, McSpadden G (2006) Biological control of plant pathogens. Plant Health Instructor. https://doi.org/10.1094/PHIA-2006-1117-02

Panth M, Hassler SC, Baysal-Gurel F (2020) Methods for management of soilborne diseases in crop production. Agriculture 10(1):16

Punja ZK, Grogan RG (1982) Effects of inorganic salts, carbonate-bicarbonate anions, ammonia, and the modifying influence of pH on sclerotial germination of Sclerotium rolfsii. Phytopathology 72(6):635–639

Rahman M, Islam T, Jett L, Kotcon J (2021) Biocontrol agent, biofumigation, and grafting with resistant rootstock suppress soil-borne disease and improve yield of tomato in West Virginia. Crop Prot 145:105630

Rani GD (2008) Advances in soil borne plant diseases. New India Publishing, New Delhi

Ray P (2015) Hi-Tech horticulture and climate change. In: Choudhary ML, Patel VB, Siddiqui MW, Mahdi SS (eds) Climate dynamics in horticultural science, principles and applications. CRC Press, Boca Raton, FL, pp 1–22

Recep K, Fikrettin S, Erkol D, Cafer E (2009) Biological control of the potato dry rot caused by Fusarium species using PGPR strains. Biol Control 50(2):194–198

Reddy PP (2010) Bacterial and viral disease and their management in horticultural crops. Scientific Publishers, Jodhpur

Reddy PP (2014) Plant growth promoting rhizobacteria for horticultural crop protection, vol 10. Springer, New Delhi, pp 978–981

Ristvey AG, Belayneh BE, Lea-Cox JD (2019) A Comparison of irrigation-water containment methods and management strategies between two ornamental production systems to minimize water security threats. Water 11(12):2558

Sachdev S, Singh RP (2017) Sustainable management of soil borne pathogens of tomato. Int J Sci Technol Soc 3:36–40

Sarolia DK, Samadia DK, Choudhary BR, Singh D (2018) Production of quality seed and planting materials

Savita GS, Nagpal A (2012) Citrus diseases caused by Phytophthora species. GERF Bull Biosci 3(1):18–27

Singh RR, Meena LK, Singh P (2017) High tech nursery management in horticultural crops: a way for enhancing income. Int J Curr Microbiol Appl Sci 6(6):3162–3172

Smyth TJ, Cassel DK (1995) Synthesis of long-term soil management research on Ultisols and Oxisols in the Amazon. In: Lal R, Stewert BA (eds) Soil management - experimentlal basis for sustainability and environmental quality. Advances in Soil Science, pp 13–60

Souza R, Ambrosini A, Passaglia LMP (2015) Plant growth-promoting bacteria as inoculants in agricultural soils. Genet Mol Biol 38(4):401–419

Spadaro D, Gullino ML (2019) Sustainable management of plant diseases. Innov Sustain Agric:337–359

Stoffella PJ, Kahn BA (eds) (2001) Compost utilization in horticultural cropping systems. CRC Press, Boca Raton, FL

Sultana N, Ghaffar A (2010) Effect of fungicides, microbial antagonists and oilcakes in the control of Fusarium solani, the cause of seed rot, seedling and root infection of bottle gourd, bitter gourd and cucumber. Pak J Bot 42(4):2921–2934

Tariq M, Khan A, Asif M, Khan F, Ansari T, Shariq M et al (2020) Biological control: a sustainable and practical approach for plant disease management. Acta Agric Scand Sect B Soil Plant Sci 70(6):507–524

Tigrel A, Arslan M, Arıcı B, Yücesan B (2022) Selection and preparation of explants for the clonal propagation of horticultural plants in plant factory systems. In: Commercial scale tissue culture for horticulture and plantation crops. Springer, Cham, pp 23–48

Tiwari RK, Lal MK, Naga KC, Kumar R, Chourasia KN, Subhash S et al (2020) Emerging roles of melatonin in mitigating abiotic and biotic stresses of horticultural crops. Sci Hortic 272:109592

Velasco L, Ruiz L, Galipienso L, Rubio L, Janssen D (2020) A historical account of viruses in intensive horticultural crops in the Spanish mediterranean arc: new challenges for a sustainable agriculture. Agronomy 10(6):860

Verma LR, Sharma RC (1999) Diseases of horticultural crops: vegetables, ornamentals, and mushrooms. Indus Publishing, New Delhi

Verma M, Brar SK, Tyagi RD, Surampalli RN, Valero JR (2007) Antagonistic fungi, Trichoderma spp.: panoply of biological control. Biochem Eng J 37(1):1–20

Vikram S, Vijaya K, Afaque Q (2018) Viral elimination strategies for Musa spp. Research & Reviews. J Microbiol Virol 8(1):7–14

Viswanath HS, Bhat KA, Bhat NA, Wani TA, Mughal MN (2018) Antibacterial efficacy of aqueous plant extracts against storage soft rot of potato caused by Erwinia carotovora. Int J Curr Microbiol Appl Sci 7:2630–2639

Welbaum GE (2015) Vegetable production and practices. CABI, Wallingford

Wezel A, Casagrande M, Celette F, Vian JF, Ferrer A, Peigné J (2014) Agroecological practices for sustainable agriculture. A review. Agron Sustain Dev 34(1):1–20

Wisler GC, Norris RF (2005) Interactions between weeds and cultivated plants as related to management of plant pathogens. Weed Sci 53(6):914–917

Yang K-T (2020) Identification of bldA as a Major Regulator of Virulence in the Sweetpotato Soil Rot Pathogen Streptomyces ipomoeae and Characterization of Distinct Clostridium Sweetpotato Soft Rot Isolates. LSU Doctoral Dissert 5438. https://digitalcommons.lsu.edu/gradschool_dissertations/5438

Zhao J, Wang L, Wang Z, Chen X, Zhang H, Yao J et al (2013) Identification of eighteen Berberis species as alternate hosts of Puccinia striiformis f. sp. tritici and virulence variation in the pathogen isolates from natural infection of barberry plants in China. Phytopathology 103(9):927–934

GPSR Compliance

The European Union's (EU) General Product Safety Regulation (GPSR) is a set of rules that requires consumer products to be safe and our obligations to ensure this.

If you have any concerns about our products, you can contact us on ProductSafety@springernature.com

In case Publisher is established outside the EU, the EU authorized representative is:

Springer Nature Customer Service Center GmbH
Europaplatz 3
69115 Heidelberg, Germany

Batch number: 10756376

Printed by Printforce, the Netherlands